Structure of the Stratosphere and Mesosphere

International Geophysics Series

Edited by

J. VAN MIEGHEM

Royal Belgian Meteorological Institute
Uccle, Belgium

olume 1	BENO GUTENBERG. Physics of the Earth's Interior. 1959
'olume 2	JOSEPH W. CHAMBERLAIN. Physics of the Aurora and Airglow. 1961
'olume 3	S. K. RUNCORN (ed.). Continental Drift. 1962
/olume 4	C. E. JUNGE. Air Chemistry and Radioactivity. 1963
Volume 5	ROBERT C. FLEAGLE AND JOOST A. BUSINGER. An Introduction to Atmospheric Physics. 1963
Volume 6	L. DUFOUR AND R. DEFAY. Thermodynamics of Clouds. 1963
Volume 7	H. U. ROLL. Physics of the Marine Atmosphere. 1965
Volume 8	RICHARD A. CRAIG. The Upper Atmosphere: Meteorology and Physics. 1965
Volume 9	WILLIS L. WEBB. Structure of the Stratosphere and Mesosphere. 1966

In preparation

S. MATUSHITA AND WALLACE H. CAMPBELL. Physics of Geomagnetic Phenomena.

Structure of the Stratosphere and Mesosphere

BY

WILLIS L. WEBB

Atmospheric Sciences Laboratory
U. S. Army Electronics Command
White Sands Missile Range, New Mexico

and

Physics Department
University of Texas at El Paso
El Paso, Texas

ACADEMIC PRESS · 1966
New York and London

ACADEMIC PRESS INC.
111 Fifth Avenue, New York, New York 10003

United Kingdom Edition published by
ACADEMIC PRESS INC. (LONDON) LTD.
Berkeley Square House, London W.1

Library of Congress Catalog Card Number: 66-16446

PRINTED IN THE UNITED STATES OF AMERICA

Preface

Man's view of the structure and dynamics of the atmosphere has always been limited by the extent of available data. Until 1960, synoptic analysis of the atmosphere was confined to the region below approximately 30 km where balloon sounding systems could operate effectively. Development of the Meteorological Rocket Network has effectively doubled the vertical dimension of the atmosphere which can be subjected to synoptic analysis. This volume is devoted to a summary of the development of this new observational network and a first comprehensive analysis of the resulting data. The relatively small amount of data yet available in stratospheric and mesospheric regions makes it likely that the current picture is incomplete, although there has already emerged sufficient new information to show that synoptic exploration of the upper atmosphere is not just desirable but is a necessity. The tools are now available, and the next phase of upper atmospheric exploration should be the expansion of the network into a global synoptic system. This Global Meteorological Rocket Network will provide the data required to test certain assumptions which are necessarily made herein, and will undoubtedly clarify many features of upper atmospheric structure which current data do not adequately portray.

Assumptions relative to static conditions where radiational processes dominate have played a leading role in past theoretical and experimental investigations of the upper atmosphere. A fundamental deduction from these initial phases of exploration is that such assumptions are subject to gross error. Clearly, kinetic processes over the entire spectrum from molecular motions to global circulations must be incorporated into any description before a reasonably accurate picture of upper atmospheric structure will be obtained. Data are required, and so more sophisticated concepts of upper atmospheric physical processes are dependent upon implementation and the production of the Global Meteorological Rocket Network.

Meteorological Rocket Network investigations of the stratospheric circulation have forcefully raised important questions relative to the global distribution of composition and physical processes. Wind and temperature profiles which are now obtained efficiently must be supplemented by additional measurements at least on a sporadic basis, and possibly in a synoptic program. Of prime interest are ozone and water vapor in the composition area, and surely the scale and activity of eddy motions which can support fluxes of energy or minor contaminants must be established. The current Meteorological Rocket Network is not the answer to our need for synoptic data in the upper atmosphere but is only a beginning.

Electromagnetic interactions between the ionized circulation of the upper atmosphere and the earth's magnetic environment are just now beginning to be understood. Our current understanding of the stratospheric circulation indicates very strongly that a meteorological analysis of the synoptic situation in the ionosphere will be most productive, both to the meteorologist studying the total atmospheric structure and to the communications engineer attempting to use the ionosphere in his communications link. This means upward extension of the Meteorological Rocket Network, which must necessarily be preceded by development of new sensors and techniques which will facilitate synoptic exploration in that region. Both the lateral expansion of the current Meteorological Rocket Network and the upward extension into the ionosphere will probably occur in the cooperative spirit which has prevailed since Lloyd Berkner and Sidney Chapman proposed the International Geophysical Year experiment. Most of the new data contained in this volume are the result of just such cooperation, with missile range meteorologists spearheading the move to get maximum results from their project efforts through cooperation in the Meteorological Rocket Network.

The Meteorological Rocket Network does not obtain synoptic data in the upper portions of the atmospheric regions studied here. Data from individual rocket soundings of the ionosphere, from systematic studies of phenomena such as noctilucent clouds, and from drift and diffusion of ionization trails have been used to extend upward our knowledge of upper atmospheric structure. Numerous other sources of information relative to the circulation and structure of the stratosphere and mesosphere have been applied to obtain the picture of that region of the earth's atmosphere which is presented here. More specifically, this work could not have been completed at this time without the excellent labors of the scientific personnel of the Schellenger Research Laboratories of Texas Western College, El Paso, Texas, who accomplished the assimilation and publication of all MRN data, and who also provided special analysis of certain data used in preparation of this manuscript. In addition, the author is indebted to his wife, Lanice, for the transcription of the manuscript; his son, Michael, for the preparation of the figures; and Mrs. Annette Noland for detailed editing of the manuscript. The author is also indebted to all of a large number of other individuals who furnished specific information in certain areas of this book which were beyond the scope of background of the author.

May 1966 WILLIS L. WEBB

Table of Contents

1

Introduction

The earth's atmosphere is a complex zone of interaction between the global surface and the solar environment. This insulating role is of major service to mankind, but it introduces certain complications into physical processes which govern the interaction. Complications become apparent at several earth's radii distance in the near space and generally increase in intensity with decreasing altitude to a maximum in the atmosphere or at the earth's surface. Not all of the many phenomena follow such simple rules; however, it is a fair approximation that the atmosphere is roughly spherically stratified into several layers in which particular characteristics predominate. The significance of certain of these stratified regions to human activities has set them apart, and our scientific exploration and technological application have proceeded along independent lines, as though they were not a part of an entity which exerts a controlling influence on all aspects of the system. Certain important corners of our atmosphere have glowed in pointed investigation for several decades, and it is now that the sophistication of our society permits the welding of these fragmentary analyses into a unified whole.

The stratosphere and mesosphere are two of the neglected layers that have been relegated to secondary roles in the atmospheric system. Lack of knowledge initially resulted in adoption of assumptions of quiescence which would tend to forever confine these intervening layers to a simple supporting function. Our expanding technology has forced inspection of these abandoned regions because of their incorporation into our active sphere of influence; and the resulting new information has pointed up the error of omitting such important components of the system. Possibly the most important advance in atmospheric sciences to result from stratospheric and mesospheric investigations has been the precipitous awareness of the significance of the atmospheric system in its entirety.

Data obtained through balloon sounding of the first few kilometers of the atmosphere during the nineteenth century founded the concept of a decreasing pressure and temperature structure. There was little evidence to alter the obvious conclusion that these parameters would continue to decrease with increasing height, and the obvious activity of precipitation and particulate matter in the lower levels led to preliminary separation of the lower and upper atmospheres. Typical of thinking during that period is the following quotation from a meteorology text by William M. Davis published by Ginn and Company, Boston, in 1899:

> The upper air, pure and dry, free from clouds and dust, far from the surface of the earth and out of reach of ordinary convective action, must possess a low temperature and must change its temperature slowly and by small amounts.

At the time the above was being written, the Frenchman Teisserenc de Bort was learning that the temperature structure over Europe exhibited a decidedly different characteristic above about the 11-km altitude level, with the tropospheric positive temperature lapse rate (negative gradient) sharply broken, and becoming nearly isothermal above. Objections to Teisserenc de Bort's data, such as errors resulting from radiation warming of the temperature sensor, were quickly dispelled by exploratory flights at night and under varying experimental conditions to the general satisfaction of the scientific community of that day. Realization that the structure of the atmosphere might be far more complicated than had been assumed sparked renewed interest in upper atmospheric characteristics among atmospheric research scientists, principally as a result of the facets of interaction of atmospheric processes with traversing energy which this detailed structure might introduce. Most of the immediate interest in investigation of this upper atmospheric region came from those scientists who were studying solar radiant energy flux at the earth's surface and propagation of sound waves over long distances through the earth's atmosphere. Discovery of this new atmospheric layer was of only passing interest to the meteorologist of the era because of a rudimentary knowledge of synoptic processes in the atmosphere, the considerable difficulties involved in obtaining timely data in these altitude ranges, and lack of adequate communications to effect timely compilation of the information required to obtain a comprehensive picture of the atmospheric structure.

Even so, meteorologists did begin to consider these new facts as a part of their sphere of interest, as is evidenced by the following quotation from Volume I of Sir Napier Shaw's *Manual of Meteorology* which was published in 1926:

The use of these balloons has resulted in the most surprising discovery in the whole history of meteorology. Contrary to all expectation the thermal structure of the atmosphere in the upper regions is found to be such that isothermal surfaces are vertical surfaces succeeding one another with diminishing temperature from the pole outwards towards the equator, which underneath this remarkable structure, which shows the lowest temperatures of the atmosphere to be at very high levels over the equator, lies the structure with which we are ordinarily familiar consisting of approximately horizontal layers warmer in the lower latitudes and colder in the higher and, with some interesting exceptions, diminishing in temperature with height until the upper layer of vertical columns is reached. The lower region of the atmosphere in which the temperature is arranged in horizontal layers was called by Teisserenc de Bort the "troposphere" and the upper region of so-called "isothermal columns" the "stratosphere." The name "tropopause" has been coined to indicate the level at which the troposphere terminates.

The investigations carried out by de Bort extended only a few kilometers above that level at which the temperature lapse rate stopped its steady fall and, since these first observations were made in upper midlatitudes where there is characteristically little change of temperature with height in these upper levels, the concept of an isothermal upper atmosphere became common. Association of the word "stratosphere" with constancy in temperature was very firmly established during the early years of the twentieth century, and a variety of nomenclature was developed to designate those atmospheric spherical shells which fell outside the limits of that definition. With mounting evidence that the isothermal concept did not apply to a significant portion of the atmosphere other than at high latitudes, the World Meteorological Organization (WMO) adopted the following definition at its third session in Rome in 1961:

Stratosphere: A region situated between the tropopause and the stratopause in which the temperature generally increases with height.

As is evidenced by the data presented in Fig. 1.1 the latter definition is more satisfying in that it is more accurately descriptive of the thermal structure which is found over a major portion of the global surface area. Even so, when one excludes equatorial regions where it is essentially satisfactory, the adequacy of the description leaves much to be desired. There is, in fact, an isothermal layer in the lower portions of the stratosphere in middle and high latitudes, and at more poleward locations the thickness and importance of this segment of the stratosphere become increasingly obvious. The net result is that most atmospheric scientists now follow the above definition in general, with a further subdivision of this principal atmospheric layer into an upper and lower stratosphere.

Vertical temperature structure data for rather typical summer and winter cases are presented in Fig. 1.1 to illustrate the layering which typifies atmospheric thermal equilibrium. These data were obtained by the three principal observation systems which currently may be employed to obtain a complete vertical temperature profile in the atmosphere from the surface to the mesopause. Since present observational practice does not generally include the mesosphere in synoptic analysis, these profiles were formed by incorporating data obtained in 1961 and 1962 at Wallops Island, Virginia from the grenade experiment (W.

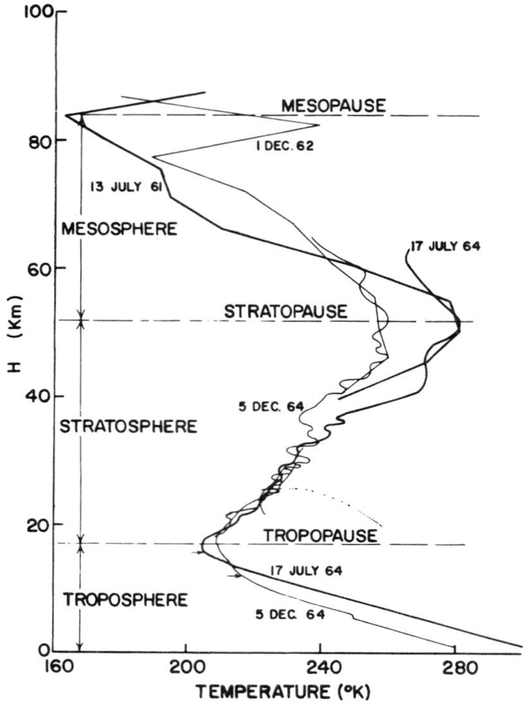

Fig. 1.1. Typical midlatitude temperature profiles for the summer and winter seasons from balloonsonde, rocketsonde, and grenadesonde sensing systems. Data from the surface to above 60-km altitude were obtained at White Sands Missile Range, New Mexico, by the U. S. Army Electronics Research and Development Activity and those from 40 to 80 km were obtained over Wallops Island, Virginia, by the National Aeronautics and Space Administration.

Smith *et al.*, 1964) in the mesosphere, and in tropospheric and stratospheric regions by the Meteorological Rocket Network station at White Sands Missile Range, New Mexico in 1964.

While these are individual profiles, they illustrate clearly the considerable degree of consistency with which the atmospheric vertical temperature structure is observed by the several systems. The annual variations in temperature in midlatitudes are indicated to be similar in sign in the troposphere and the upper stratosphere, with warmer values occurring in the summer season and colder temperatures in the winter. The high-temperature layer centered near 50 km exhibits an annual temperature variability of the same order as that observed at the earth's surface. The upper portion of the mesosphere, on the other hand, has an annual variation which is approximately 180° different in phase, with minimum temperatures observed in the summer and maximum temperatures observed in the winter season. It is clear that a significantly different mechanism must be involved in establishing the vertical thermal structure of the mesosphere than is important at lower levels.

A most important point to note in the data illustrated in Fig. 1.1 is the detail structure which characterizes the stratospheric region. This information is observed to be a persistent feature of the stratosphere by the very sensitive sensors of the Meteorological Rocket Network, and is probably also a feature of the mesospheric structure which has been filtered out as a result of inadequate sensitivity of observational techniques. These data suggest a far more active region of the atmosphere than had been assumed, and require a new insight into the physical processes which can be instrumental in producing the observed results. It is probable that the picture thus far obtained lacks clarity as a result of the limitations of sensitivity and accuracy of the current upper atmospheric observational systems, and it is rather certain that new approaches to the observational problem will be required before a comprehensive understanding of the dynamics of the stratosphere and mesosphere can be obtained.

Considerable conflict is still evident in the literature over the point at which a division of the stratosphere should be effected, and a principal point of this treatise will be the advancement of a well-defined separation based on the horizontal temperature field that in turn prescribes the seasonal vertical wind profiles. As illustrated in Fig. 1.2 the proposed term and its definition are as follows:

> *Stratonull:* A surface of minimum meridional temperature gradient located in the central stratosphere, which is usually evidenced by a minimum or change in gradient in the vertical wind profile.

Observational data obtained during the past few years indicate the stratonull surface to be evident in all except polar regions and during spring and fall reversal periods, and to be located at an altitude of approximately 24 km. The stratonull is ill-defined in polar regions, as is the case with most atmospheric structural separations, owing to lack of significant horizontal stratification. There are also indications of considerable variations in middle latitudes, possibly even to the extent of fracturing of the surface, similar to that which has been found on detailed inspection of the tropopause surface (Danielsen, 1964). A principal attribute of the adoption of the stratonull surface as an informative division of the upper atmosphere is the fact that it represents the separation of two principal circulation systems which serve to establish the dynamic equilibrium of all except the electrified ionospheric

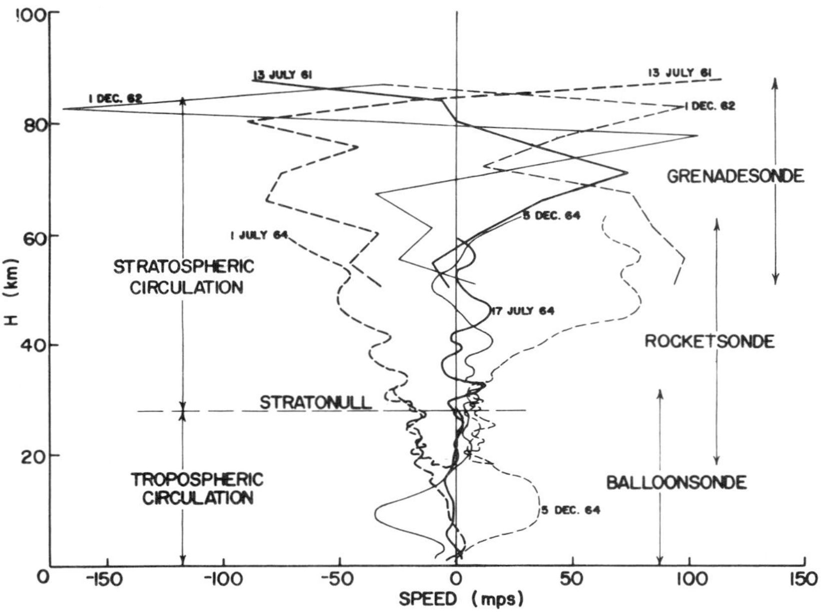

Fig. 1.2. Wind data from the soundings plotted in Fig. 1.1 are presented in components, with zonal curves dashed (west positive) and meridional curves solid (south positive), indicating the direction from which the wind is blowing.

upper layers of the atmosphere. Sample wind profiles illustrating features of the vertical structures which are typical of summer and winter seasons are presented in Fig. 1.2. These data were obtained as companion measurements with the temperature data reported in Fig. 1.1. The dashed curves represent the zonal components, plotted as positive when winds are from the west. Solid curves depict the meridional components, with south plotted as positive. The altitude of the stratonull is indicated for these particular cases, and the vertical extents of the tropospheric and stratospheric circulation systems are illustrated.

Tropospheric winds in the winter case exhibit a strong meridional component, with a maximum speed of 35 meters/sec from the north at 10-km altitude. The stratonull surface is a bit high in this case, lying near 28 km when compared with a more usual height of 24 km. The stratosphere displays its usual midlatitude, dominantly zonal, wind structure here, with light meridional winds and a strong, well-developed zonal flow. Peak westerly winds of the stratospheric winter season are characteristically found near the stratopause, with a positive wind lapse rate the rule in the mesosphere; in the summer season the easterly winds increase rather steadily with height to near the mesopause. Summer easterly winds are generally weaker than winter westerlies in the stratosphere, but this trend is reversed in the mesosphere where the easterlies increase and the westerlies generally decrease, both as a function of height. In effect, then, the two circulation systems exemplified by these data have much the same maximum speeds, although the momentum of the winter flow can be expected to be significantly greater owing to the large density lapse rate, which provides little mass for the strong winds of the summer easterlies in the mesosphere.

A most striking feature of the data of Fig. 1.2 is the dramatic increase in meridional wind components in the mesosphere. In these examples meridional winds, which are generally quite light in the stratosphere, become as strong as the zonal components in the middle mesosphere, and even exceed the zonal flows in the upper mesosphere. This gross change in the mode of circulation heat exchange is probably attributable to frictional effects, which should negate the validity of our usual assumptions of geostrophic equilibrium in the flow of the upper atmosphere. This gradual transition of the principally zonal stratospheric circulation defines the top of that circulation system and the beginning of the ionospheric circulation system. At this point, structural breakdown of the atmosphere along kinetic lines has three major subdivisions: the tropospheric circulation from the surface to 24 km; the stratospheric circulation from 24 to 80 km; and the ionospheric circulation above 80-km altitude.

The troposphere and the lower stratosphere are very intimately related in the circulation system which governs heat exchange in the lower atmosphere. The troposphere is the lower portion of a composite wind system that responds to thermodynamic stimuli situated principally at its lower boundary (the earth's surface), and the lower stratosphere normally represents the upper portion of that circulation pattern. Just as the troposphere is divided along latitude lines into longitudinal bands which are dominated by different circulation processes, the lower stratosphere exhibits entirely different thermal and wind structures in equatorial, midlatitude, and polar regions. In general, lower stratospheric wind profiles are characterized by positive lapse rates at all latitudes, with this decrease in wind speed with height occurring in temperature lapse rates which are negative in tropical regions and essentially isothermal in middle latitudes and polar regions. These positive lapse rates in the flow have their maximum strength and variability in midlatitudes where peak speeds of the tropospheric jet streams are of the order of 50 meters/sec or more at the tropopause level. The data show that even over the strongest tropospheric jet streams the wind speed drops to very low values or the wind even reverses direction in the upper portions of the lower stratosphere.

Evidence is mounting that the tropopause is not characterized by the strong physical separation of two different atmospheric layers as had been visualized earlier. It is known to be quite variable in altitude, particularly in midlatitudes, and is, on occasion, broken into more than one surface over local areas. The tropopause is now suspected to be a level of dynamic equilibrium between the troposphere and lower stratosphere with an active mass transfer across it. The atmospheric layer which was initially labeled the stratosphere and is now known as the lower stratosphere is really an integral part of the lower atmosphere the dynamics of which are controlled by the tropospheric circulation system, although this control is a negative process in that the input energies which activate the tropospheric circulation are farthest removed from the lower stratospheric shell. The physical characteristics of the lower stratosphere (great vertical stability) in effect place a lid on the tropospheric scene of turbulent interchange and establish a level of minimum coupling between these two main circulation systems of the earth's lower and middle atmosphere.

Our accepted definition of the stratosphere implies an upper boundary to limit the increase in temperature with height. This limit is to be found in an inflection in the temperature profile near 50-km altitude in all except polar regions where the change is generally observed at higher altitudes. As a companion definition to the WMO stratospheric statement above, the following upper boundary definition is offered:

Stratopause: The top of the inversion layer in the upper atmosphere, usually about 50–55 kilometers.

The stratopause is thus the level of a maximum in the vertical temperature structure which is produced by absorption of solar ultraviolet energy by the substance ozone (O_3), a minor constituent of the stratosphere. The altitude at which this maximum occurs (50 km) is well above the level of maximum ozone concentration (approximately 25 km) and represents the equilibrium altitude where heat input, capacity, and assorted loss mechanisms combine to produce the maximum temperature.

As a result of its vertical thermal structure, the upper stratosphere is a very stable layer. A 30-km-thick stratum with a temperature rise of 80° in tropical regions and 50° in midlatitudes effectively prohibits convective-type activity with which we are familiar in the troposphere. This extreme stability of the upper stratosphere is less pronounced in polar regions, but even there the negative lapse rates of the order of 1°C per kilometer altitude preclude development of convective activity similar to tropospheric thunderstorms under equilibrium conditions. This is not to say that there will be no vertical motion. The vertical motions which do occur, however, will be damped by nonlinear processes to restricted amplitudes and may represent a strong sink for energy from the general circulation (that is, friction).

As a result of the physical situation described above, the stratopause represents the upper surface of a very stable volume of atmosphere. It has characteristics relative to the atmosphere above it which must be remarkably similar to those of the earth's surface. The stratopause represents a diffuse heat source which, due to the positive lapse rate above it, must result in convective interchange in the layers above and in the development of a large-scale circulation system unique to that region. Some of the most important advances which have come from the past several years' exploration of stratospheric structure concern the stratopause and the nature of the circulation which has its origin in the stratopause area.

The stratopause exhibits great stability in time and space. Only in the winter polar region is the situation significantly altered. Here turbulent motions in the troposphere are minimal, and no pronounced tropopause is apparent. Little or no solar energy is available for stratospheric heating, so that relatively small deviations away from an isothermal vertical temperature distribution are to be expected. The winter polar atmospheric structure represents the weak point in the stability of the upper atmosphere.

The upper atmospheric layer bounded at its base by the stratopause is the mesosphere. Again there is a need for careful definition since this term has also been used to represent alternate regions of the upper atmosphere in the past. The currently accepted definition is:

Mesosphere: A region (situated between the stratopause and the mesopause) in which the temperature generally decreases with height.

The mesosphere enjoys a physical situation which is quite similar to the troposphere, with the exception that in the winter, at least, a positive lapse rate in the wind profile dominates the region. Both the thermal and kinetic structure are oriented toward instability relative to convective activity, so considerable turbulent activity can reasonably be expected in this atmospheric region. A maximum in this convective activity is to be expected in the summer hemisphere. An interhemispheric circulation of considerable magnitude may well exist in view of the large meridional wind components which appear to represent friction effects and thus result in nongeostrophic flow. A possibly important diagnostic tool for analysis of this region is the analysis of pressure perturbations which may originate in turbulent eddies and propagate away with the speed of sound.

The final atmospheric surface which must be defined for our purposes is the mesopause, which is the cap on the convective mesosphere. The nomenclature which we will adopt here is:

Mesopause: The base of the inversion at the top of the mesosphere (usually found about 80–85 kilometers).

The mesopause is a surface which is the locus of the lowest temperatures to be found in the earth's atmosphere. The magnitude of the temperature decrease from the bottom to the top of the mesosphere is only slightly greater than that observed across the troposphere in equatorial regions. The base temperature, on the other hand, is of the order of 30° cooler in the case of the mesosphere. The result is a cold mesopause located at an altitude of about 80 km, above which the ionospheric temperature increases rapidly with height to values well above any observed in the lower atmosphere.

As is evident in the above discussion, most of the nomenclature in these upper atmospheric layers is based on the thermal structure. This is not the only available criterion, or even the most lucid. For instance, the term "ozonosphere" is not uncommon in the literature. The ozonosphere refers to the spherical shell bounded by the tropopause and the mesopause, encompassing all of the atmosphere in which the ozone reaction dominates other atmospheric chemical reactions.

Considerable emphasis is placed on the flow as an analysis tool in this study. This results primarily from the hard fact that vertical profiles of wind are the most plentiful data available. In addition, throughout the stratosphere a considerable stability makes the geostrophic approximations reasonably acceptable, and thus the wind data can provide a measure of the horizontal temperature and pressure fields. Measurement of vertical wind profiles has proved to be an inexpensive probing tool, using simple flight equipments and available ground tracking devices. This emphasis on circulation could well lead to an additional set of nomenclature, but in the interest of simplicity we will tie the two major circulation systems of interest here to the atmospheric regions in which they originate.

Thus, the "tropospheric circulation" (Fig. 1.2) refers to that wind system which dominates the troposphere and lower stratosphere. In like manner, the "stratospheric circulation" refers to that wind system which dominates the upper stratosphere and most of the mesosphere. This technique permits the development of a meaningful group of terms to describe the kinetic properties of the middle atmosphere with the addition of only one new term (stratonull) and only mild distortions of previously established definitions. Adherence to this limitation may well prove to be an important accomplishment.

Our knowledge of the lower stratosphere has resulted from probings from beneath with balloon systems. Experimental difficulties associated with measurements on a balloon platform at low pressures after tropospheric winds have carried the sensor far from the ground measuring equipments limited the utility of such data in the lower stratosphere until the middle 1900's. Improvements in techniques produced a steady increase in mean peak altitudes achieved with balloon radiosondes by the United States Weather Bureau until the stratonull surface was exceeded in the 1950's and routine analysis of the extreme lower portion of the stratospheric circulation system became feasible. Interest in these data was nominal until the discovery of gross disturbances in stratospheric structure by Richard Scherhag (1952) at Berlin in the winter of 1951–1952. The apparent downward propagation of these phenomena raised fundamental questions as to the validity of assumptions concerning negligibility of stratospheric influences on the tropospheric weather systems. Such influences could be promulgated by highly significant alterations in scale heights which might accompany these "explosive warmings" of the lower stratosphere.

Shaw's enthusiasm for his "most surprising discovery" would probably have known no bounds had he had available a synoptic picture of the lower stratosphere. It was in an atmosphere of expectant discovery that the Meteorological Rocket Network (MRN) was initiated over North America in October of 1959. The observed volume of atmosphere was effectively doubled in one step, and a new dimension in meteorology was realized with the addition of a second quasi-independent layer for circulation equalization of asymmetric energy inputs.

Actually, a general knowledge of the existence of a second circulation in the upper atmosphere was obtained from sporadic probings before the advent of the MRN. The first indication came in the early 1900's through observation of sound propagation from the surface through the upper atmosphere and back to the surface again. It was quickly deduced that this special mode of propagation resulted from a warm layer at approximately 50-km altitude. Investigators also concluded that there must be a strong seasonal variation in winds at high levels to produce the observed lack of symmetry in refraction of the sounds with direction. Except for some unfortunate overestimates of the temperature of the stratopause and an almost complete lack of sensitivity, the technique provided for an excellent first look at the stratosphere and was applied at various locales and times.

Our next look at the stratosphere and mesosphere came with the development of rocket vehicles. Numerous experimental techniques were used during the exploratory phase of applying rocketry to the upper atmospheric research problem. In general, the principal purpose of these early rocket firings was to test equipment, so the atmospheric experimental techniques usually turned out to be less than optimum. Most of the stratospheric and mesospheric data acquired during this period were lacking in sensitivity, in large part owing to the high speed with which the region was traversed and the nature of the sensors. It was from these first attempts that the Rocket Panel (1952) drew up the first estimate of the vertical atmospheric structure over White Sands Missile Range, New Mexico.

There was considerable pressure for independent rocket systems which could accommodate atmospheric experiments in an optimum configuration. The Navy's Viking was a first step in this direction followed by the Aerobee, which became the work horse for upper atmospheric observation during the 1950's. The grenade experiment for temperature and wind sounding was developed for use in the Aerobee, and became one of the leading experimental techniques for stratospheric data acquisition during the 1950's. The stratosphere and mesosphere above White Sands Missile Range became the best observed region of the atmosphere during this early experimental phase.

A new trend began with preparations for the International Geophysical Year. Efforts were made to package the experiments in smaller rocket vehicles such as the Nike Cajun. Costs could be reduced in flight hardware, ground equipments, and personnel, and the experimental restraints could be reduced to great advantage where special firing schedules were required. Use of this rocket system at remote sites became feasible, and during the late 1950's the first direct measurements of the high-latitude upper stratosphere were made.

The rocket grenade experiment for probing the upper atmosphere involved the ejection and detonation of small explosive charges at predetermined intervals along the trajectory. The pressure perturbations generated by these explosions could then be observed on arrival at the ground and the differences in the arrival signals were then related to the propagation parameters in the atmospheric region between the bursts. Thus, mean values for the temperature and wind can be obtained through altitude intervals which are a function of the payload characteristics of the rocket vehicle and its trajectory. The maximum altitude at which the experiment can generally be conducted is approximately 100 km, so the system thus lends itself to upper atmospheric observation programs extending to above the mesopause. The grenade experiment does not lend itself to synoptic use in the sense of balloon soundings, but does provide for representative samplings at particular points in the pursuit of solutions to particular atmospheric problems. Several major investigations have utilized this technique, particularly in the area of the latitudinal variability of the mesosphere.

The need for a rocket system which had a utility equivalent to the balloon technique with an observational capability sufficient for description of the stratospheric circulation precipitated the development of smaller rocket and sensor systems. Experimentation during the late 1950's quickly demonstrated that the observational data could be enhanced, particularly with respect to sensitivity, through the use of very simple, inexpensive, and easily applied sensors. Chaff was the first of these new sensors, with different configurations employed in different altitude ranges to obtain wind measurements to altitudes as high as 80 km. While this technique was satisfactory for many problems, it had disadvantages in that the chaff represented a diffuse target for radar observation, which led to certain errors in observation. The magnitude of these errors increased with time after chaff deployment, making it difficult to acquire complete wind profiles throughout the stratosphere and mesosphere.

The need for measurement of at least one thermodynamic parameter forced the development of a parachute system which could carry a telemetry system. With no experience in the dynamics of parachutes in the thin air of the mesosphere, the first experiments in parachute deployment from a rocket constituted a bold step forward. Initial models proved highly effective when deployed from the Arcas rocket, with only minor adjustments required to achieve satisfactory reliability. Possibly this initial success was unfortunate, since our knowledge of the flight characteristics of our stratospheric observational platforms has advanced slowly in the absence of a strong motivation. In any case, it was soon apparent that the 15-ft diameter metalized silk parachute of the Arcas system did a satisfactory job of wind sensing as well as providing a suitable descent pattern for other observations from above 60-km altitude down to the peak altitudes reached by balloons. A major part of the stratospheric wind data which have thus far been obtained has been acquired by radar tracking of a descending parachute.

Measurement of temperature by contact methods in the lower mesosphere and the entire stratosphere was first accomplished on the Arcas parachute system as it performed the wind sensing function. The temperature was sensed by a small semiconductor sphere which was placed in the front of the transmitter assembly to insure that the air sampled by the bead was uncontaminated. Standard meteorological sounding equipments were used to sense, telemeter, and record upper stratospheric temperature information. A balloon system employed the same equipments and techniques for observation in the troposphere and lower stratosphere as the rocket system employed in the upper stratosphere. Numerous additional sensors have been experimented with, but none have evidenced the utility of the wind and temperature sensors mentioned above, so thus far these systems have provided most of the data which have been accumulated in the stratospheric and mesospheric exploration program.

Development of small rocket vehicles of the Loki and Arcas types and sensors which were suitable for routine use led directly to the formation of a Meteorological Rocket Network (MRN) (Webb *et al.*, 1961). The initial observation program called for workday firings at noontime for one month each season, with the first firings beginning in Ocotber, 1959. The MRN expanded steadily, and in May, 1961, routine year-around observations on each Monday, Wednesday, and Friday were begun at six stations (Webb *et al.*, 1962). The MRN has continued to grow until at this writing some 23 sites are equipped for stratospheric observation, and about a dozen are firing routinely. The result has been the accumulation of more than 4000 profiles of stratospheric structure during the first five years of MRN operation.

Data from the MRN have revealed a complex circulation structure in the stratosphere and mesosphere regions. A distinct monsoonal regime dominates the gross picture, with weaker easterly winds in the summer season of about four and one-half months replaced by strong westerlies during the long winter season. The center of activity in these hemispheric flows is at lower latitudes in the summer case and in midlatitudes in the winter case. The middle of the winter season is sometimes highlighted by gross disturbances in the zonal circulation, with a resulting occasional division of the circumpolar flow into two or more cells. Adjustment of the stratospheric and mesospheric structure in a given locale to these hemispheric changes causes rapid variations which can be described adequately only by an increased firing rate at the MRN station in that locale. Certain of these events are of such restricted geographical extent that the spatial distribution of MRN observation points is inadequate, and even an internationalization of the MRN will be hard pressed to provide adequate coverage. For a large part of the year, however, the current firing schedule of each Monday, Wednesday, and Friday at locations roughly 1000 km apart is a rather satisfactory, though not optimum, program for the MRN.

One surprising result of the MRN experimentation was the discovery of a large amount of detailed structure in the vertical profiles of all meteorological parameters which could be measured with adequate sensitivity. The wind profiles were not the smooth curves which one could reasonably expect in a flow this far removed from terrestrial obstructions. Temperature data also showed a marked variability, and this information was quickly followed by similar data from new sensitive ozone sensors. It was natural to assume that the atmosphere would become more quiescent in its upper reaches, although observation of the sun's atmosphere did not point in that direction, but it was probably the smoothed profiles obtained by early sensing systems which had the most telling influence in developing this misconception. The wind, temperature, and ozone profiles presented in Figs. 1.1 to 1.3 illustrate the magnitude of error in that assumption. One must remember that the sensors used to obtain these data have some response problems and the actual gradients are probably greater than indicated in the figures.

The ozone sounding presented in Fig. 1.3 is an example of the type of profiles which are to be observed in the central United States during the late winter season. These data provide evidence of advective transport of air across middle latitudes from ozone-rich high latitudes and the relatively "clean" upper tropospheric air of low latitudes. It is to be expected that synoptic analysis of this type of data will provide a new insight into the dynamics of exchange of air between the troposphere and lower stratosphere across the tropopause barrier.

Atmospheric ozone concentration is altitude dependent since the several reaction rates involved in establishing the concentration vary rapidly with height. Principal reactive activity occurs in the upper stratosphere, so that the ozone which is formed in or penetrates the lower stratosphere by diffusion or advection enjoys a protected existence, and is thus a very conservative quantity. It is valuable as a tracer material in this mixing zone between the stratosphere and troposphere.

Variations of the type exhibited in the ozone profile (Fig. 1.3) are characteristic of advective motions in the scale of synoptic pressure systems in the midlatitude troposphere. Currents of air from the high-latitude, ozone-rich, lower stratosphere are interchanged with upper tropospheric air from lower latitudes to develop layered strata of marked difference in ozone concentration in a fashion that would be very unlikely to develop through other processes. The scene of this large-scale mixing is the upper portion of the tropospheric circulation system in lower midlatitudes, which had been noted earlier to be controlled by physical processes which are characteristics of the troposphere.

Even more striking are the small-scale variations which appear in all of the records. The amount of this detail structure and the amplitude

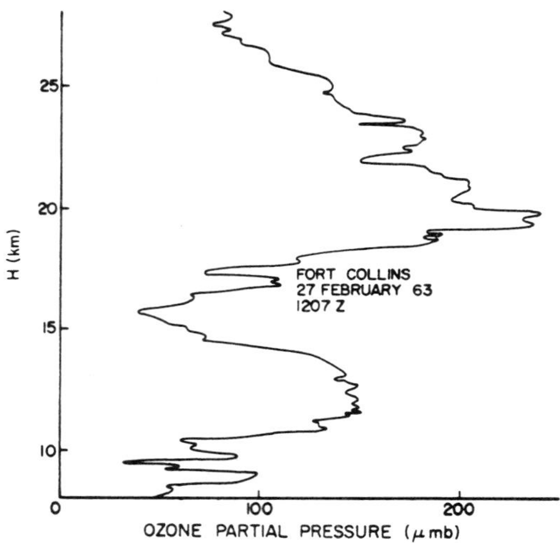

Fig. 1.3. An example of the type of vertical ozone structure observed with a balloon-borne Regener ozonesonde observation at a midlatitude station in late winter by the U. S. Air Force ozonesonde network.

are functions of time and space, but are invariably present and constitute a form of turbulent motion which was not expected in the stratosphere and remains a point of considerable speculation. In view of the great stability of the stratospheric region it is evident that the motions are forced, and the possibility exists that the motions which produce these complex profiles are themselves internal waves initiated by interaction between the flow and obstructions such as mountains, cumulus clouds, and turbulent eddies. Theories along these lines are controversial at this time, but the data are not. For the experimenters at least, initial disbelief of the new data was largely set aside by detailed inspection of the sensing systems. Parachute systems were found to execute motions which could only be the result of interactions with a variable flow environment, and the observed temperature data simply could not be ascribed to the various possible sources of error.

A most important supplement to direct parametric measurements in the upper atmosphere is the observation of natural phenomena such as clouds which provide information on the motions and physical state of the air at the level of formation as well as in adjacent regions. Excellent examples of such clouds are the noctilucent and nacreous clouds, which have been identified as occurring in the vicinity of the mesopause and stratonull, respectively. Noctilucent clouds are observed principally in the summer months at high latitudes, while nacreous clouds are a winter phenomenon of middle latitudes. It is possible through simple observation of these clouds to obtain information on similarities between hemispheres as well as to discern likely physical processes attending their formation which then provide guidance on the observational program that should be conducted by rocket systems.

Unless otherwise stated the upper atmosphere will be assumed to be in hydrostatic equilibrium. Thus the usual relationship

$$dp = -\rho g\, dh, \tag{1.1}$$

where p is pressure, ρ is density, g is acceleration of gravity, and h is altitude, will be employed to characterize the vertical structure, giving an exponential form to the pressure and density distributions such that

$$P = P_0 \exp\left(-\frac{mgh}{KT}\right), \qquad \rho = \rho_0 \exp\left(-\frac{mgh}{KT}\right), \tag{1.2}$$

where P_0, ρ_0 are the pressure and density at the base of the layer, m is average particle mass, K is Boltzmann constant, and T is temperature, which are dependent on the temperature and average particle mass distribution. Since eddy mixing involves accelerations on atmospheric elements smaller than the dimensions of the eddies, it is likely that the hydrostatic relationship is subject to some error, particularly in the low-density and high-wind-speed environment of portions of the upper atmosphere. These assumptions also imply effectiveness of diffusive separation of the various atmospheric gases, which is generally known not to occur at low levels. It should be remembered, therefore, that the applicability of such assumptions may be limited and that these assumptions may be invalid in special cases even in the troposphere, such as in the vicinity of tornadoes or thunderstorms.

In addition to the above assumptions relating to dynamic parameters, the observed winds will be assumed to be a good approximation to the geostrophic wind as expressed for our purpose by

$$f\mathbf{v}_g \times k = g\nabla_p h, \tag{1.3}$$

where $f = 2|\boldsymbol{\omega}| \sin \varphi$ is the Coriolis parameter, ω is earth's angular velocity, φ is latitude, k is the vertical unit vector, $\mathbf{v}_g$ is the geostrophic wind on a constant-pressure surface, and $\nabla_p h$ is the pressure gradient ($ih_x + jh_y$); that is, frictional forces are assumed to be negligible and the pressure gradient and coriolis "forces" are in equilibrium. Since the kinematic viscosity (η) is related to the temperature and mean free path of the fluid by the equation

$$\eta = \frac{\beta T^{3/2}}{\rho(T + S)}, \tag{1.4}$$

where $\beta = 1.458 \times 10^{-6}$, S is Sutherland's constant, and ρ is density, it is clear that this parameter will increase markedly with height as the density decreases. Frictional forces will therefore increase with altitude if eddy mixing occurs, so that a given size eddy will be far more efficient in dissipating the kinetic energy of the general upper atmospheric circulation into heat than will that eddy in the troposphere.

It would be premature to attempt a comprehensive survey of stratospheric and mesospheric physical processes at this time. While the data obtained by the MRN may be of sufficient quality, they are inadequate in time and space to permit detailed analysis of many of the phenomena which they reveal. Our purpose here, then, will be to formulate a structure on which a complete picture may be built as the required data are acquired. It is possible that the limited data will result in misinterpretations, but even these may provide the desired impetus to achieve an adequate understanding of physical processes in the upper atmosphere.

For clarity our discussion is divided into four principal subject areas. The first (Chapter 2) is a brief review of the systems which have produced the data for our analysis. Since the MRN is the principal data source for our consideration, the tools for that effort are emphasized, with only passing mention of associated instrumentation systems. Subsequent chapters deal with the layered regions which we have defined as the lower stratosphere, the upper stratosphere, and the mesosphere, in that order. Each layer has been treated independently according to the priority of interest and available data, and certain subject areas have been omitted because of their more comprehensive treatment elsewhere.

Balloon systems have been used in the exploration of the lower stratosphere and because of their efficiency it is probable that this mode of observation will dominate that region in the future. The upper stratosphere is the domain of the meteorological rocket, and, in general, investigators in this area are a separate breed, although in the majority of the cases they derived their training from balloon observation systems. While meteorological rocket systems are currently hard pressed to sample the mesosphere adequately, it is clear that they must be extended so that synoptic coverage of the complete stratospheric circulation can be achieved. Data obtained by more sophisticated systems form the heart of Chapter 5, however, and the investigators who perform these experiments are a third group of investigators of different heritage. Because the reference material will be most useful to these scientists in their separate corners, a bibliography is incorporated in each chapter. In some cases there is an overlap which requires a particular listing in more than one chapter, but in general the listings are made only once, in that chapter to which the information is unique.

For instance, references on the rocket grenade experimental results are found mostly in the Chapter 5 bibliography, MRN data references are found in Chapter 4, and ozone reports are largely located in Chapter 3. In each case there is material dealing with other chapters, but it is located in the subject area of its greatest utility. A good general rule is that a reference will follow the chapter concerning the highest layer to which it is applicable. It is, of course, impossible to include all publications of interest, but an attempt has been made to provide reference to an adequate body of material for initial efforts of a serious upper atmospheric investigator.

REFERENCES

Aleskseev, P. P. (1957). Raketnye issledovaniia atmosfery. *Meteorol. i Gidrol.* **8**, 3–13.

Arkhangel'skii, V. N., and E. I. Sukhotskii. (1958). Novaia standartnaia atmosfera. *Meteorol. i Gidrol.* **5**, 55–58.

Bates, D. R. (1957). Composition and structure of the atmosphere. "The Earth and Its Atmosphere," pp. 97–112. Basic Books, New York.

Bellamy, J. E. (1961). Requirements for high-altitude meteorology. *Navigation* **8**, 1965–1975.

Benton, M. (1958). Literature of space science and exploration. U. S. Naval Research Laboratory, Washington, D. C., Bibliography, Report No. 13.

Berkner, L. V. (1958). Manual on rockets and satellites. *Ann. Intern. Geophys. Yr.* **6**, 508 pp.

Boyd, R. L. F., and M. J. Seaton, eds. (1954). "Rocket Exploration of the Upper Atmosphere," 307 pp. Pergamon Press, Oxford.

Chapman, S. (1950). Upper atmospheric nomenclature. *J. Atmospheric Terrest. Phys.* **1**, 121.

Cole, A. E., A. Court, and A. J. Kantor. (1962). Supplemental atmospheres. U. S. Air Force, Cambridge Research Laboratories, L. G. Hanscom Field, Massachusetts, Research Note 62-899.

Danielsen, E. F. (1964). Radioactivity transport from stratosphere to troposphere. *Mineral Ind. J.* **33**, No. 6, 1–7.

Faust, H. (1960a). Raketen, Satelliten und Meteorologie. *Meteorol. Rundschau* **13**, 130–134.

Faust, H. (1960b). Raketenmeteorologie. *Naturw. Rundschau* **13**, 95–99.

Faust, H. (1961a). Das nordamerikanische Forschungsraketennetz. *Meteorol. Rundschau* **14**, 126.

Faust, H. (1961b). The significance of European and world-wide rocket research network. *Weltraumfahrt* **12**, 84–89.

Faust, H., and W. Attmannspacher. (1959). Cell structure of the atmosphere. German Weather Service, Final Contract Report DA-91-508-EUC-387, AD 231 502, and AD 231 503.

Godson, W. L. (1963). Meteorological rocket plans-recommendations for the IQSY. *In* "First International Symposium on Rocket and Satellite Meteorology" (H. Wexler and J. E. Caskey, Jr., eds.), pp. 173–186. Wiley, New York.

Goody, R. M. (1954). "The Physics of the Stratosphere," 187 pp. Cambridge Univ. Press, London and New York.

Gossard, E., and W. Munk. (1954). On gravity waves in the atmosphere. *J. Meteorol.* **11**, 259–269.

Greenfield, S. M. (1956). Synoptic weather observations from extreme altitudes. Rand Corp., Santa Monica, California, P-761, 12 pp.

Hare, F. K. (1962). The stratosphere. *Geograph. Rev.* **52**, 525–547.

Harris, I., and R. Jastrow. (1959). An interim atmosphere derived from rocket and satellite data. *Planetary Space Sci.* **1**, 20–26.

Heninger, S. K., Jr. (1960). "A Handbook of Renaissance Meteorology," 254 pp. Duke Univ. Press, Durham, North Carolina.

Izakov, M. N., and G. A. Kokin. (1957). Izmerenie temperatury, davleniia i plotnosti atmosfery S Pomoshch'iu raket. U.S.S.R. Glavnoe Upravlenie Gidrometeorologicheskoi Sluzhby. Komitet po Provedeniia Mezhdunarodnogo Geofizicheskogo Goda. *Inform. Sbornik.* **4**, 57–61.

Kallmann, H. F. (1959). A preliminary model atmosphere based on rocket and satellite data. *J. Geophys. Res.* **64**, 615–623.

Kellogg, W. W. (1958). IGY rockets and satellites: A report on Moscow meetings, August, 1958. *Planetary Space Sci.* **1**, 71–84.

Kellogg, W. W. (1959). Review of IGY upper air results. Rand Corp., Santa Monica, California, P-1717, 21 pp.

Kellogg, W. W. (1964). Pollution of the upper atmosphere by rockets. Rand Corp., Santa Monica, California, RM-3961-PR, 86 pp.

Kellogg, W. W., and G. F. Schilling. (1951). A proposed model of the circulation in the upper stratosphere. *J. Meteorol.* **8**, 222–230.

Khrgian, A. K. (1958). Fizika atmosfery. Gos. Izdat. Fiziko-Matematicheskoi Literatury, Moscow, 472 pp.

Kiss, E. (1960). Annotated bibliography on rocket meteorology. *Meteorol. Geoastrophys. Abst.* **11**, 1480–1535.

Kiss, E. (1961). Annotated bibliography on upper atmosphere structure. *Meteorol. & Geoastrophys. Abstr.* **12**, 776–826.

Kuiper, G. P., ed. (1954). "The Earth as a Planet," 751 pp. Univ. of Chicago Press, Chicago, Illinois.

Leak, W. M. (1954). Rocket research at NRL. U. S. Naval Research Laboratory, Washington, D. C., Report No. 4441.

Lejay, P. (1954). Méthodes modernes de recherches sur la haute atmosphère. Conference at the Palais de la Decouverte, Paris.

Lequeux, J. (1958). Les études geophysiques au moyen des fusées et des satellites. *Nature (Paris)* **3273,** 24–29.

Link, F., and L. Neuzil. (1957). Raketové lety a výzkum vysoké atmosfeŕy. Prague, Ceskoslovenská Akademie Ved., 235 pp.

Maeda, K., and K. Hiaro. (1962). A review of upper atmosphere rocket research in Japan. *Planetary Space Sci.* **9,** 355–369.

Massey, H. S. W. (1958). Rockets and satellites in scientific research. *Endeavour* **17,** 85–89.

Massey, H. S. W. (1959). Scientific applications of rockets and satellites. *In* "Symposium on High Altitude and Satellite Rockets, Cranfield, England," pp. 7–14. New York Philosophical Library, New York.

Massey, H. S. W., and R. L. F. Boyd. (1958). "The Upper Atmosphere," 333 pp. Hutchinson, London.

Mitra, S. K. (1952). "The Upper Atmosphere," 2nd ed., 713 pp. Asiatic Society, Calcutta.

Newell, H. E., Jr. (1950). A review of upper atmosphere research from rockets. *Trans. Am. Geophys. Union* **31,** 25–34.

Newell, H. E., Jr. (1953). "High Altitude Rocket Research," 298 pp. Academic Press, New York.

Newell, H. E., Jr. (1954). Rockets and the upper atmosphere. *Sci. Monthly* **78,** 30–36.

Newell, H. E., Jr., ed. (1959). "Sounding Rockets," 334 pp. McGraw-Hill, New York.

Petrov, B. M. (1959). U. S. S. R. rocket and earth satellite programme for the IGY. *In* "Symposium on High Altitude and Satellite Rockets, Cranfield, England," pp. 56–57. New York Philosophical Library, New York.

Poloskov, S. M., and B. A. Mirtov. (1957). The study of the upper atmosphere by means of rockets, at the Academy of Sciences, U. S. S. R. *J. Brit. Interplanet. Soc.* **16,** 95–100.

Ratcliffe, J. A., ed. (1960). "Physics of the Upper Atmosphere," 586 pp. Academic Press, New York.

Rocket Panel, Harvard College Observatory. (1952). Pressures, densities, and temperatures in the upper atmosphere. *Phys. Rev.* **88,** 1027–1032.

Scherhag, R. (1952). Die Explosionsartigen Stratospharenerwarmungen des Spatwinters 1951/52. *Ber. Deut. Wetterdienstes U. S. Zone* **38,** 51–63.

Singer, S. F. (1955). Research in the upper atmosphere with high altitude sounding rockets. *Vistas Astron.* **2,** 878–912.

Smith, C. P., Jr. (1954). Summary of upper atmosphere rocket research firings. U. S. Naval Research Laboratory, Washington, D. C., Report 4276.

Smith, W., L. Katchen, P. Sacher, P. Swartz, and J. Theon. (1964). Temperature, pressure, density and wind measurements with the rocket grenade experiment, 1960–1963. *NASA (Natl. Aeron. Space Admin.)* Tech. Rept. **TR-R-211.**

Stroud, W. G. (1958). Meteorological rocket soundings in the Arctic. *Trans. Am. Geophys. Union* **39,** 789–794.

Stuhlinger, E. (1959). Raketen und Satelliten im Dienste der Geophysik. *Naturwissenschaften* **46,** 303–309.

Surtees, W. J. (1964). An outline of some characteristics of the upper atmosphere. Canadian Armament Research and Development Establishment, Valcartier, Quebec, Report T.R. 470/64, 124 pp.

Sutton, O. G. (1960). High atmosphere research in the Meteorological Office. *Meteorol. Mag.* **89,** 97–98.

Teisserenc de Bort, L. P. (1902). Variations de la temperature de l'air libre dans la zone entre 8-km et 13-km d'altitude. *Compt. Rend.* **134,** 987–989.

Thompson, P. D. (1949). The propagation of small surface disturbances through rotational flow. *Ann. N. Y. Acad. Sci.* **51,** 463–475.

Townsend, J. W., Jr., H. Friedman, and R. Tousey. (1957). History of the upper-air-rocket research program at the Naval Research Laboratory. U. S. Naval Research Laboratory, Washington, D. C., Report No. 5087, 44pp.

Vassy, A., and E. Vassy. (1954). Absorption of solar radiation by the atmosphere. U. S. Air Force, Cambridge Research Center, Geophysical Research Papers No. 30, pp. 427–446.

Vassy, E. (1954a). La haute atmosphére. *Meteorologie* [4] **35,** 185–194.

Vassy, E. (1954b). Notes on French Ministry of Supply rocket research programme. *In* "Rocket Exploration of the Upper Atmosphere" (R. L. F. Boyd and M. J. Seaton, eds.), pp. 212–213. Pergamon Press, Oxford.

Webb, W. L., W. E. Hubert, R. Miller, and J. F. Spurling. (1960a). Initiation of the Meteorological Rocket Network. Inter-Range Instrumentation Group, Meteorological Working Group, Meteorological Rocket Network Committee, White Sands Missile Range, New Mexico, Document No. 105–60, AD 246 950, 240 pp.

Webb, W. L., W. E. Hubert, R. Miller, and J. F. Spurling. (1960b). Initiation of the Meteorological Rocket Network. *Bull. Am. Meteorol. Soc.* **41,** 261–268.

Webb, W. L., W. E. Hubert, R. L. Miller, and J. F. Spurling. (1961). The first Meteorological Rocket Network. *Bull. Am. Meteorol. Soc.* **42,** 482–494.

Webb, W. L., W. I. Christensen, E. P. Varner, and J. F. Spurling. (1962). Inter-Range Instrumentation Group participation in the Meteorological Rocket Network. *Bull. Am. Meteorol. Soc.* **43,** 640–649.

Webb, W. L., W. I. Christensen, D. W. Hansen, H. B. Tolefson, W. O. Banks, and H. R. Rippy. (1965). The Meteorological Rocket Network. Inter-Range Instrumentation Group, Meteorological Working Group, Meteorological Rocket Network Committee, White Sands Missile Range, New Mexico, Document No. 111–64, 324 pp.

Wilckens, F. (1962). Bibliographe über Raketenmesserfahren in der Stratosphare und Mesophare. Flugwissenschaftliche Forschungsanstalt, Deutsche Luftwaffen Wetter Nachrichten Dienst, Munchen, Bericht N. 11, 56 pp.

Wyckoff, P. H. (1958). Rocket as a research vehicle. *Am. Geophys. Union, Geophys. Monograph* **2,** 102–107.

2

Data Acquisition Systems

Introduction

The stratosphere and mesosphere were largely inaccessible to man's probing inspection until the turn of the century. In part this resulted from a lack of interest due to a seeming remoteness. The data deficiency was masked by assumptions of uniformity which assigned the upper regions of the atmosphere to a negligible role in the total system. Small densities and energy storage capacities were cited as proofs that the atmospheric scientist need not bother with these uninteresting atmospheric layers. Development of radio techniques and discovery of the ionosphere precipitated intensive study of that region through use of ground-based experimental equipments which effectively bypassed the stratosphere and mesosphere. Recurring suggestions that many phenomena observed in the ionosphere might well be closely related to physical processes in the lower atmosphere were given little attention. The resulting division and separation of ionospheric scientists from the general meteorological study of the atmosphere have proved detrimental to all concerned. A survey of modern literature indicates that this yen for division has not abated, but predictably occurs again with the development of each new technique for atmospheric analysis.

Meteorologists must bear the brunt of the fault since the basic error is failure to seize a new observational tool and exploit it for maximum gain in knowledge concerning the atmosphere. In most cases of record a user technology, exemplified by the communications industry in the case of ionospheric exploration, supports the development of experimental techniques and in so doing founds a new scientific discipline. After unified application of the data becomes quite apparent some attempts at reconciliation are made, usually with relatively little success. The pursuit of such an approach to scientific study in our sophisticated society is lamentable. Correction of the error can only be achieved by a broadening of the viewpoint of atmospheric scientists to permit assimilation of all methods of atmospheric analysis. This broadening of concepts must necessarily start with instrumentation techniques if the atmosphere is to be studied in a unified manner.

Considerable literature exists relative to indirect techniques for sampling of the stratosphere and mesosphere. These systems principally concern the use of sound waves to probe the stratosphere and radio waves to probe the mesosphere. While these techniques have produced notable advances in our knowledge of these regions, they suffer from a common defect in that they produce data of little sensitivity, and thus are limited to gross observation of the upper atmosphere. An important result of the application of these systems has been the development of concepts of smooth stratospheric and mesospheric structure. Man's active invasion of the upper atmosphere in the rocket age has placed emphasis on the more detailed aspects of upper atmospheric structure, and exploration of the detail features has brought the realization that small-scale phenomena exercise an important influence on the dynamics of that region.

Detailed exploration of the stratosphere and mesosphere awaited the development of suitable vehicles for the transport of instrumentation into this hitherto relatively inaccessible region. Such mobile observation platforms made their appearance in the form of balloons and aircraft which could penetrate into the lower stratosphere and rockets which offered access to all of the 15- to 85-km altitude region of interest. The wide range of environmental conditions which characterize the stratosphere and mesosphere precludes the application of any one measuring system over the entire range. Our initial observational knowledge came from a variety of instrumentation systems applied by various investigators at numerous locations. Individual experimental requirements generally dictated the observational plan, and a comprehensive atmospheric analysis received only scant attention. Again, much was learned about specific items, but progress in the unified exploration of the atmosphere was slow. One thing is clear from an over-all look at the data obtained. There remains today a steady and pronounced decrease in the amount of available data with increasing height.

Scarcity of upper stratospheric and mesospheric data may be attributed principally to economic considerations. A vertical balloon sounding of temperature and wind in the lower stratosphere may be obtained for an order-of-magnitude expenditure of 100 dollars, while minimum sampling to mesospheric altitudes requires an order of magnitude greater economic outlay. Possibly of greater impact than monetary considerations has been the relative lack of interest displayed by the scientific community concerning this intriguing portion of the earth's atmosphere. It is significant that the capabilities of available low-cost probing systems such as balloon soundings were not extended to their maximum until the data were required for flight purposes. It would appear obvious that expansion of man's knowledge of the atmospheric environment should precede incorporation of that environment into his operations sphere. In most cases the impetus toward expansion of our observational systems has come from the requiring projects, and thus meteorological knowledge has lagged the actual need for new scientific information on physical processes in the upper atmosphere.

Availability of balloons, aircraft, and rockets which can transport instruments into the upper atmosphere is only one aspect of a complex observation problem. Suitable sensing techniques must be available and an efficient system for telemetering acquired data must be employed. Initial data on stratospheric structure were obtained through *in situ* recording of the measurements and recovery of the flight equipment by parachute after balloon burst. Many uncertainties attend the operation of such an observational system. Radio techniques provided the necessary avenue for a significant intensification of stratospheric probings with balloon carriers during the early 1900's. Radio telemetry frequencies which have been applied to the atmospheric observational problem center at 72.2, 403, and 1680 Mc. A steady improvement in balloon characteristics and sensor techniques as well as telemetry and tracking systems resulted in an accumulation of significant amounts of lower stratospheric data. Thus, only the most modern systems, which can be expected to provide our principal data source in the future, will be discussed here.

A typical ground station for reception and processing of meteorological data from stratospheric and mesospheric observation systems is illustrated in Figs. 2.1 and 2.2. This is the basic Model 4000R radiotheodolite receptor and direction finder which operates at a carrier frequency of 1660 to 1700 Mc and is employed by the United States Weather Bureau at upper air sounding stations. A 403-Mc transmitter antenna (Fig. 2.1) is mounted on the direction finder scanner assembly to provide a communications circuit for measurement of slant range between the ground station and the flight instrument. Meteorological data obtained by instruments on the flight units are telemetered to the ground station over the 1680-Mc data link, and slant range is determined by phase comparison of the outgoing 403-Mc carrier signal as it returned to the ground station modulation of the 1680 carrier frequency. The ground system illustrated here is the result of over 20 years of development, starting with the AN/GMD-1 of the World War II days. The military version of this Weather Bureau system is the AN/GMD-2, which is quite similar in most operating characteristics, but is significantly different in physical configuration. The 4000R represents an engineering

Fig. 2.1. Front view of the U. S. Weather Bureau Model 4000R radiotheodolite receptor and direction finder used at upper air sounding stations. The dish and central scanner system operate at 1660–1700 Mc in a receiving mode and the superimposed antenna system transmits a 403-Mc ranging signal. (*Courtesy U. S. Weather Bureau*)

optimization of the GMD systems to the Weather Bureau's particular upper air observation problems.

The above ground system was developed specifically to provide for reception of meteorological data from balloon systems which sample near-vertical profiles from the surface to nominal balloon burst altitudes of 30 km. Most of the data thus far available have been acquired in the simpler mode of operation, using only the 1680 channel of telemetry and employing hydrostatic relationships to evaluate altitude and wind displacements as the balloon ascends. This same mode of operation, supplemented by radar height measurements, has been successfully applied to meteorological rocket observation systems to extend the available observational sphere to an altitude of 70 km. An objective of the meteorological rocket system development includes the use of a single telemetry and tracking system to obtain complete vertical profiles

Fig. 2.2. View of the instrumentation panel and pedestal of the U. S. Weather Bureau 4000R receptor system. (Courtesy U. S. Weather Bureau)

from a parachute-borne instrument deployed from a rocket vehicle. The Model 4000R radiotheodolite (and the AN/GMD-2) thus provides a comprehensive ground system which, with a compatible flight unit, can accomplish all of the tracking and telemetry operations required for a complete vertical profile observation of the free atmosphere. The flight systems that are currently in use are limited in their application, particularly in the effective altitude range over which they can be used. They will, therefore, be treated separately in the following pages.

2.1 Balloon Soundings

The sampling configuration employed for balloon probing of the stratosphere is illustrated in Fig. 2.3. Trains usually consist of a 20-meter separation of the balloon and parachute and an additional 20-meter separation between the recovery parachute and instrumentation. On release, the balloon is approximately 2 meters in diameter and has a

Fig. 2.3. A balloon-borne radiosounding system ready for release. The 4000R receptor is roof mounted and weather shielded above the balloon inflation shelter and the data recording and processing areas are adjacent. (Courtesy U. S. Weather Bureau)

free lift of 1200 gm in addition to its own weight of 1.2 kg. A nominal burst size of 10 meters occurs at an altitude of approximately 30 km, and variations in balloon size and air density serve to maintain relatively constant drag and produce a uniform ascent rate. A vertical speed of 400 meters/min is common, and thus total observation times of the order of one hour are the rule. A standard Weather Bureau upper atmosphere observation station provides for inflation and release of the balloon as well as optimizing the observational point of the radiotheodolite in a plastic shelter on the roof. Calibration, recording, and analysis equipment are housed in the building so that a two-man crew can perform an upper atmospheric observation and transmit the data in a two-hour period.

A sounding balloon has a nearly vertical velocity relative to its environment as a result of the lift it possesses. It also has a lateral velocity relative to the ground due to entrainment in the windstream. In general, the balloon responds quite rapidly when it enters a layer of different wind velocity, so the flow about the balloon and its train is largely vertical except for pendulum-type motions which the system may, and generally does execute. Any modification of the environment resulting from the passage of the balloon will be distributed in a nearly vertical column and thus may contaminate the sample. Actual observation of the system in flight indicates that pendulum-type oscillations induced at launch and by environmental wind shears cause the sensors at the lower end of the train to spend most of their sampling time well away from the vertical centerline of the balloon's trajectory. While some modification of the atmospheric environment must be assumed to occur as the balloon ascends, the over-all sampling technique appears to yield representative data, at least in the troposphere. These points are less well understood in the lower stratospheric environment, in part owing to our limited knowledge of the region, and it appears likely that the probability of significant contamination of the sample would increase with increasing altitude.

Descent of the instrument package after balloon bursts is held in safe limits by a parachute which has an inflated diameter of 1 meter, reducing the fall rate of the system to a value of about three times the balloon's ascent rate. Except in special circumstances, the flight units are considered expendable, and thus recovery of the instruments is not a major issue. The uncontrolled drift of these sampling systems over distances of the order of hundreds of miles from the launch sites constitutes one of the more pressing problems of the current upper atmospheric observations systems. This, along with the need for more timely and adequately distributed data, will maintain incentive to perfect observational techniques which will minimize these negative factors. Until the problems are solved, value of the data is expected to outweigh the disadvantages of current systems so that the observational program will continue.

During certain periods of the year the balloon sounding system may be borne away from the point of launch as it ascends, sometimes with considerable speed. Under these conditions the stratospheric portion of the observation is made at distances of the order of 50 to 100 km from the ground equipment. The resulting low-elevation-angle observations, particularly when used with computed altitudes, provide wind measurements of low quality. Application of the ranging system in the Model 4000R radiotheodolite and the AN/GMD-2 has done much to

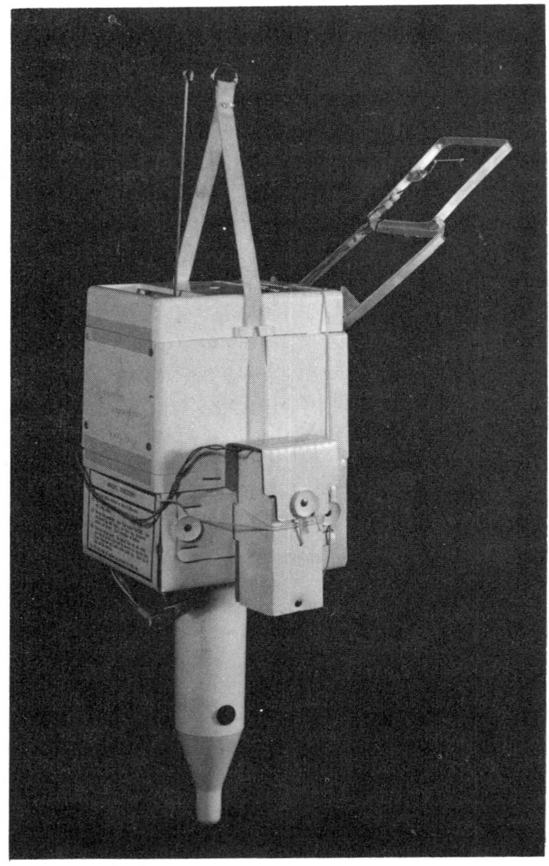

Fig. 2.4. A balloon-borne flight unit for use with the Model 4000R equipment. The 1660–1700-Mc transmitter antenna system and associated electronics are housed in the plastic bottle underneath and the 403-Mc receiver antenna is the hooked wire above. Water-activated batteries are enclosed, as is a ducted humidity sensor. The rod thermistor temperature element is carried in a bracket mount and a hypsometric pressure-measuring device is strapped on the side. (*Courtesy U. S. Weather Bureau*)

reduce this source of error, although the considerable ranges over which these measurements must be made present formidable obstacles to the acquisition of accurate data. A most important contribution to this highly disadvantageous observational geometry is the rather long observational period which is currently required. Attempts to speed up the ascent rate of the balloon have proved fruitful, but resulting restrictions on other desirable attributes such as a high peak altitude have limited acceptance of these balloon systems. This long period of observation and the rather fluid remote site of high-level observation constitute real problems in applicability of the data to atmospheric problems.

An illustration of the flight instrumentation used in balloon soundings is presented in Fig. 2.4. A 1680-Mc transmitter with associated electronics and batteries is housed in the central package. Also included are an aneroid pressure cell for measurement in the troposphere, a hypsometric pressure sensor for use in the stratosphere, and a humidity sensor which varies in electrical resistance with moisture content of the sensing surface. In addition, an exposed rod thermistor (2 mm in diameter and 3.8 cm in length) with a coating is used to sense the temperature of the ambient air as the system rises.

Pressure measurements are made in the lower stratosphere (100 to 10 mb) with a quoted accuracy of 1 mb below the 20-mb level, 0.5 mb up to the 10-mb level, and 0.1 mb at high altitudes. Temperature profile measurements with the rod thermistor result in smoothed data in the lower stratosphere owing to the comparatively large heat capacity of the thermistor and the rapid ascent of the balloon system.

These considerations point toward the 4000R system's being an extremely efficient tropospheric measuring system, with somewhat less desirable characteristics for measurement in the lower stratosphere. Economical considerations are also quite favorable for the balloon-borne vertical sounding system, and clearly it is the first choice in an observational system when the required measurements are within its capability.

Since our principal interest is in observation systems which are capable of producing data throughout the stratosphere and mesosphere, it is essential that more adequate techniques be employed. Such systems necessitated the application of new concepts, and the advent of rocket techniques provided at least an interim solution to the problem. To this date the balloon and rocket systems are complementary, and both must be applied to get atmospheric profiles from the surface to the mesopause. Techniques which are in evolution have the characteristics which can be expected to insure the capability of sampling the entire atmosphere in a far more efficient manner than is now possible.

2.2 Rocket Systems

Measurement of atmospheric parameters was one of the tasks to which rocket vehicles were applied during the later 1940's. In most cases the meteorological observations were a secondary consideration to the basic purpose of the rocket firing, and inevitably the conditions for observation were less than satisfactory. These early investigations contributed much to our knowledge of the upper atmosphere, but they failed to discern the wealth of detail structure which has proved to be one of the most important characteristics of the stratosphere and mesosphere. Shortcomings of the early investigations centered largely on the attention to on-board observation systems on vehicles, which traversed the region of interest at rather high speed, and on unit costs which precluded the acquisition of adequate quantities of data for proper meteorological analysis. Several of the more basic difficulties in rocket observa-

Fig. 2.5. A Nike Cajun sounding rocket system being prepared for launch at White Sands Missile Range, New Mexico. A Nike Ajax booster provides initial thrust and, after a 15-second coast phase, the Cajun rocket carries the payload to peak altitude. (Courtesy U. S. Army)

tion of the stratosphere and mesosphere were negated by the adoption of progressively smaller vehicles and the application of deployed sensors which carry out the measuring program under conditions which are most favorable for accurate observation.

The Nike Cajun and Nike Asp series of sounding rocket systems provide for deployment of 22- to 220-kg payloads over an altitude range of 180 to 200 km. An example of one of these systems and its launcher complex is illustrated in Fig. 2.5. A 15-sec time delay between stages is employed in this research rocket system to achieve maximum altitudes. The system has a gross weight of approximately 600 kg at lift-off, and thus requires considerable ground support. Typical trajectory data for such a firing at White Sands Missile Range are presented in Fig. 2.6.

These vehicles exhibit a wind sensitivity which requires a large impact area for safe return of the flight units. The deviation which such a system experiences results from the weather-cocking of the rocket as it enters a new wind environment, as well as from other deviating influences. When the rocket is under thrust it picks up momentum along the new trajectory after any deviation, and thus significantly alters the point at which it impacts. These effects are generally compensated by measuring the wind field, estimating the trajectory deviations, and incorporating

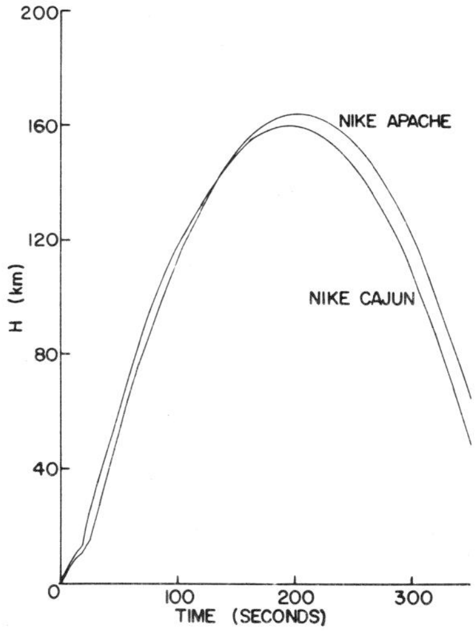

Fig. 2.6. Nike Cajun and Nike Apache trajectories for a firing at White Sands Missile Range.

corrective angles in the launcher to cause impact at the desired point. These factors are assimilated in the following equations (Walter, 1962), which can be used to calculate the trajectory from observed data to arrive at the proper launching settings:

$$\ddot{x} = A_a\xi_a + A_r\xi_r + B_t\xi_r + C_w x + C_x, \tag{2.1}$$

$$\ddot{y} = A_a\eta_a + A_r\eta_r + B_t\eta_r + C_w y + C_y, \tag{2.2}$$

$$\ddot{z} = A_a\zeta_a + A_r\zeta_r + B_t\zeta_r + C_w z + C_z, \tag{2.3}$$

$$\ddot{\theta}' = A_m(-\xi_a \sin\theta' + \eta_a \cos\theta' \cos\alpha' + \zeta_a \cos\theta' \sin\alpha') + B_j\dot{\theta}'(\dot{\alpha}' \cos\theta' + C_n\dot{\beta})\dot{\alpha}' \sin\theta', \tag{2.4}$$

$$\ddot{\alpha}' = A_m(-\eta_a \sin\alpha' + \zeta_a \cos\alpha')/\sin\theta' + B_j\dot{\alpha}' - (2\dot{\alpha}' \cos\theta' + C_n\dot{\beta})\dot{\theta}'/\sin\theta', \tag{2.5}$$

$$\ddot{\beta} = A_f + B_r\dot{\beta} + B_s, \tag{2.6}$$

where x, y, z are position coordinates of the rocket; A_a, A_r, A_m, A_f are aerodynamic coefficients; ξ_a, η_a, ζ_a are direction cosines of aerodynamic velocity; ξ_r, η_r, ζ_r are direction cosines of the rocket axis; B_t, B_j, B_r, B_s are thrust coefficients; C_w, C_x, C_y, C_z, C_n are trajectory coefficients; θ', α' are modified attitude angles; and $\dot{\beta}$ is the roll rate of the rocket.

The corrected launcher angles derived in this way can be only as accurate as the input data. Clearly there will be errors, and their magnitude will vary with the meteorological situation. A given wind speed will have maximum effect as the rocket leaves the launcher where its speed is least. On the other hand, the wind speed may increase with height and result in large deviations at considerable altitudes. Continuously indicating wind systems are usually employed on towers in the launch area for low-level observations, as well as special balloon probes, but there will invariably be time-variability-induced errors in addition to the usual observational errors. As a result, there will be instances in which the rocket system may not be used because of safety considerations. It is also obvious that a considerable expenditure of effort is required to perform the above functions. At this time the use of such systems is required if data are to be obtained to the mesopause and above.

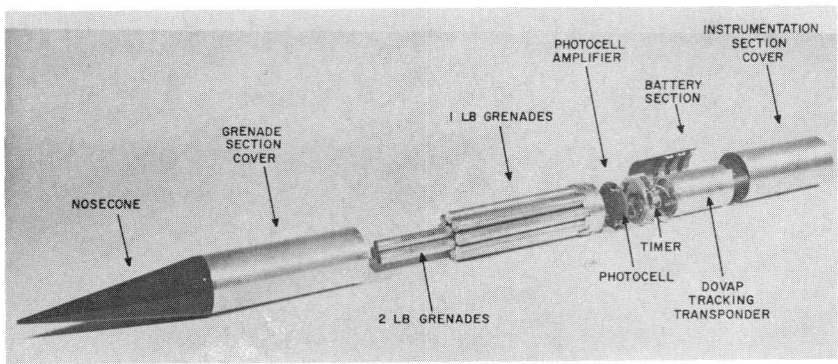

Fig. 2.7. A grenade payload used in the Nike Cajun for upper atmospheric temperature and wind observations. Individual grenades are ejected at intervals and detonated by a lanyard system at a distance of 15 feet from the rocket. (*Courtesy National Aeronautics and Space Administration*)

For our purposes, the most useful experiments of the many which have been conducted in this class are the grenade and falling sphere techniques. Detonation of explosive charges of from 1 to 4 lb at intervals along the trajectory permits the evaluation of mean temperatures and winds between explosions by observation and measurement of the resulting pressure perturbations at the surface. An example of the flight hardware employed in the grenade experiment is illustrated in Fig. 2.7. Ground-based instrumentation provides data on the location and time of initiation of each event as well as the time and angles of arrival of the expanding pressure perturbations. Mean data of considerable accuracy ($\pm 10°C$ and ± 5 meters/sec) can be obtained over altitude intervals of 5 to 10 km with this system at mesospheric altitudes.

Density of the mesospheric region has been probed by a falling

Fig. 2.8. A falling sphere used to measure mesospheric density through atmospheric drag measurements. The sphere is carried to altitude by a Nike Cajun type of rocket system and separated for the measuring phase. (Courtesy U. S. Army)

sphere technique (Jones and Bartman, 1956), which is illustrated in Fig. 2.8. An omnidirectional accelerometer which is centrally mounted in the sphere is used to observe the drag of the environment on the system as it falls after separation from the rocket carrier. With certain information relative to aerodynamic drag on the sphere, the density of the ambient air can be calculated. Mesospheric density data are obtained to an accuracy of 5% with this system.

As is illustrated in Fig. 2.8, all components of the transmitter and associated electronics, power supplies, and mechanical parts of the accelerometer are enclosed in the 6-inch diameter sphere. Slot antennas are used to radiate the telemetry signals, and at the same time maintain the desired spherical shape of the system. In practice, the sphere assembly is ejected from the rocket carrier on ascent soon after 50-km altitude is attained. This provides for comparison of the data on the up and down legs of the sphere's trajectory. Weight of the sensing system is 4.2 kg.

Knowledge acquired by the grenade, falling sphere, and other techniques during the IGY strongly indicated that there was much to be

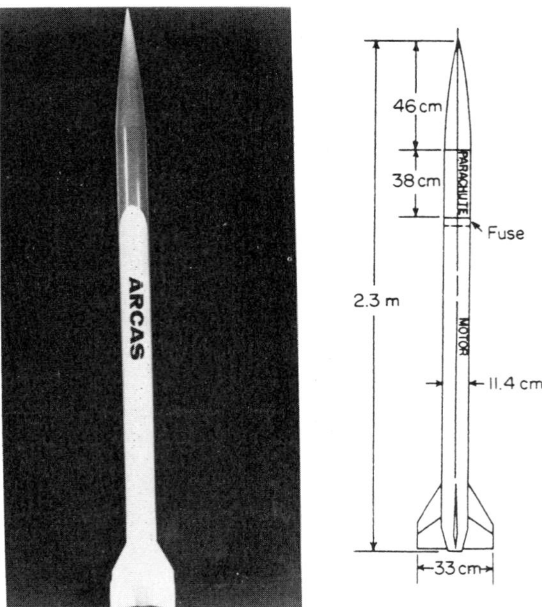

Fig. 2.9. The Arcas sounding rocket system used for meteorological research studies in the upper atmosphere and as a principal vehicle in the MRN. (Courtesy Atlantic Research Corporation)

learned by intensified temporal and spatial observation of the stratosphere and mesosphere. Synoptic observations with rockets required rather drastic reductions in costs of individual flight units, ground equipments, and supporting personnel and facilities. Reductions in all of the factors were implemented in large part by adoption of small rockets, generally falling in the weight range of less than 100 lb. Numerous vehicles have been tried but only the Arcas- and Loki-type systems will be discussed here, since they have provided for most of the soundings thus far accomplished. These rocket systems provide the currently available vehicles which are applicable to synoptic observation of the stratosphere and mesosphere. Each vehicle has its own limitations: both fall short of an optimum technique, but they are highly useful at this time and are paving the way toward more efficient systems.

The Arcas rocket system illustrated in Fig. 2.9 was developed by the Atlantic Research Corporation under the sponsorship of the Office of Naval Research (NOnr-2477(100)) to fill the need for a small rocket system for special stratospheric research studies. The primary objective was that the rocket should deliver at least 12-lb payloads to above 60-km altitude. The Arcas system has done this successfully and, in addition, has contributed more measured stratospheric wind and temperature profiles to the Meteorological Rocket Network than any other rocket system. Its capability exceeds that required for the meteorological payloads which have proved efficient up to this time and thus has not been used at its peak efficiency. It provides for expansion as new sensors become available, and vehicles of this type can be expected to play a strong role in continued improvement of meteorological rocket techniques. The routine application of developed systems in a synoptic configuration may well lead to less sophisticated systems for economic reasons.

The Arcas rocket system was test flown in the fall of 1958 at White Sands Missile Range, New Mexico. It was placed in service during the first series of MRN firings in October 1959. A gross weight of 29 kg at lift-off is reduced to 9 kg after expulsion of the fuel over a burning time of 28 sec. At burnout the rocket is nominally above 15-km altitude with a velocity of approximately 130 meters/sec. Subsequently, the vehicle coasts to its peak altitude and expells its payload at about 130 sec after launch. The expended rocket then falls back to impact at approximately 300 sec after launch.

A principal attribute of the Arcas rocket system is the low acceleration imposed on the payload. Instrumentation can be assembled and flown successfully with minimum attention to hardening to withstand the rigorous environment which characterizes most rocket flights. The rocket emerges from the closed-breech launcher with a velocity of approximately 80 meters/sec, and follows the trajectory program illustrated in Fig. 2.10. A following piston confines the rocket motor exhaust gases to the launcher for additional acceleration, and styrofoam guides are used to keep the rocket centered in the tube during the launch. The rocket weather-cocks as it encounters new winds, turning into the wind in a predictable manner. Inaccuracies in available information on the wind field which the rocket will encounter result in uncertainties in the impact point, and as a result a rather large safe impact area is required. As a consequence, the Arcas has been used principally at established missile ranges where suitable impact areas are available. In general, an impact circle of 20-km radius is desirable to assure an ability to accomplish a sounding on a majority of available dates.

As is indicated in Fig. 2.9, a gas generator is used to deploy the payload. A pyrotechnic delay train located in the top of the motor sections is ignited at rocket motor burnout, burns for 100 sec, and ignites the gas generator to expel the payload. All components of the meteorological sensing system are thrust forward from the rocket tube with an ejection

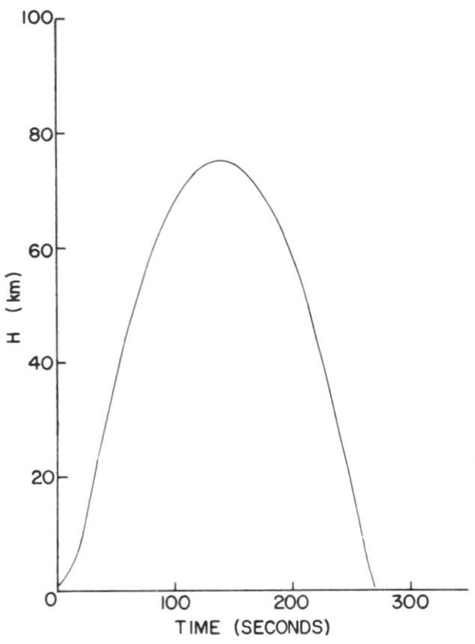

Fig. 2.10. Time-height trajectory curve for the Arcas rocket system when fired at White Sands Missile Range.

velocity of approximately 5 meters/sec. The gas generator piston is stopped by a cable approximately 30 cm out of the rocket, and frees the parachute from its container. Momentum of the instrumentation package unfurls the parachute until breaking tension is applied to the weak cord attaching the top of the parachute to the gas generator piston. The parachute-instrumentation system then swings into a near-vertical position for the descent. Successful completion of the ejection maneuver requires some instability in flight of the rocket vehicle to preclude damaging of the parachute as it slows down in the path of the rocket. Experience has demonstrated that the coning motions of the yawing-spinning free-flight rocket in the weak aerodynamics of the upper atmosphere are adequate to effect a clean ejection of the payload.

The heat shield-payload assembly is fixed to the rocket structure with shear pins which are broken by shearing forces at ignition of the gas generator. The heat shield is freed from the instrumentation base as it leaves the rocket, and separates from the system as soon as the parachute drag becomes appreciable. The sensors are thus uncovered and are the first item to enter the new environment as the system falls. The type of instrument which has evolved for use in synoptic studies

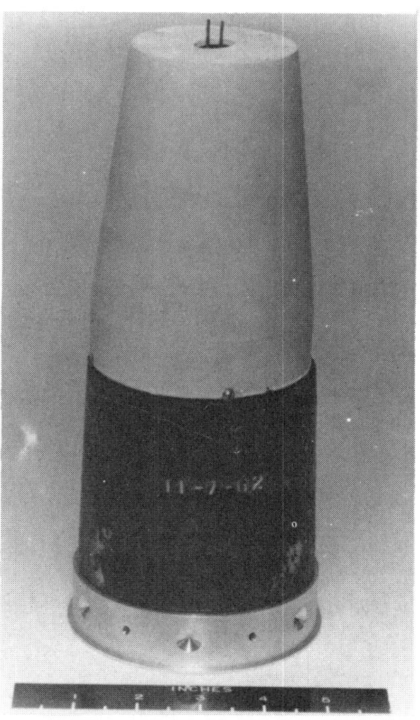

Fig. 2.11. The Delta payload developed for the Arcas rocket system to measure stratospheric temperatures. (Courtesy U. S. Army)

of the stratospheric temperature is illustrated in Fig. 2.11. This is the Delta system developed by Clark and McCoy (1962) to use a 10-mil diameter spherical bead thermistor and the basic 1680-Mc telemetry system employed in balloon observational systems (Fig. 2.4). Principal problems of such an observation system concern the adaptation of the payload to the upper atmospheric environment after its deployment from the hot rocket nosecone and the suitability of the falling parachute as an adequate temperature observing platform.

The forward 12 inches of the Arcas cylindrical section is packaged with a 15-ft diameter silk parachute housed in a split cylinder which permits deployment when freed from the confines of the rocket. Development of the Arcas parachute system provided the first experience available relative to the physics of such a system in the upper atmosphere, and remarkably little difficulty has been experienced in its field application. Fall rates obtained with a 3-lb payload are illustrated in Fig. 2.12. The observational time between 70- and 30-km altitude is approximately 30 min, which is generally acceptable from the view of available power in the flight unit and desirable geometry in the tracking system for wind measurement.

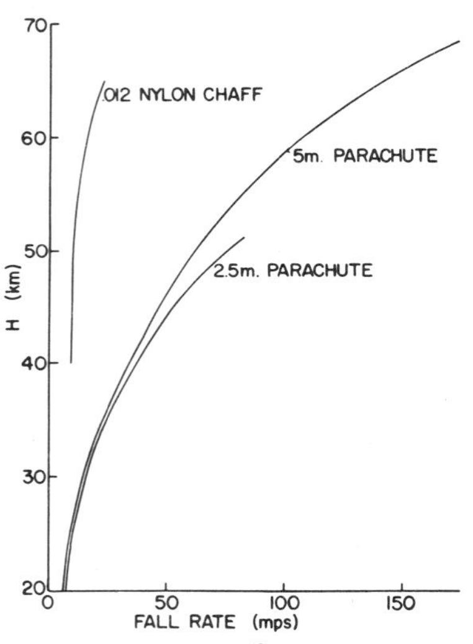

Fig. 2.12. Fall rates of the 5-meter diameter Arcas parachute and the 2.5-meter diameter Loki parachute systems and 0.012 nylon chaff used in the Loki, Judi, and Hasp systems.

Deployment of the Arcas parachute at higher altitudes is undesirable because of the increased terminal velocities which occur. Response to wind shears is reduced at these higher fall rates, and temperature measurements are deteriorated owing to aerodynamic heating of the sensor on contact with the high-speed flow. Some improvement in the fall characteristics could be achieved by reducing the drag below about 50-km altitude, but no efforts have been directed along this line thus far. Improvements in fall characteristics will have to be made before the system can be used in the lower stratosphere and the troposphere.

Extensive analyses of the sensitivities and accuracies of the wind and temperature measuring systems have been conducted by several authors. The wind measurements obtained by radar tracking of the metalized parachutes with the usual radars are generally considered to be accurate to within 5 meters/sec over resolution intervals of a few

Fig. 2.13. Launch of an Arcas rocket system is illustrated. Separation of the piston and styrofoam hardware used to pressurize the launcher and keep the rocket centered in the barrel is effected as the rocket leaves the launcher. (Courtesy U. S. Army)

hundred meters without dynamic corrections over the entire altitude range of 70 to 30 km. Analysis of the bead thermistor sampling and observing system has produced correction factors which result in the acquisition of temperature profiles of an estimated accuracy of $\pm 2°C$ between 30- and 50-km altitude and $\pm 5°C$ at 65-km altitude. Sensitivity of both wind and temperature measurements to small-scale variations deteriorates from the order of 100 meters in the 30-km altitude range to the order of a kilometer at 60 km. There is considerable evidence for the need for improved sensitivities of these sensing systems in the upper stratosphere and lower mesosphere. The launch of an Arcas rocket system is illustrated in Fig. 2.13.

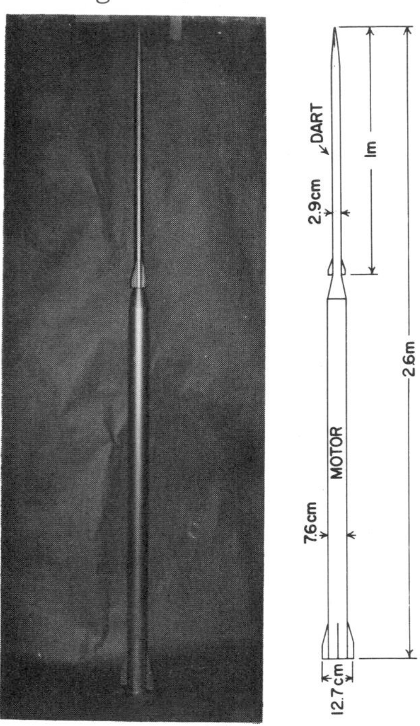

Fig. 2.14. The Loki rocket system for use in upper atmospheric measurements. (Courtesy U. S. Army)

The Loki rocket vehicle was first developed by the Jet Propulsion Laboratories primarily as an antiaircraft weapon. It is a small, free-flight, high-speed rocket system designed to elevate a 7-lb dart to altitudes as high as 30 km. Its original application was to be in large numbers against aircraft. More recently it was found that through certain redesign procedures the vehicle performance could be increased to the extent that it could deliver the same payload to above 60-km altitude. This caused the vehicle to be of considerable interest to research groups engaged in upper atmospheric studies. Only slightly different versions of the Loki are the Hasp II and the Judi systems. If adequate for a particular experiment, the Loki system is highly efficient owing to the ease with which it can be handled and launched (total weight only 15 kg) and the comparatively low cost of this small rocket system. All of these systems are generally illustrated by the photograph and drawing presented in Fig. 2.14. Propulsion is provided by a rocket motor which includes stabilizing fins and an attachment for carrying the dart payload. At the end of propulsion the rocket motor separates from the payload-carrying dart and comes to a stop at a relatively low altitude owing to increased drag and poor stability. This portion of the system then falls back to earth within a few kilometers of the launch point. In general, this portion of the system is completely unstable upon reaching the surface and thus falls back with low enough velocity to be of little danger to anything except the most sensitive facilities.

One principal attribute of the Loki rocket concerns its speed. A burning time of only 1.8 second results in accelerations of over 200 G's during the launch phase. These quick transitions to high speed make the rocket relatively insensitive to low-level winds. Thus the need for large impact areas for the rocket body is eliminated and the system can be used in many applications for which other systems are not suited. The high-G forces produce problems relative to instrumentation for the Loki system, but there is evidence that this problem is not of paramount importance. Possibly of greater importance are the rather severe constraints on size and weight of the payload. In any case, the small-size and relatively economical Loki system is a strong contender for a desirable MRN-type vehicle.

The Loki dart carries the payload to peak altitude and expels it at this point. An electrically initiated fuse delay train ignites an expulsion charge to deploy the atmospheric sensing system. All of the above components are located in the ogive portion of the dart vehicle. The payload of the dart system, which is characterized by a weight of approximately 0.8 kg and a volume of approximately 18 inch^3, is housed in the central portion of the rocket along with adequate heat-shielding and mechanical-deployment systems. On ignition of the expulsion charge at peak altitude, the tail portion of the dart is sheared free from its mounting, allowing the separation of the payload container from the dart. The payload container consists of a split cylinder of insulating material which is spring loaded to assure opening on removal from the confines of the rocket body.

In general, the payloads available in the Loki Dart were inadequate for research sounding purposes in the atmosphere, with most early applications directed toward the measurement of winds, first with chaff materials and later with a 2-meter diameter parachute system. The Loki system has proved to be highly effective and efficient in obtaining wind data on a synoptic basis throughout the stratosphere and mesosphere. Recent improvements and miniaturization of electronics have greatly increased the likelihood that this vehicle will be able to fulfill many of the routine observational requirements in the stratosphere and mesosphere. The 2-meter-diameter parachute is illustrated in Fig. 2.15. It has a fall rate characteristic, with a 340-gm payload, of the type illustrated in Fig. 2.12.

The rather exceptional flight characteristics of the Loki system include an average thrust of 2033 lb lasting for approximately 1.8 sec

Fig. 2.15. Parachute system used for wind measurements for Loki-type vehicles. (Courtesy U. S. Army)

which deploys the dart vehicle at an altitude of 1.5 km with a velocity of 1700 meters/sec. The low drag of the dart system permits an efficient coast phase, with expulsion of the payload at peak altitude the final event in the rocket portion of the observation. The dart body then falls back to earth at the prescribed impact point, and since the forward portion is not altogether unstable, it attains relatively high velocity on re-entry, constituting a hazard. The high speed with which the dart traverses the lower atmosphere during the upleg of the trajectory results in considerable heating of the outer surface of the rocket vehicle. Suitable heat-lagging materials are required to protect the payload from excessive temperatures.

The Loki rocket system is a spin-stabilized free-flight rocket. A spiral launcher 3 meters in length is used to impart an initial spin to the vehicle. Canted fins on the dart and rocket motor maintain a spin rate of approximately 15–18 rps at booster burnout, increasing to 65 rps during dart coast phase. As a result of the relatively small dispersion experienced by this rocket vehicle, the launcher is simply aimed at the point in space at which deployment is desired. The AN/FPS-16 radar has proved capable of tracking the dart throughout its flight and thus provides for immediate acquisition of the inert sensor payloads when they are expelled from the rocket. Other radar systems which have not proved capable of skin tracking the dart during its high-speed flight usually experience little trouble acquiring the target after it is ejected at high altitude, principally because of the small dispersion between flights.

Several applications have been made of the original basic Loki rocket configuration to attain new sounding rocket capability. Most notable of these are the Hasp system developed by the Naval Ordnance Laboratories and the Judi rocket system developed by Rocket Power, Inc., under the Army's sponsorship. The Hasp system incorporates a 2-meter diameter parachute from which is suspended a radio transmitter broadcasting at 403-Mc frequency. Measurements of temperature and pressure are included. Winds are observed by radar tracking of the metalized parachute. The 403-Mc transmitter, a power supply, and associated components have a total weight of approximately 620 gm.

Principal innovations in the Judi system relative to the previous models of the Loki rocket include an ejection of the payload from the front end of the Judi dart. This makes it possible to place the transmitter system in the upper portion of the dart body and to use an isolated nosecone ogive as a transmitting antenna during the upward portion of the flight. AN/GMD-1 tracking of the dart up to expulsion altitude can thus be accomplished and the acquisition of the payload by radar and direction finder on ejection from the dart body is facilitated. The expulsion procedure then includes a separation of the ogive and forward ejection of the payload through the action of a gas generator located in the rear of the dart body. A pyrotechnic delay train is used which provides for ignition of the delay timing mechanism prior to launch of the rocket.

The system also provides space for future incorporation of a receiver for ranging, both on the up and down legs of the observation, which makes it compatible with the AN/GMD-2 system and thus eliminates the need for radar tracking to obtain winds as the parachute descends. Some improvements have been achieved in fuels and in techniques of loading the Loki booster rocket motor so that altitudes above 75 km are consistently reached with this vehicle. The Loki vehicle system is very attractive for meteorological purposes owing to its low over-all weight, the small amount of supporting facilities required for launch, and the low unit cost compared with other rocket sounding systems. One small difficulty associated with the Loki system concerns the rather limited payload capabilities coupled with the very high stresses imposed on payloads during the launch phase.

A new contender for the MRN-type application has developed in the form of a gun probing system. The Army's Ballistic Research Labora-

Fig. 2.16. Five-inch gun system and projectile used to deploy meteorological sensors at altitudes above 70 km. (Courtesy U. S. Army)

tories at Aberdeen, Maryland, have perfected spin-stabilized subcaliber projectiles that can be launched from smooth-bore 5- and 7-inch diameter guns to reach peak altitudes of 70 and 90 km, respectively. The 5-inch gun with an extended barrel, and a close-up of the projectile to be flown in this equipment, are illustrated in Fig. 2.16. Propulsion gases are confined to the barrel by a midsection sabot which falls away from the projectile on emergence from the barrel. The projectile then flies to peak altitude where an expulsion charge deploys the payload. Chaff, parachute, and superpressure sphere sensors have been successfully deployed from this equipment as well as certain chemical releases and stratospheric smoke trails. Principal virtues of the gun system include the great accuracy with which the payload can be deployed and the flight equipments recovered, and the low unit cost that can be expected under extensive usage.

Limitations are placed on the payload design as a result of the large accelerations imposed on the flight vehicle during the launch phase. Acceleration loadings in excess of 40,000 G's are to be expected in the 5-inch system. Electronic instrumentation is not yet available for use in this system, but efforts are under way to produce solid-state telemetry equipments that will operate successfully in the gun probe system. If successful, the gun probe will be particularly attractive for the MRN application in view of the small impact areas needed for operation. It is expected that a range cone from the launcher to a circle of one-mile radius will be adequate for operational deployment of the system, even under the most adverse meteorological conditions.

2.3 Summary

The rocket, balloon and aircraft systems and sensor techniques briefly described in this chapter effectively represent the state of the art in stratospheric and mesospheric exploration at the beginning of 1965. Developments are under way which will, and necessarily must, revolutionize the facility with which synoptic observation of the upper atmosphere may be conducted. These improvements will be aimed at lower costs and greater utility in obtaining data at the required time and place. Rapid progress is to be expected during the next several years as the MRN becomes firmly established as an integral part of our atmospheric observation system.

Certain of the unknowns which exerted a negative influence during the early development of the MRN are being clarified as a result of the data obtained with these experimental systems. Actual needs for upper atmospheric data are better defined in space and time as are the most informative measurements. Difficulties relative to application of experimental techniques have been evaluated for several sensing systems so that it is possible to assess the relative costs of the various elements of data. Thus, the changing aspect of the atmosphere can be incorporated into the planning of a synoptic network, rather than simply force an upward extension of familiar techniques for observation of phenomena with which we have had experience at lower altitudes.

These comments point toward the fact that firm groundwork has been laid for an efficient MRN during these first six years. Expanding needs of our advancing technology can be counted on to produce an increasing demand for environmental data. At the same time, improvements in MRN techniques will produce an enhanced capability. A bright future is in prospect for the MRN, hinging only on the productive efforts of the atmospheric scientists associated with the network.

REFERENCES

Ainsworth, J. E., and H. E. LaGow. (1956). Vacuum gauge chamber response time. *Rev. Sci. Instr.* **27,** 653–654.

Ainsworth, J. E., D. F. Fox, and H. E. LaGow. (1961). Measurement of upper atmosphere structure by means of the pilot-static-tube. *NASA* (*Natl. Aeron. Space Admin.*), *Tech. Note* **TN D-670.**

Aleksejev, P. P. (1957). Rocket investigation of the upper atmosphere. *Meteorol. i Gidrol.* **8,** 3–13.

Allen, H. F. (1963). Proposed Keweenaw Research Range. College of Engineering, Report No. 0537-1-P, Univ. of Michigan Press, Ann Arbor, Michigan.

Anderson, A. D. (1957). Comments on "Upper air wind measurements with small rockets." *J. Meteorol.* **14,** 473–474.

Anderson, A. D. (1960). A wind-measuring system for the tactical prediction of fallout. *Naval Res. Rev.* pp. 7–14.

Anderson, A. D., and W. E. Hoehne. (1956). Experiments using window to measure high altitude winds. *Bull. Am. Meteorol. Soc.* **37,** 454–457.

Atlantic Research Corp., Alexandria, Va. (1960). Arcas rocketsonde system: High altitude and meteorological sounding rocket. Final Report, 8 pp.

aufm Kampe, H. J. (1956). Upper-air wind measurements with small rockets. *J. Meteorol.* **13,** 601–602.

aufm Kampe, H. J. (1960). Meteorological rocket network for measuring atmospheric parameters up to 250,000 ft. *Weatherwise* **13,** 192–195.

Auld, C. O. (1957). Feasibility of a meteorological rocket for synoptic measurements to 300,000 ft. U. S. Naval Ordnance Laboratory, White Oak, Maryland, Report 5761.

Badgley, F. I. (1957). Response of radiosonde thermistors. *Rev. Sci. Instr.* **28,** 1079–1084.

Ballard, H. N. (1961). Response time of and effects of radiation on the VECO bead thermistor. *Instr. Soc. Am. Conf. Preprint* **169-LA-61.**

Ballard, J. C. (1941). On the interpretation of temperature measurements made at high levels. *Monthly Weather Rev.* **69,** 33–40.

Bandeen, W. R. (1959). The recording of acoustic waves from high-altitude explosions in the rocket-grenade experiment and certain other related topics. U. S. Army Signal Research and Development Laboratories, TR 2056, AD 231 943.

Bandeen, W. R., and J. Otterman. (1960). Temperature correction on the rocket-grenade experiment due to the finite-amplitude-propagation effect. *J. Geophys. Res.* **65,** 851–855.

Bandeen, W. R., R. M. Griffith, and W. Nordberg. (1959). Measurements of temperatures, densities, pressures and winds over Fort Churchill, Canada, by means of the rocket grenade experiment. U. S. Army Signal Research and Development Laboratories, TR 2076, 71 pp.

Barr, W. C. (1958). Heat transfer considerations in the design of high-altitude temperature sensing elements. *Bull. Am. Meteorol. Soc.* **39,** 435.

Barr, W. C. (1960). Theoretical evaluation of cylindrical chaff as a wind sensor at high altitudes. U. S. Army Signal Research and Development Laboratories, TR 2138, AD 241 876.

Barr, W. C. (1961). Theoretical considerations in the design of atmospheric temperature sensing elements. U. S. Army Signal Research and Development Laboratories, TR 2195.

Bartman, F. L., L. W. Chaney, L. M. Jones, and V. C. Liu. (1956). Upper-air density and temperature by the falling-sphere method. *J. Appl. Phys.* **27,** 706–712.

Battan, L. J. (1958). Use of chaff for wind measurements. *Bull. Am. Meteorol. Soc.* **39,** 258–260.

Baynton, H. W. (1961a). Rocketsonde temperature profiles over WSMR. *Bull. Am. Meteorol. Soc.* **42,** 11–16.

Baynton, H. W. (1961b). AN/ANQ-15 rocketsonde tests at White Sands Proving Ground. *Bull. Am. Meteorol. Soc.* **42,** 34–41.

Bedinger, J. F., E. R. Manring, and S. N. Ghosh. (1958). Study of sodium vapor ejected into the upper atmosphere. *J. Geophys. Res.* **63,** 19–29.

Belmont, A., R. Peterson, and W. Shen. (1964). Evaluation of meteorological rocket data. General Mills, Inc., Minneapolis, Report 2567, Contract NASw-558, 96 pp.

Berkner, L. V. (1958). Manual on rockets and satellites. *Ann. Intern. Geophys. Yr.* **6,** 508 pp.

Berning, W. W. (1954). Trajectory measurements for rockets used in upper atmosphere research. *In* "Rocket Exploration of the Upper Atmosphere" (R. L. F. Boyd and M. J. Seaton, eds.), pp. 65–72. Pergamon Press, Oxford.

Bettinger, R. T. (1958). The exploration of the high atmosphere with rockets. *Bull. Am. Meteorol. Soc.* **39,** 435.

Beyers, N. J. (1960). Preliminary radar performance data on passive rocket-borne wind sensors. *IRE, Trans. Military Electron.* **4,** 230–233.

Beyers, N. J., and O. W. Thiele. (1960). Meteorological rocket wind sensors. U. S. Army Signal Missile Support Agency, White Sands Missile Range, New Mexico, Special Report 41, AD 242 764.

Beyers, N. J., and O. W. Thiele. (1961). Performance characteristics of systems for high altitude wind and temperature observations. *Instr. Soc. Am. Conf. Preprint* **170-LA-61-1.**

Beyers, N. J., O. W. Thiele, and N. K. Wagner. (1962). Performance characteristics of meteorological rocket wind and temperature sensors. U. S. Army Electronics Research and Development Activity, White Sands Missile Range, New Mexico, TR SELWS-M-4.

Blagonravov, A. A. (1957). Issledovanie verkhnikh sloev atmosfery pri pomoshchi vysotnykh raket. *Vestn. Akad. Nauk SSSR* **27,** 25–32.

Boyd, R. L. F., M. J. Seaton, and H. S. W. Massey. (1954). *In* "Rocket Exploration of the Upper Atmosphere" (R. L. F. Boyd and M. J. Seaton, eds.), p. 376. Pergamon Press, Oxford.

Boyer, E. D. (1963). Five-inch Harp tests at Wallops Island. U. S. Army Ballistics Research Laboratories, Aberdeen, Maryland, Memo. Report No. 1532.

Bradford, W. C. (1961). Theoretical and experimental dispersions for the Arcas-Robin rocket. Eglin Air Proving Ground Center, Florida, Tech. Document APGC-TDR-61-62.

Bradford, W. C., and P. W. Myers. (1962). Dispersion study for the Arcas-Robin with 230 feet/second launch velocity. Eglin Air Proving Ground Center, Florida, Tech. Document APGC AIDR 62-2.

Brockman, W. E. (1963). Computer programs in Fortran and Balgol for determining winds, density, pressure and temperature from the Robin-falling balloon. University of Dayton Research Institute, Report AF19(604)-7450.

Broglio, L. (1963). Review of Italian meteorological activities and results. *In* "First International Symposium on Rocket and Satellite Meteorology" (H. Wexler and J. E. Caskey, Jr., eds.), pp. 94–118. Wiley, New York.

Brown, J. A. (1960). Arcas high-altitude meteorological rocketsonde system. *Trans. N. Y. Acad. Sci.* [2] **22,** 275–283.

Bruch, A., and G. M. Morgan, Jr. (1961). Wind variability in the mesosphere as determined by the tracking of falling objects. New York University, Final Report AF19 (604)-6193, 104 pp.

Burgess, E. (1952). High-altitude research. *Engineer, London* **194,** 338–340 and 370–373.

Chubb, T. A., E. T. Byram, H. Friedman, and J. E. Kupperian. (1958). The use of radiation, absorption and luminescence in upper air density measurements. *Ann. Geophys.* **14,** 109–116.

Clark, G. Q. (1961). Development of a rocket telemetry package for the Meteorological Rocket Network. Inter-Range Instrumentation Group, White Sands Missile Range, New Mexico, Document No. 105-60, AD 246 950, pp. 147–162.

Clark, G. Q., and J. G. McCoy. (1962). Meteorological rocket thermometry. U. S. Army Signal Missile Support Agency, White Sands Missile Range, New Mexico, Tech. Report MM-460.

Clark, G. Q., W. L. Webb, and K. R. Jenkins. (1960). Rocket sounding of high atmosphere meteorological parameters. *IRE, Trans. Military Electron.* **4,** 238–243.

Cline, D. E. (1957). Rocket beacon wind sensing system. U. S. Army Signal Research and Development Laboratories, Engr. Report E-1205.

Cline, D. E. (1958). Double-direction-finding rocket meteorological sensing system. U. S. Army Signal Research and Development Laboratories, TR 1981, AD 212 478.

Cochran, V. C., and J. K. Hansen. (1963). Theoretical performance of the Arcas and boosted Arcas. U. S. Army Signal Missile Support Agency, White Sands Missile Range, New Mexico, Tech. Report MM-432.

Dorling, E. B. (1959). The first six skylark firings. Royal Aircraft Establishment, Farnborough, England, 69 pp.

Dow, W. G., and N. W. Spencer. (1953). Measures of ambient pressure and temperature of the upper atmosphere. University of Michigan, Final Report AF19(122)-55.

Duncan, L. D. (1960a). Theoretical performance of the Arcas. U. S. Army Signal Missile Support Agency, White Sands Missile Range, New Mexico, Special Report 34.

Duncan, L. D. (1960b). Automatic rocket impact predictor. *IRE, Trans. Military Electron.* **4,** 243–245.

Facy, L. (1963). Les voilures de parachutes destinées aux hautes altitudes. *In* "First International Symposium on Rocket and Satellite Meteorology" (H. Wexler and J. E. Caskey, Jr., eds.), pp. 187–198. Wiley, New York.

Faust, H. (1960). Raketenmeteorologie. *Naturw. Rundschau* **13,** 95–99.

Ference, M., Jr., W. G. Stroud, and J. R. Walsh. (1956). Measurement of temperatures at elevations of 30 to 80 kilometers by the rocket-grenade experiment. *J. Meteorol.* **13,** 5–12.

Force, C. T., and W. E. Walker. (1958). Design characteristics and performance of Sandia's Deacon-Arrow chaff rocket system. Sandia Corp., Albuquerque, TR 4229.

Godson, W. L. (1963). Meteorological rocket plans-recommendations for the IQSY. *In* "First International Symposium on Rocket and Satellite Meteorology" (H. Wexler and J. E. Caskey, Jr., eds.), pp. 173–186. Wiley, New York.

Green, C. F. (1954). Utilization of the V-2 (A-4) rocket in upper atmosphere research. *In* "Rocket Exploration of the Upper Atmosphere" (R. L. F. Boyd and M. J. Seaton, eds.), pp. 28–45. Pergamon Press, Oxford.

Griner, G. N., and T. A. Mahar. (1963). Sparrow (c-8) -HV Arcas dispersion analysis. Atlantic Research Corp., Alexandria, Va., Report No. ARC-FA1-2D1.

Groves, G. V. (1956a). Introductory theory for upper atmosphere wind and sonic velocity determination by sound propagation. *J. Atmospheric Terrest. Phys.* **8**, 24–38.

Groves, G. V. (1956b). Theory of the rocket-grenade method of determining upper atmospheric properties by sound propagation. *J. Atmospheric Terrest. Phys.* **8**, 189–203.

Groves, G. V. (1956c). Effect of experimental errors on determinations of wind, velocity, speed of sound and atmospheric pressure in the rocket-grenade experiment. *J. Atmospheric Terrest. Phys.* **9**, 237–261.

Groves, G. V. (1956d). A rigorous method of analysing data of the rocket-grenade experiment. *J. Atmospheric Terrest. Phys.* **9**, 349–351.

Groves, G. V. (1963). High atmosphere wind studies by skylark rocket cloud releases. *In* "First International Symposium on Rocket and Satellite Meteorology" (H. Wexler and J. E. Caskey, Jr., eds.), pp. 60–69. Wiley, New York.

Hammond, R. (1955). Rockets for atmospheric research. *Instr. Pract.* **9**, 427–430.

Hansen, W. H., and F. F. Fischbach. (1958). The Exos sounding rocket. Michigan University Research Institute, Final Report AF19(604)-1943.

Harp Project. (1962). Description and status. McGill University, Montreal, Canada, Report No. 625.

Harp Project. (1963). First twelve firings and status as of July 30, 1963. McGill University, Montreal, Canada, Report No. 635.

Hart, P. J., and I. Gluttmacher. (1959). Flight summaries for the U. S. Rocketry Program for the International Geophysical Year, Part 1, July 5, 1956–June 30. 1958. IGY Rocket Report Series, No. 2, 193 pp.

Haycock, O. C. (1960). Improvements in the falling sphere instrumentation. Final Report AF19(604)-4956, AD 236 470.

Heinrich, H. G. (1956). Drag and stability of parachutes. *Aeron. Eng. Rev.* **15**, 73–81.

Heinrich, H. G. (1963). The effective porosity of parachute cloth. *Flugwiss*, **11**, 389–397.

Heinrich, H. G., E. L. Haak, and R. J. Niccum. (1963). Research and development of Robin meteorological rocket balloons. University of Minnesota Report, Vol. 2, Contract No. AF19(604)-803L.

Israel, G. (1963). Mesure de la pression atmosphérique par une méthode aerodynamique. *In* "First International Symposium on Rocket and Satellite Meteorology" (H. Wexler and J. E. Caskey, Jr., eds.), pp. 76–85. Wiley, New York.

Itokawa, H. (1959). Japanese sounding rockets—Kappa and Sigma. *In* "Sounding Rockets" (H. E. Newell, Jr., ed.), pp. 273–286. McGraw-Hill, New York.

Izakov, M. N., and G. A. Kokin. (1957). Izmerenie temperatury, davleniia i plotnosit atmosfery s pomoshch'iu raket. U.S.S.R. Glavnoe Upravlenie Gidrometeorologicheskoi Sluzhby. *Inform. Sb.* **4**, 54–61.

Jenkins, K. R. (1958). Measurements of high altitude winds with Loki. U. S. Army Signal Missile Support Agency, White Sands Missile Range, New Mexico, TM 544, AD 200 001.

Jenkins, K. R. (1962). Empirical comparisons of meteorological rocket wind sensors. *J. Appl. Meteorol.* **1**, 196–202.

Jenkins, K. R., and W. L. Webb. (1958). Rocket wind measurements. *Bull. Am. Meteorol. Soc.* **39**, 436.

Jenkins, K. R., and W. L. Webb. (1959). High-altitude wind measurements. *J. Meteorol.* **16**, 511–515.

Joint Task Force Seven. (1959). Meteorological ordance report Operation Hardtack. Meteorological Center, University of Hawaii, JTFMC-TP-10, AD 217 464.

Jones, L. M. (1956). Transit-time accelerometer. *Rev. Sci. Instr.* **27**, 374–377.

Jones, L. M. (1959). Measuring upper air structure. *Astronautics* **2**, 29–50.

Jones, L. M., and F. L. Bartman. (1956). A simplified falling sphere method for upper air density. University of Michigan, TR AF19(604)-999, AD 101 328.

Jones, L. M., J. W. Peterson, and E. J. Schaefer. (1959). Upper air density and temperature: Some variations and an abrupt warming in the mesosphere. *J. Geophys. Res.* **64**, 2331–2340.

Keegan, T. J. (1961). Meteorological rocketsonde equipment and techniques. *Bull. Am. Meteorol. Soc.* **42**, 715–721.

LaGow, H. E., and J. Ainsworth. (1956). Arctic upper atmosphere pressure and density measurements with rockets. *J. Geophys. Res.* **61**, 77–92.

LaGow, H. E., R. Horowitz, and J. Ainsworth. (1958). Rocket measurements of the Arctic upper atmosphere. World Data Center A, IGY Rocket Report No. 1, pp. 26–37.

Lally, V. E., and R. Leviton. (1958). Accuracy of wind determination from the track of a falling object. U. S. Air Force, Cambridge Research Laboratories, Surveys in Geophysics, No. 93.

Langer, G., and J. Stockham. (1960). High-altitude tracking by chemical smokes. *J. Geophys. Res.* **65**, 3331–3338.

Lenhard, R. W., Jr., and M. P. Doody. (1963). Accuracy of meteorological data obtained by tracking the Robin with MPS-19 radar. U. S. Air Force, Cambridge Research Laboratories, L. G. Hanscom Field, Massachusetts, Instrumentation for Geophysics and Astrophysics, No. 35.

Leviton, R. (1961). The Robin-falling-sphere. U. S. Air Force, Cambridge Research Laboratories, L. G. Hanscom Field, Massachusetts, Research Note 73.

Leviton, R., and W. E. Palmquist. (1959). A rocket balloon instrument. *Bull. Am. Meteorol. Soc.* **40**, 374.

Liu, V. C. (1956). On a pitot-tube method of upper atmosphere measurements. *J. Geophys. Res.* **61**, 171–178.

Liu, V. C. (1957). On the motion of a projectile in the atmosphere. *Z. Angew. Math. Phys.* **8**, 76–82.

Maeda, Ken-ichi. (1963). Japanese rocket sounding for meteorology. *In* "First International Symposium on Rocket and Satellite Meteorology" (H. Wexler and J. E. Caskey, Jr., eds.), pp. 86–93. Wiley, New York.

Marks, S. T., and E. D. Boyer. (1963). A second test of an upper atmosphere gun probe system. U. S. Army Ballistics Research Laboratories, Aberdeen, Maryland, Report No. 1464.

Marks, S. T., L. G. MacAllister, J. W. Gehring, H. D. Vitagliano, and B. T. Bently. (1961). Feasibility test of an upper atmosphere gun probe system. U. S. Army Ballistics Research Laboratories, Aberdeen, Maryland, Report No. 1368.

Mason, H. P., and W. N. Gardner. (1957). A limited correlation of atmospheric sounding data and turbulence experienced by rocket-powered models. *Natl. Advisory Comm. Aeron., Tech Notes* **3953**.

Massey, H. S. W., and R. L. F. Boyd. (1958). Probing with sound waves. "The Upper Atmosphere," pp. 84–94. Hutchinson, London.

Massey, H. S. W., and R. L. F. Boyd. (1958). Research by balloons and rockets. "The Upper Atmosphere," pp. 49–73. Hutchinson, London.

Masterson, J. E. (1959a). A review of meteorological sounding rockets. *Proc. 1st Intern. Symp. Rockets Astronautics, Tokyo, 1959,* pp. 216–223.

Masterson, J. E. (1959b). Loki-Wasp. *In* "Sounding Rockets" (H. E. Newell, Jr., ed.), pp. 124–142. McGraw-Hill, New York.

Minzner, R. A., M. Dubin, and E. W. Beth. (1954). Ambient temperature determinations from yawing rockets in the presence of winds. *In* "Rocket Exploration of the Upper Atmosphere" (R. L. F. Boyd and M. J. Seaton, eds.), pp. 17–25. Pergamon Press, Oxford.

Newell, H. E., Jr. (1954). Rockets and the upper atmosphere. *Sci. Monthly* **78,** 30–36.

Newell, H. E., Jr. (1955). Temperatures in the upper atmosphere. *Temp., Meas. Control Sci. Ind., Vol. II, 3rd Symp. Temp., Wash., D. C., 1954* pp. 429–444. Reinhold, New York.

Newell, H. E., Jr. (1957). The high altitude sounding rocket. *Jet Propulsion* **27,** 261–262.

Ney, E. P. (1958). Air temperature measurement. Atmos. Phys. Program, University of Minnesota, Contract Nonr-710(22).

Ney, E. P. (1961). The measurement of atmospheric temperature. *J. Meteorol.* **18,** 60–68.

Nordberg, W. (1959). Upper atmosphere rocket sounding on the Island of Guam. U. S. Army Signal Research and Development Laboratories, TR 2078, AD 228 443.

Ordway, F. E., and R. C. Wakeford. (1957). 1957 Research rocket roundup. *Missiles Rockets* **2,** 39–48.

Otterman, J. (1958). Effect of finite-amplitude propagation in the rocket-grenade experiment for upper atmosphere temperature and winds. University of Michigan, Report DA-36-039-SC-64659, AD 162 059.

Otterman, J. (1959). Finite-amplitude propagation effect on shock-wave travel times from explosions at high altitudes. *J. Acoust. Soc. Am.* **31,** 470–474.

Otterman, J., I. J. Sattinger, and D. F. Smith. (1961). Analysis of a falling sphere experiment for measurement of upper atmosphere density and wind velocity. *J. Geophys. Res.* **66,** 819–822.

Parker, M. J. (1958). The Naval Ordnance Lab. Weather-Rocket Hasp. *Bull. Am. Meteorol. Soc.* **39,** 435.

Peterson, J. W., H. F. Schulte, and E. J. Schaefer. (1959). A simplified falling sphere method for upper air density, Part II; density and temperature results from eight flights. University of Michigan, Report AF19(604), AD 219 983.

Peterson, J. W., D. A. Robinson, and H. F. Schulte. (1960). Falling sphere instrumentation development. University of Michigan, Report AF19(604) 2415, AD 237 085.

Poetzschke, H. (1960). Wind determination from an Aerobee firing. *J. Geophys. Res.* **65**, 368–369.

Poloskov, S. M., and B. A. Mirtov. (1957). The study of the upper atmosphere by means of rockets, at the Academy of Sciences, U. S. S. R. *J. Brit. Interplanet. Soc.* **16**, 95–100.

Poppoff, I. G. (1958). Low-cost meteorological rocket. *Astronautics* **26**, 44–46.

Poppoff, I. G. (1959). Rocketsondes for high-altitude soundings. *Stanford Res. Inst. J.* **3**, 56–60.

Prince, M. J., and A. L. Quirk. (1955). Rocket instruments for the measurement of total solar radiation in the upper atmosphere. University of Rhode Island, Upper Atmosphere Research Laboratories, Report No. 1, 54 pp.

Raff, S. J. (1959). Report of Phase I of the Feasibility Committee for 200,000 ft altitude instrumented Hasp. U. S. Naval Ordnance Laboratory, White Oak, Maryland, NAVORD Report 6763.

Rapp, R. R. (1960). The accuracy of winds derived by the radar tracking of chaff at high altitudes. *J. Meteorol.* **17**, 507–514.

Regener, V. H. (1959). Automatic chemical determination of atmospheric ozone. *Advan. Chem. Ser.* **21**, 124–127.

Regener, V. H. (1960). On a sensitive method for the recording of atmospheric ozone. *J. Geophys. Res.* **65**, 3975–3977.

Reisig, G. H. R. (1956). Instantaneous and continuous wind measurements up to the higher stratosphere. *J. Meteorol.* **13**, 448–455.

Rolt, H. (1955). Rockets for atmospheric research. *Instr. Pract.* **9**, 427–230.

Rosen, M. W., and R. B. Snodgrass. (1954). The high altitude sounding rocket. *In* "Rocket Exploration of the Upper Atmosphere" (R. L. F. Boyd and M. J. Seaton, eds.), pp. 46–52. Pergamon Press, Oxford.

Shvidkovaskii, E. G. (1958). A rocket flies into the stratosphere. Sovetskaia Aviatsiia, 4 pp.

Sicinski, H. S., N. W. Spencer, and G. W. Dow. (1954). Rocket measurements of upper atmosphere ambient temperature and pressure in the 30-75 km region. *J. Appl. Phys.* **25**, 161–168.

Sieden, P. E. (1957). Wide range thermistor vacuum gauge. *Rev. Sci. Instr.* **28**, 657–658.

Singer, S. F. (1953). Synoptic rocket observations of the upper atmosphere. *Nature* **171**, 1108–1109.

Singer, S. F. (1956). Research in the upper atmosphere with high altitude sounding rockets. *J. Atmospheric Terrest. Phys.* **4**, 878–912.

Singer, S. F., and A. L. Lawrence. (1957). Terrapin: an upper atmosphere research vehicle. *Jet Propulsion* **27**, 281–284.

Smith, L. B. (1960). The measurement of winds between 100,000 and 300,000 ft by use of chaff rockets. *J. Meteorol.* **17**, 296–310.

Snavely, B. L. (1956). Proposed Instrumentation of the Loki rocket for meteorological measurements. U. S. Naval Ordnance Laboratory, White Oak, Maryland, NAVORD Report 4430.

Soberman, R. K., S. A. Christ, J. J. Manning, L. Rey, T. G. Ryan, R. S. Skrivanek, and N. Wilhelm. (1964). Techniques for rocket sampling of noctilucent cloud particles. *Tellus* **16**, 89–95.

Spencer, N. W. (1958). Research in the measurements of ambient pressure, temperature and density of the upper atmosphere by means of rockets. University of Michigan, Engineering Research Institute, Final Report, 40 pp.

Spencer, N. W., and W. G. Dow. (1954). Density-gauge methods for measuring upper-air temperature, pressure and winds. *In* "Rocket Exploration of the Upper Atmosphere" (R. L. F. Boyd and M. J. Seaton, eds.), pp. 82–97. Pergamon Press, Oxford.

Spencer, N. W., H. F. Schulte, and H. S. Sicinski. (1954). Rocket instrumentation for reliable upper atmosphere temperature determination. *Proc. IRE* **42**, 1104–1108.

Spencer, N. W., R. L. Boggess, L. R. Brace, and M. A. El-Moslimany. (1958). Radioactive-ionization-gauge pressure-measurement system. University of Michigan, Engineering Research Institute, Contract AF19(604)-545.

Stephens, W. H. (1959). British upper atmosphere sounding rocket. *In* "Symposium on High Altitude and Satellite Rockets, Cranfield, England," pp. 23–28. New York Philosophical Library, New York.

Stroud, W. G., E. A. Terhune, J. H. Venner, J. R. Walsh, and S. Weiland. (1955). Instrumentation of the rocket-grenade experiment for measuring atmospheric temperature and winds. *Rev. Sci. Instr.* **26**, 427–432.

Temple University Research Institute. (1952). Research in the physical properties of the upper atmosphere with V-2 rockets. Final Report, 101 pp.

Thaler, W. J., and J. E. Masterson. (1956). A rapid high altitude wind determining system. *Bull. Am. Meteorol. Soc.* **37**, 177.

Townsend, J. W., Jr., and R. M. Slavin. (1957). Aerobee-Hi development program. *Jet Propulsion* **27**, 263–265.

University of Michigan. (1954). Falling sphere method for upper air density and temperatures. *In* "Rocket Exploration of the Upper Atmosphere" (R. L. F. Boyd and M. J. Seaton, eds.), pp. 98–107. Pergamon Press, Oxford.

Van Allen, J. A., and M. B. Gottlieb. (1954). The inexpensive attainment of high altitude with balloon-launched rockets. *In* "Rocket Exploration of the Upper Atmosphere" (R. L. F. Boyd and M. J. Seaton, eds.), pp. 53–64. Pergamon Press, Oxford.

Wagner, N. K. (1961). Theoretical time constant and radiation error of a rocketsonde thermistor. *J. Meteorol.* **18**, 606–614.

Wagner, N. K. (1963). Theoretical accuracy of the meteorological rocketsonde thermistor. *Proc. Intern. Symp. Stratospheric Mesospheric Circulation,* pp. 527–536. Inst. Meteorol. Geopyhs., Freien Univ., Berlin.

Walter, E. L. (1962). Six variable ballistic model for a rocket. U. S. Army Signal Missile Support Agency, Missile Meteorology Division, White Sands Missile Range, New Mexico, TR 445.

Webb, W. L., and K. R. Jenkins. (1959a). Application of meteorological rocket systems. *J. Geophys. Res.* **64**, 1855–1863.

Webb, W. L., K. R. Jenkins, and G. Q. Clark (1959b). Flight testing of the Arcas. U. S. Army Signal Missile Support Agency, White Sands Missile Range, New Mexico, Report 1.1.60.

Webster, R. C., W. C. Roberts, Jr., and W. P. Donnell. (1960). Development of the Arcas rocketsonde system. Atlantic Research Corp., Alexandria, Va., Final Report, Nonr-2477-(00), AD 235 341.

Weisner, A. G. (1954). The determination of temperatures and winds above thirty kilometers. *In* "Rocket Exploration of the Upper Atmosphere" (R. L. F. Boyd and M. J. Seaton, eds.), pp. 133–142. Pergamon Press, Oxford.

Welinski, B. R. (1960). Robin, meteorological balloon development. University of Dayton, Final Report AF 19(604)-6653, 31 pp.

Whitlock, C. H., and H. N. Murrow. (1964). Performance characteristics of a preformed elliptical parachute at altitudes between 200,000 and 100,000 feet obtained by in-flight photography. *NASA (Natl. Aeron. Space Admin.) Tech. Note,* **TND--2183.**

Williams, D. R. (1957). Feasibility of a meteorological rocket for synoptic measurement to 300,000 ft. U. S. Naval Ordnance Laboratory, White Oak, Maryland, NAVORD Report 5761.

Wyckoff, P. H. (1958). Rocket as a research vehicle. *Am. Geophys. Union, Geophys. Monograph* **2,** 102–107.

3

The Lower Stratosphere

Introduction

The lower stratosphere was defined in the introduction as that volume of the upper atmosphere which is bounded beneath by the tropopause and above by the stratonull surfaces. The stratonull surface has a nominal height of 24 km over the entire globe, with the exception of a restricted region around the poles, while the tropopause is high (approximately 17 km) in equatorial regions and low (approximately 10 km) in high latitudes. Thus, the lower stratosphere is a thin layer in equatorial regions and is a comparatively thick stratum at high latitudes. Both the upper and lower boundaries exhibit considerable irregularities in middle latitudes, with breaks and multiple stratifications not uncommon.

Tropospheric circulations dominate the advective structure of the lower stratosphere, and since these circulations have long been the subject of detailed consideration they will not be discussed here. It is adequate to note that the intensity of this tropospheric circulation is a maximum in middle latitudes and is relatively weak in equatorial and polar regions. Since the maximum wind speeds of the tropospheric circulation are generally to be found near the tropopause and a minimum in wind speed is characteristic of the stratonull, the lower stratosphere is typified by a decreasing speed in the vertical wind profiles. The instabilities associated with this sharp positive lapse rate in the wind structure are, in the mean, canceled by the rather stable thermal structure, with the result that the lower stratosphere may be stable or unstable relative to mass interchange with the troposphere across the tropopause depending upon the relative importance of these parameters.

The same cyclonic and anticyclonic systems which produce the large latitudinal excursions of subtropical and subpolar air out of their normal habitats into the mixing grounds of the midlatitudes generate similar activity in the lower stratosphere. The asymmetric meridional profile of the lower stratosphere complicates the procedure, however, with the source of circulation clearly stratospheric in high latitudes and more generally tropospheric in the tropical case. Thus, advective motions will result in intense stratifications of the more conservative properties of these decidedly different air mass sources.

The lower stratosphere may be viewed as a rather clean region of the atmosphere. On occasion, significant amounts of particulate matter may be introduced by intrusion of meteoroid material from above or volcanic debris from below. These are rather special events, however, and their appearance is usually limited in spatial extent and time to a small fraction of the possible observational instances. Generally, they can only be observed under the optimum conditions at sunrise and sunset, but when significant quantities of particulate material are present in the lower stratosphere the optical effects are spectacular. Fall rates are such that even the smallest particles which can produce optical effects are soon precipitated and the normal clean lower stratosphere is restored.

Very low temperatures are a general characteristic of the lower stratosphere, with a range from —40° to —80°C. These low values of temperature preclude the residence of significant amounts of water vapor and thus minimize the very important contribution which liquid water or frost attachment sublimation to nuclei makes toward rendering atmospheric particulate matter detectable optically. Even in the relatively dirty troposphere, particulate material is usually noticeable only in restricted areas, usually near the ground, except when water vapor deposits enhance the particulate sizes. While the lower stratosphere is surely dry in the sense that a small amount of water is present, the significance of water in the thermodynamic and dynamic process could well be of great consequence.

A dominant feature in the chemical characteristics of the lower stratosphere is the presence of ozone. This triatomic oxygen molecule is formed principally in the upper stratosphere, but is shielded from destructive processes in the lower stratosphere so that it exhibits a maximum concentration there. Diffusion and subsidence abet the reaction processes to form a reservoir of the highly toxic ozone molecules in the lower stratosphere, at the base of which the lifetime of an ozone molecule may be months or years. The tropopause is a very effective base on which the "ozonosphere" rests, with a very sharp gradient in ozone concentration the rule wherever a tropopause surface is located. The troposphere acts as a gross sink for ozone molecules, probably as a result of oxidation with the numerous particulate materials which are a common feature of tropospheric composition. In addition, turbulent mixing, which dominates detail transport in the troposphere, dilutes the ozone which comes under its influence.

While the first penetrations of the lower stratosphere were accomplished by balloon systems before the end of the nineteenth century, it is only since World War II that intensive *in situ* exploration has been accomplished. Improved balloon systems make it possible to sample vertical structure in the lower stratosphere in a very efficient fashion. Development of constant-altitude balloons capable of maintaining large payloads in the lower stratosphere for extended periods has made possible the conduct of complex experiments under reasonably favorable conditions. Restrictions are imposed on experimental techniques by the presence of the balloon system, but in general the available observational tools for exploration of the lower stratosphere have not been exploited.

The need for a knowledge of the structural characteristics of the lower stratosphere has gradually come into focus in the past two decades. Conduct of operations in this region with high-speed vehicles has necessitated a new look at those parameters which might influence the efficiency of these systems. The concept of a homogeneous upper atmosphere in which high-speed flight would be smooth has already fallen by the wayside in the upper troposphere with the advent of clear air turbulence. Meager data from sensitive instrumentation available on lower stratospheric structure do not indicate uniformity, although the size of these disturbances is not necessarily of dimensions which will produce significant interaction with an aircraft.

A more encompassing pressure for information on physical processes in the lower stratosphere is the need for a comprehensive knowledge of the dynamics of the earth's atmosphere in its entirety.

3.1 Tropopause

The concept of a tropopause surface separating the troposphere and the stratosphere has become one of the very meaningful terms which is applied to the upper atmosphere. The concept of a tropopause found its origin in the discovery that the temperature lapse rate which is so clearly characteristic of a troposphere experienced a significant change at the levels immediately above 10-km altitude. These early observations of temperature structure in the upper troposphere and the lower stratosphere were obtained in midlatitude Europe and were thus biased toward the considerable stability of the tropopause in this region. It was understood very early that the height and the temperature of the tropopause varied with synoptic situations, but the belief that a tropopause was an effective separating layer dividing the upper atmosphere from the troposphere became firmly rooted in these early observations obtained at strictly midlatitude stations. Subsequent observations of the temperature structure and the vertical structure of other atmospheric parameters have led to some questions as to the advisability of placing too much importance on a single parameter such as the tropopause. On the other hand, the tropopause has become one of the more definitive indicators for meteorologists and thus deserves considerable attention in its own right.

Considerable controversy has persisted relative to a proper definition of the tropopause, principally as a result of the considerable variations which have been observed at various locales. A most generally accepted definition of the tropopause was sanctioned by the World Meteorological Organization in 1957 in the following words:

> 1. The "first tropopause" is defined as the lowest level at which the lapse rate decreases to 2°C per kilometer or less, provided also the average lapse rate between this level and all higher levels within 2 kilometers does not exceed 2°C per kilometer.
>
> 2. If, above the first tropopause, the average lapse rate at any level and at all higher levels within one kilometer exceeds 3°C per kilometer, then a "second tropopause" is defined by the same criteria as the first tropopause. This tropopause may be either within or above the one kilometer layer.
>
> The first tropopause determined by this definition is generally accepted as the tropopause for a given situation in general usage. Only in more detailed studies is attention directed toward subsequent tropopause levels. In the latter category falls some of the more interesting aspects of the problem of vertical temperature structure of the atmosphere. Structure of the tropopause surface in midlatitudes during the stormy winter season is generally complex and may locally be ill-defined to the extent that several tropopause occurrences are noted on a single sounding. In general the cold winter polar region offers a considerable problem relative to the analysis for tropopause height and on occasion defies explicit measurement.

The definition of a tropopause layer plays a major role in atmospheric analysis in that it provides for definite separation of the regions of the atmosphere which receive their driving energy from significantly different sources. The region below the tropopause receives its energy principally from surface heating, which is then distributed throughout the troposphere by convective motions of the atmosphere. Above the tropopause, energy is absorbed in a bulk fashion by atmospheric interaction with solar and terrestrial radiant flux. Special spatial and time characteristics of the heat input generated by this diffuse absorption process can be expected to be closely related to the aspect and intensity of the incident radiation as well as any variability of the atmospheric absorbing parameters which may be incurred.

Actual original data have demonstrated that tropical regions, which are subjected to the almost perpendicular rays of the sun throughout the year, characteristically have a high cold tropopause. Characteristic mean values are altitudes of the order of 17 km and temperatures of the order of −80°C. The tropical tropopause is a very constant factor in the meteorological inventory and offers a very special region for study, well removed from the harassing effects of midlatitude storms. Certain complicating effects of earth's rotation are minimized at this point, although the tropical regions are recognized as having their own special environmental characteristics.

There is a general downward slope of the tropopause with increasing latitude. This down slope is quite gentle through the first 25 degrees of latitude throughout the year. During the less vigorous circulation season a smooth steadily increasing slope is exhibited by the tropopause surface into polar regions. The altitude of the tropopause in polar regions is generally 10 km or less. Although the above prescribed meridional distribution of tropopause height is a general feature most of the year, on occasion strong circulation systems may result in significant modification of this rather simple picture. Strong invasions of midlatitude regions by polar masses may frequently introduce quite low and relatively warm tropopause formations deep into the midlatitude belts. The high cold tropopause over the tropics is more generally representative of midlatitudes, and this configuration usually remains firm in the face of the polar invasion. Under these conditions a rather severe disruption of the normally smooth tropopause surface results. These very strong gradients in the tropopause are invariably associated with jet stream locations in which large amounts of kinetic energy are available.

A most disturbing influence on the selection of the tropopause level in the polar regions concerns the fact that, during the long polar winter night, surface layers on occasion represent the coldest temperature available in the lower atmosphere and thus preclude the attainment of the necessary 3 degrees per kilometer lapse rate specified in our definition. This means, in essence, that we have no troposphere over the polar regions under these conditions. Such a result leaves much to be desired, however, in view of our stated reason for determining the tropopause altitude; in other words, the failure of the temperature to decrease with height is not the result of absorbed radiant energy in this case. So this region does not fall into the category we had previously defined as the stratosphere. Our definition of the tropopause simply fails to meet the criteria for which it was set up.

An interesting difference exists between the temperature structure over the arctic and antarctic polar regions. In the mean, the tropopause over the arctic is high and cold in the summer with representative values of 260 mb and —50°C, and low and warm in the winter with representative values of 20 mb and —52°C. The antarctic tropopause is high and cold in the winter with representative values of 230 mb and —70°C, and low and warm in the summer with representative values of 300 mb and —50°C. The reason for these differences is not at all clear, although there is evidence that the antarctic tropopause regime is the more stable case with less variability exhibited. While variations with synoptic situations in the antarctic region are noted, they attain nowhere near the intensity observed occasionally in the arctic regions.

During the first half of the twentieth century the tropopause became accepted as an impervious surface separating the physical processes

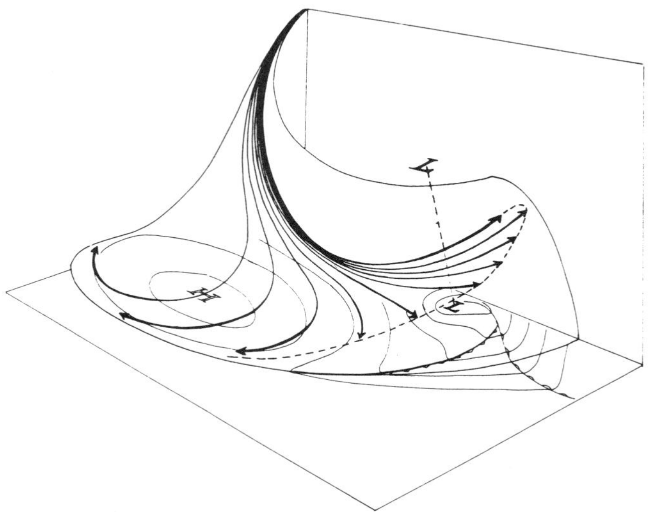

Fig. 3.1. Danielsen's view of stratospheric intrusion into the troposphere in the circulation systems established by tropospheric cyclones. (Courtesy Journal of Mineral Industries)

which dominate the troposphere and stratosphere. Since a dynamic concept would be more complex, it is more or less natural that a static picture of no mass transfer across this boundary would develop. While a case can be made for considerable separation of the tropospheric and stratospheric air masses, certain important exceptions appear to occur. Analyses of spatial distributions of radioactive materials with time (Danielsen, 1964) have indicated significant invasions of stratospheric air into the tropospheric storm systems as illustrated in Fig. 3.1. If one accepts these downflows of stratospheric air across the tropopause as a characteristic feature of cyclonic activity, it is obvious that a depletion of stratospheric air will result in the winter hemisphere. There will also necessarily be a return flow, possibly in the form of surges of convective activity, but more likely through slow migration of tropospheric air across the tropopause boundary. In most cases the measurements required to discuss these events are unavailable.

These considerations led to the characterization of the tropopause as a leaky barrier, which may be considered as a perfect seal for certain purposes, but as a rather poor surface in other cases. The processes which form the tropopause are dynamic in their application and appear to be capable of adjustment to certain vertical motions. Evaluations of residence times for stratospheric air which assumed a closed tropopause would appear to be highly questionable, with significantly shorter residence time the more likely case.

3.2 Stratonull

A principal unifying aspect of stratospheric and mesospheric regions of the earth's atmosphere concerns the circulation that dominates them. Their highly organized circulation system is characterized by very light winds at a base of approximately 24-km altitude, increasing comparatively steadily to a peak profile speed at altitudes from 50 to 65 km, and decreasing with altitude throughout the mesosphere. It is particularly of note that the region of increasing wind speed with height occurs in a very stable region of increasing temperature with height. This is opposite to the situation in the troposphere, where the increasing wind speeds with height are embedded in a layer which has a decreasing temperature with height. In addition, the decreasing wind speeds with height which are characteristic of the mesosphere occur in a region which shows a decreasing temperature with height, whereas the upper side of the tropospheric jet streams in the lower stratosphere is accompanied by an isothermal or slightly increasing temperature profile. There is adequate reason therefore to expect the stratospheric and mesospheric wind system to exhibit significantly different structural features than its lower atmospheric counterparts. In general, the gross hemispheric circulations which dominate the stratospheric and mesospheric regions of the earth's atmosphere represent the best observed fixtures and most obvious starting point for any analysis of the characteristics of this region.

The boundary between the tropospheric and the stratospheric-mesospheric circulation systems has only recently been defined. This is principally because the level of this separation is located near the upper portions of balloon soundings; that is, at about 25-km altitude. In fact, only recently have balloon soundings been able to attain the 25 km required to reach this minimum wind level. The accuracy of wind measurements by many of the early techniques for exploring the lower stratosphere decreased with altitude, particularly in this region above 25 km. As a result, most attention was focused on the decay with height of winds of the tropospheric jet stream regions until near zero wind speeds were attained in the lower stratospheric regions near the termination point of balloon soundings. The features of another strong independent circulation system in the upper stratosphere became clear only when systematic sounding of this atmospheric region became possible through high-altitude balloon systems and rocket vehicles. The rocket techniques have inherently favorable characteristics for exploration of the stratosphere in a short sampling period and rather accurate systems for obtaining wind and temperature measurements.

Observations in recent years point clearly to the existence of a well-defined separating layer between the tropospheric and stratospheric circulations, particularly at middle and low latitudes. It is true that the separation of the lower and upper atmospheric circulation systems is less clearly defined in polar regions, and the analysis is as yet inadequate in a restricted region near the equator. It is also true that little is known relative to mass transport across this interesting boundary layer. The criteria with which we established the level of the stratonull become invalid during the spring and fall reversal periods of the stratospheric zonal circulations. The stratonull level is quite difficult to establish during the summer circulation case also. In summer, easterly winds frequently increase steadily from the surface throughout the troposphere and stratosphere, so the basis for the clear-cut determination of the stratonull level disappears. However, even during this season a distinct increase in the wind lapse rate at the stratonull level is quite common, with values in the stratospheric circulation approximately double those in the troposphere.

The stratonull level is defined here as the level of minimum meridional temperature gradient in the lower stratosphere. An important feature of the lower stratosphere is the fact that temperature increases with increasing latitude throughout the region from tropopause to about 24-km altitude. The maximum meridional temperature gradient is to be found at about the level of the equatorial tropopause where temperatures in the mean of $-80°$ are common while in polar regions values of $-60°$ or even warmer are representative for this level of about 17-km altitude. This strong heat difference results in winds which must necessarily show an easterly thermal wind component with height and results in a rapid decrease in the tropospheric wind systems which have their peak near midlatitude tropopause. This process results in the almost complete destruction of these tropospheric wind systems in the lower stratosphere below an altitude of approximately 20 km. Apparently, the driving mechanisms which produce strong wind systems in the troposphere exert a controlling influence on the thermal wind intensity of the lower stratosphere, so that the thermal wind of the lower stratosphere has the correct value required to keep winds at the stratonull level at low values. Thus, when the tropospheric jets are light, the thermal easterly winds which are characteristic of the lower stratosphere throughout the year have a low value also; and only light easterly winds are developed in this region even if the winds are very light at the usual jet stream maximum of the tropopause.

While the stratonull surface offers an informative means of subdividing the atmosphere into workable regions, there is little reason to expect it to form a real barrier to the transport of mass between the upper and lower stratosphere. In fact, it hardly seems reasonable to expect these flows to slip quietly past each other. It would be far more realistic to expect gross variations in the stratonull surface to be produced by interactions between the tropospheric and stratospheric circulations. The significance of the stratonull surface, then, is only that it is a convenient measure of the level of separation of these circulations. Through analysis of its antics additional information may well be gained relative to interaction between the tropospheric and stratospheric circulations.

3.3 Equatorial Lower Stratosphere

At the level of the stratonull in equatorial regions a very curious circulation system has been observed. A biennial cycle dominates the lower stratospheric wind system above the tropopause, with a sharp maximum immediately above the equator at an altitude of approximately 24 km. This oscillation is superimposed on a mean easterly flow, and thus the westerlies are weaker and of shorter duration than the equatorial easterlies. The source of the biennial oscillation is at this time unknown, and the acquisition of an understanding of it portends real progress in analysis of the dynamics of the atmosphere.

The biennial cycle in the zonal wind observed in this region of the atmosphere is illustrated in Fig. 3.2 as prepared by Reed (1963b). These data show that the maximum development of the equatorial easterly circulation occurs between 10 and 15 degrees latitude north and south of the equator, and that an anomalous circulation must be present in the immediate vicinity of the equator to explain the considerable diminution of the easterly winds in that region, which even appear as a westerly mean flow in the extreme lower portion of the lower stratosphere. Analysis of the data available thus far indicates that there is a general easterly drift of equatorial lower stratospheric air over a mean period of the order of tens of years. On the other hand, the weakness of the mean easterly circulation, particularly in the immediate vicinity of the equator, lends credence to the concept of a varying circulation with alternate easterly and westerly flows. If this is so, it points clearly to such a circulation having its maximum in a narrow region about the equator.

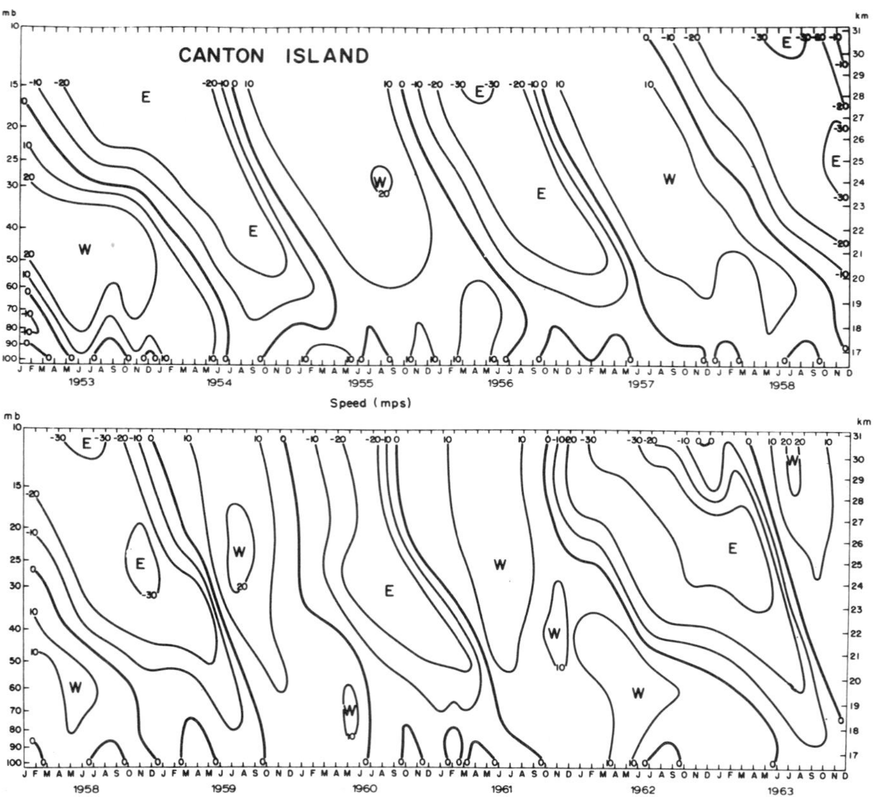

Fig. 3.2. Cyclic variation in the zonal winds of the equatorial lower stratosphere. (Courtesy U. S. Navy Weather Research Facility)

Early inspection of the data had indicated that there was general continuity with longitude of any prevailing circulation so data were included from all stations regardless of longitude location. These data, presented in Fig. 3.2, thus refer to a general circulation picture over the entire earth's equatorial region.

Reed (1963b) has shown that the amplitude of the annual cycle of wind variations in tropical regions decreases rapidly with decreasing latitude to a minimum at the equator. The biennial cycle, on the other hand, increases rapidly with decreasing latitude, attaining its maximum value over the equator at an altitude of approximately 25 km. This is illustrated in Fig. 3.3, where the dashed lines illustrate the amplitude of the annual cycle in meters per second and show practically no variation with altitude in the lower stratosphere in the immediate equatorial regions. A considerable altitude dependence of this parameter in the upper portions of the lower stratosphere is illustrated, with very little change occurring beyond the 15- to 20-degree latitude region. That is, the effect

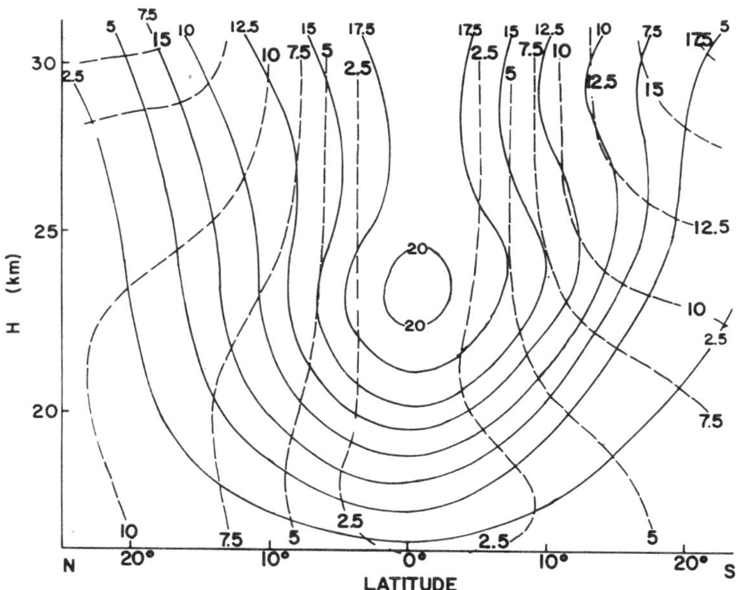

Fig. 3.3. Amplitudes of the annual (dashed) and biennial (solid) zonal wind oscillations in equatorial regions. Units are meters per second. (Courtesy U. S. Navy Weather Research Facility)

of a diminution of the annual cycle of wind variation is most obvious near the equatorial tropopause and therefore the effect extends further away from equatorial into subtropical latitudes. The solid lines in Fig. 3.3 illustrate the amplitude of the biennial cycle of the zonal winds in meters per second. One notes a dramatic difference between the biennial and annual cases in immediate equatorial latitudes. The center of activity is found in the 23- to 24-km altitude region immediately over the equator and mean biennial amplitudes of the zonal wind of 20 meters/sec are observed. The distribution of amplitude of the biennial cycle wind circulation fluctuations in the equatorial regions is essentially symmetrical about what we have already defined as the stratonull level, in all regions beneath and to either side of that level, but the biennial cycle is obviously connected into the upper stratospheric circulation since the curves of equal amplitude do not converge in that region but indicate an extension into higher latitudes as one moves into the upper stratosphere.

This biennial cycle is well illustrated by the analyzed data for Canton Island. Here we find periodic variations in the zonal winds from

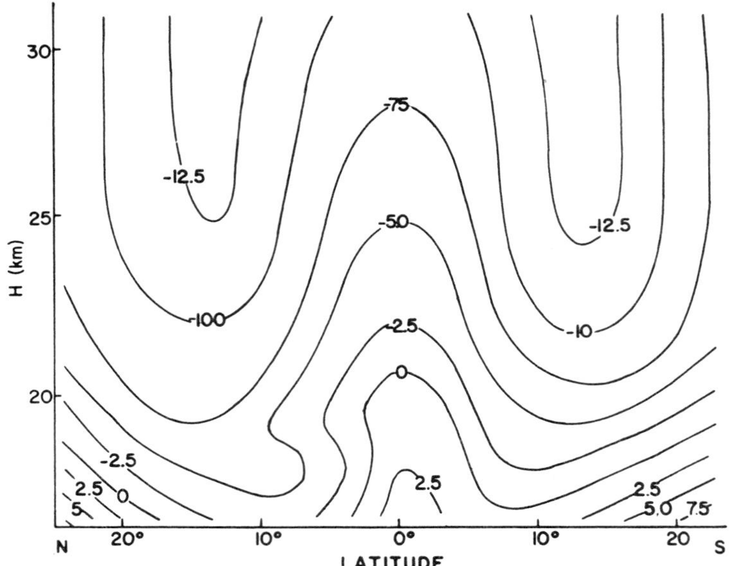

Fig. 3.4. Mean zonal wind component in meters per second of the lower stratospheric wind for the 10-year period through 1963. (Courtesy U. S. Navy Weather Research Facility)

a general westerly wind to a general easterly wind at periods of the order of a year for each of these phases. Again, the maximum amplitude of the westerly and easterly winds is to be found near the stratonull surface with peak westerly winds occurring in mid-1953, -1955, -1957, -1959, -1961, and -1963. Inspection of these data show that, in general, the easterly components are stronger than the similar westerly components and thus that the annual mean flow in the Canton Island region will average out to be easterly, as would have been predicted from the mean annual data presented in Fig. 3.4 for a location of 2° 46′ S. These intrusions of easterly winds from the upper stratospheric region down into the lower stratosphere exhibit considerable differences from one cycle to the next. For instance, the 1956 and 1960 cases would seem to indicate less development of the easterly circulation over Canton Island in the lower stratosphere. On the other hand, the 1954, 1958, and 1963 instances indicate considerably more massive invasions, particularly in the upper portions.

Spectrum analysis of 50-mb data for a 10-year period for San Juan, and for a 7-year period for Kwajalein and Canton Island, have been prepared by Dartt and Belmont (1964). The results of that work are shown in Fig. 3.5. The large amount of energy included in the biennial cycle at Canton Island and Kwajalein is evident, while a reduced amount

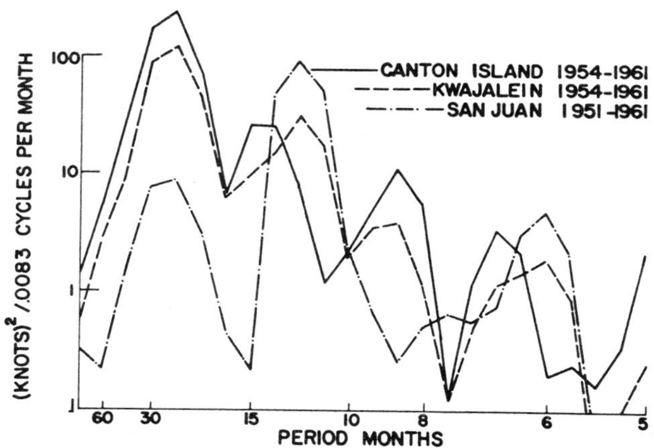

Fig. 3.5. Results of spectrum analysis of 50-mb wind data for a 10-year period at San Juan (18° 28′ N, 66° 7′ W) and a 7-year period at Kwajalein (8° 43′ N, 167° 44′ E) and Canton Island (2° 46′ S, 171° 43′ W). (Courtesy Journal of Geophysical Research)

at the more subtropical San Juan is indicated. It is clear from these data that a considerable reduction in the annual cycle occurs in the more equatorial locations. Of considerable interest in these data are the strong peaks which occur in the tropical spectrum curve at periods of approximately 8½ months. The source of energy for this period is not obvious, and the possibility exists that it might be a harmonic or subharmonic of some other input frequency.

Search for a cause of this very unusual circulation system is one of the more interesting research problems for current-day atmospheric physicists. The data imply an upper stratospheric origin and the Meteorological Rocket Network offers the opportunity to look for possible sources. A limited amount of data from equatorial MRN sites indicates that the biennial cycle decreases in intensity with height in the upper stratosphere until it is insignificant at the stratopause. The peak intensity of the biennial winds occurs at the stratonull level immediately above the equator. As will be discussed in a later section, zonal winds at the stratopause level exhibit a strong asymmetric semiannual oscillation which may well be related to the originating mechanisms which produce the biennial cycle.

3.4 Temperature Structure

The lower stratosphere has been defined as that region of the upper atmosphere which lies between the level at which the tropospheric tem-

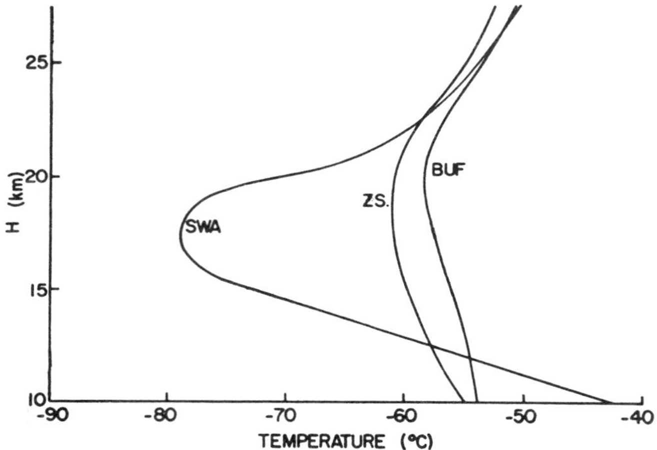

Fig. 3.6. Five-year mean vertical temperature profiles for January in the lower stratosphere obtained from 0600Z balloon soundings at Swan Island (SWA), Buffalo (BUF), and Coral Harbour (ZS).

perature lapse rate is broken and the level at which the meridional temperature gradient is a minimum or exhibits an inflection point with height with respect to space and time. Considerable ambiguity can occur in a specific situation relative to the level of these surfaces when the vertical structure of the atmosphere is quite complicated. Since both surfaces are defined by the temperature structure of the upper atmosphere it is essential that we consider the stratification of this atmospheric parameter in the lower stratosphere.

To maintain some generality in our discussion of the temperature structure we will consider 5-year average values of radiosonde temperature measurements obtained during nighttime hours at stations scattered along the 80th meridian from equatorial to polar regions for the period of 1959 to 1963. It is only during this relatively recent period that peak altitudes of routine upper air sounding systems have achieved the heights required to analyze the lower stratospheric region of the atmosphere. Figures 3.6 and 3.7 illustrate the January and July mean profiles for Swan Island (17° 24′ N), for Buffalo (42° 56′ N), and for Coral Harbour (64° 11′ N). It is apparent from these mean data that the tropopause has its maximum altitude and its lowest temperatures in equatorial regions during the winter months. It is equally obvious that the lower stratosphere is dominated in both of these major seasons by a positive meridional temperature gradient which has its peak value at about the altitude of the equatorial tropopause. During the summer season, one

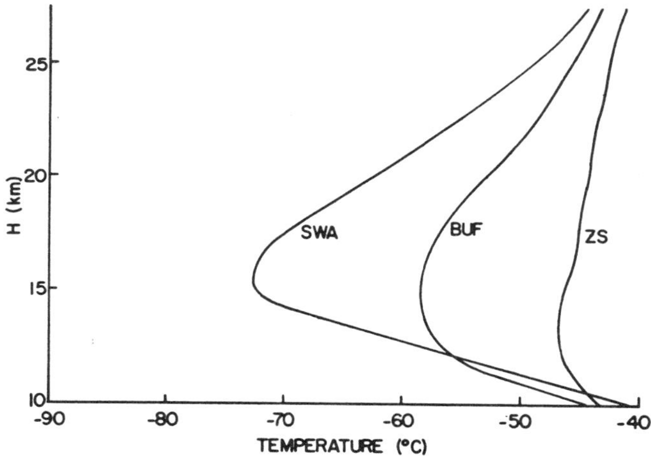

Fig. 3.7. Five-year mean vertical temperature profiles for July in the lower stratosphere obtained from 0600Z balloon soundings at Swan Island (SWA), Buffalo (BUF), and Coral Harbour (ZS).

notes a maximum temperature difference between Swan Island and Coral Harbour of approximately 28°C in these mean data, at an altitude of about 16 km. The January mean data show a more conservative temperature difference, even reversing at the higher altitudes with a maximum difference indicated by these data of approximately 21° between Swan Island and Buffalo. The heat engine which operates in the lower stratosphere can be said to be more uniform with latitude during the summer season and to achieve its maximum value during that period, while it is confined to lower latitudes in the winter season and is significantly reduced in intensity.

Further consideration of the July mean data points out the fact that the tropopause level falls with increasing latitude, ranging from values of the order of 15 km in the tropical regions to approximately 11 km in polar regions. In winter, this slope is far steeper, being of the order of 17-km altitude in the tropical regions, and ranging to below 10-km altitude in polar regions during the winter season.

Our definition of the stratonull level is based on the occurrence of a minimum in the meridional temperature gradient. During most of the year there is a reversal in the meridional temperature gradient between the lower and the upper stratosphere and therefore the minimum value of meridional temperature gradient is well defined as zero. Consideration of the January data indicates, however, that this level is not uniform with latitude. In this particular case it would appear to have an altitude of the order of 26 km in lower midlatitudes, dropping to near 22½-km altitude at high latitudes. Again one has to expect that considerable variability will be found if the resolution of our data is improved in space and time. In the summer data the level of the stratonull is less clear-cut. In fact, since all of the stratosphere exhibits a warming trend with increasing latitude, it is probable that this period will leave the stratonull generally ill-defined. However, there is indication that it is located at higher levels of the order of 28 km.

Comparison of these mean profiles for Swan Island and Buffalo point out the fact that the meridional temperature gradient has its maximum during the winter season between these two stations, indicating a very strong thermal engine in operation in the lower stratosphere during this winter season. The peak occurs at about 17-km altitude in these lower latitudes. During the summer season the meridional gradient between Swan Island and Buffalo is reduced from the 20° winter value to approximately 15° and this maximum difference is reduced in altitude to about 15 km.

Thermal winds generated by these meridional temperature gradients are thus easterly throughout the year in the lower troposphere. They will have their maximum intensity in lower latitudes during the winter season and will be relatively uniformly distributed over the entire hemisphere during the summer season. Maximum intensity of these easterly thermal winds will be at a level of approximately 17 km during the winter season. The gradients of these thermal winds will be approximately the same above and below this level of peak intensity, being initiated at about 12½-km altitude and disappearing immediately above 22½ km. During the summer season the easterly thermal winds will have their peak intensity at around 15-km altitude. The gradient of these easterly thermal winds will be approximately double beneath this level of maximum value compared with the gradient above. The lower level of this easterly thermal wind regime will have its lowest altitude in polar regions, sloping to near 10 degrees there and reaching upward toward the 15-km level in equatorial regions. The upper level will be characterized by only small values of the easterly thermal wind, the lowest of which will be achieved in the 25- to 30-km range. The considerable detailed structure evidenced even in these meager mean data indicates that a more thorough inspection of the lower stratospheric thermal field is required. The meridional temperature structure of the lower stratosphere is illustrated in Figs. 3.8 through 3.19. These charts were constructed from 5-year monthly mean values obtained from United States and Canadian chronological summaries of radiosonde observations in these regions for the years 1959 through 1963. The stations included,

TABLE 3.1

Radiosonde stations from which data were used to construct Figures 3.8 through 3.19

Station	Symbol	Number	Latitude	Longitude
Swan Island	SWA	501	17° 24′ N	83° 56′ W
Miami	MIA	202	25° 49′ N	80° 17′ W
Tampa	TPA	211	27° 58′ N	82° 32′ W
Jacksonville	JAX	206	30° 25′ N	81° 39′ W
Charleston	CHS	208	32° 54′ N	80° 2′ W
Greensboro	GSO	317	36° 5′ N	79° 57′ W
Pittsburgh	PIT	520	40° 30′ N	80° 13′ W
Buffalo	BUF	528	42° 56′ N	78° 44′ W
Maniwaki	MW	722	46° 22′ N	72° 59′ W
Moosonee	MO	836	51° 16′ N	80° 39′ W
Port Harrison	PH	907	58° 27′ N	78° 8′ W
Coral Harbour	ZS	915	64° 11′ N	83° 22′ W
Arctic Bay	AB	918	73° 00′ N	85° 18′ W
Thule	THU	202	76° 33′ N	68° 47′ W

along with their latitude, longitude, station numbers, and indicators, are listed in Table 3.1. Thus, the charts present a meridional cross section along the 80th meridian from equatorial to polar regions in the Northern Hemisphere. The contours represent lines of equal temperature gradient along the meridian with positive indicating increasing temperatures poleward.

Inspection of these charts shows a pronounced difference between the winter and summer seasons relative to the temperature structure of the lower stratosphere. In the winter case, the maximum observed gradients of meridional temperature distribution are found at lower latitudes than are observed in the summer case. Peaks in the meridional temperature gradient are observed at about 31 degrees latitude at a height of 17 km in January with a maximum strength of this lower-

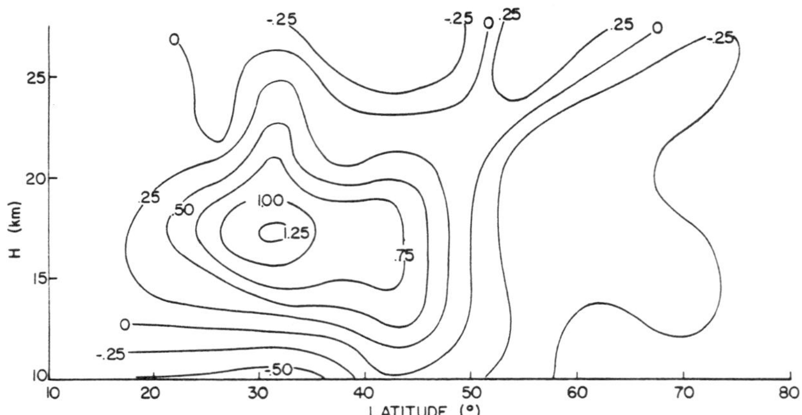

Fig. 3.8. Mean meridional temperature gradient (degrees Celsius per degree latitude) for January from radiosonde data obtained at stations near the 80° W meridian in the Northern Hemisphere during 1959–1963.

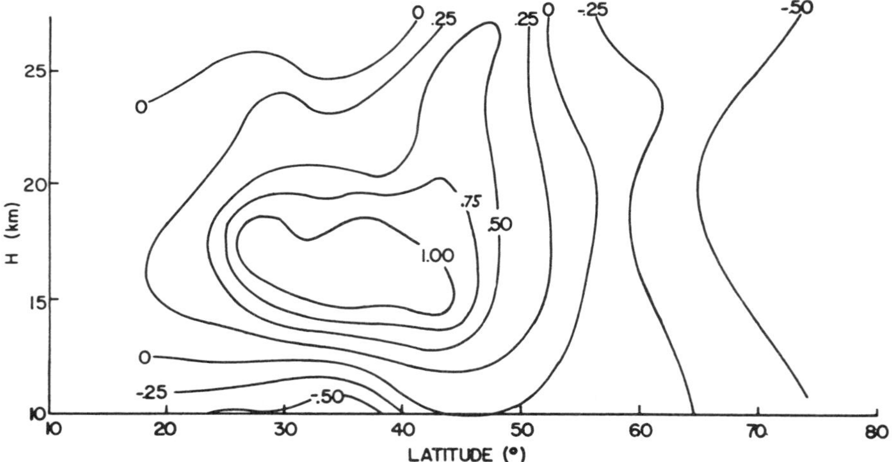

Fig. 3.9. Mean meridional temperature gradient (degrees Celsius per degree latitude) for February from radiosonde data obtained at stations near the 80° W meridian in the Northern Hemisphere during 1959–1963.

latitude thermal gradient to be found at about the same altitude and at 30 degrees latitude in April. A significant weakening of this peak gradient at low latitudes is observed immediately with the beginning of the summer season, and it is quickly supplanted by a peak meridional

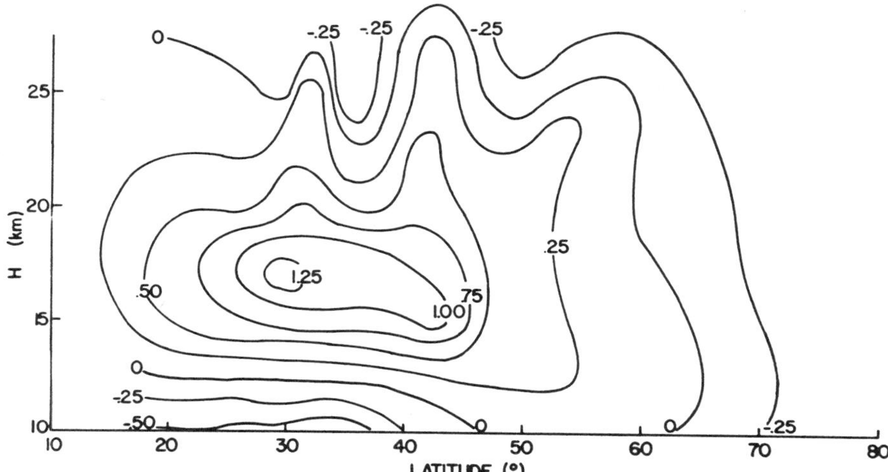

Fig. 3.10. Mean meridional temperature gradient (degrees Celsius per degree latitude) for March from radiosonde data obtained at stations near the 80° W meridian in the Northern Hemisphere during 1959–1963.

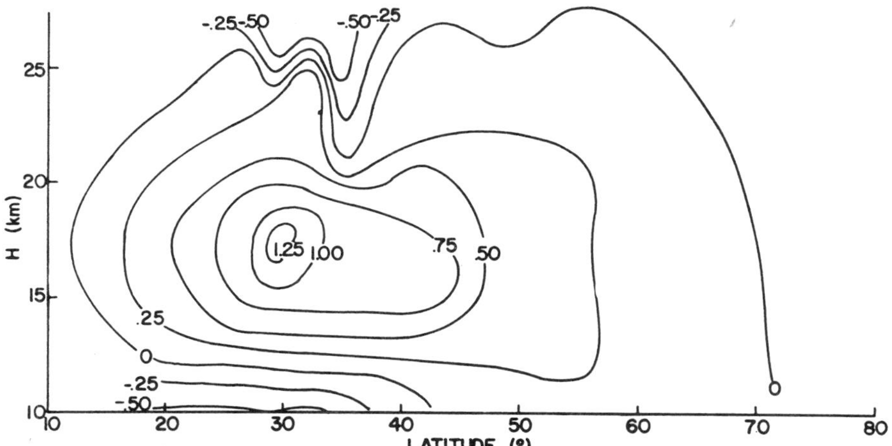

Fig. 3.11. Mean meridional temperature gradient (degrees Celsius per degree latitude) for April from radiosonde data obtained at stations near the 80° W meridian in the Northern Hemisphere during 1959–1963.

temperature gradient at the lower altitude of approximately 15 km at a latitude of 42 degrees, as is illustrated in the July and August data.

The dying out of the winter thermal regime in the lower stratosphere and the buildup of the summer structure occur at a comparatively rapid

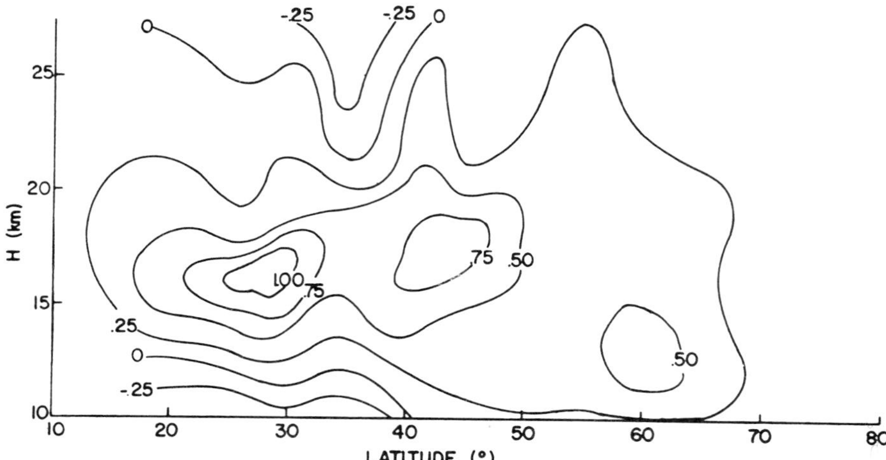

Fig. 3.12. Mean meridional temperature gradient (degrees Celsius per degree latitude) for May from radiosonde data obtained at stations near the 80° W meridian in the Northern Hemisphere during 1959–1963.

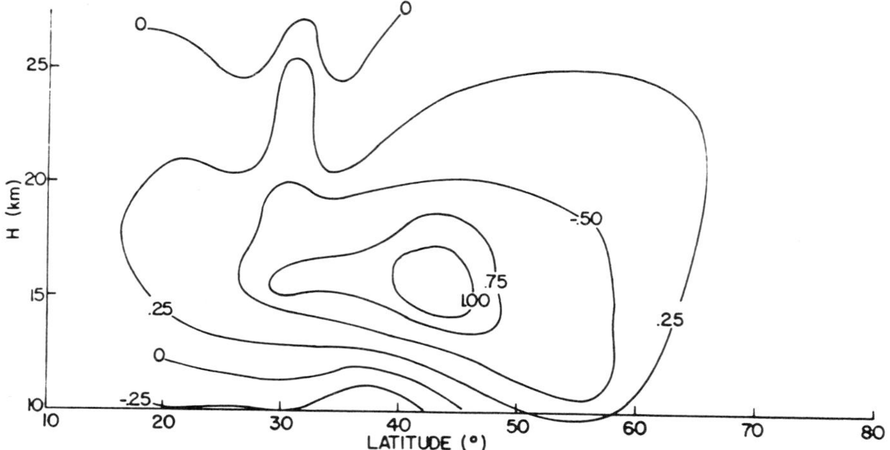

Fig. 3.13. Mean meridional temperature gradient (degrees Celsius per degree latitude) for June from radiosonde data obtained at stations near the 80° W meridian in the Northern Hemisphere during 1959–1963.

pace. The strong center of the order of +1.25°C per degree latitude, which appears in the April cross section at about 30 degrees latitude, is weakened significantly in the May data and is almost entirely absent in the June cross section. The first appearance of a peak in the meridional

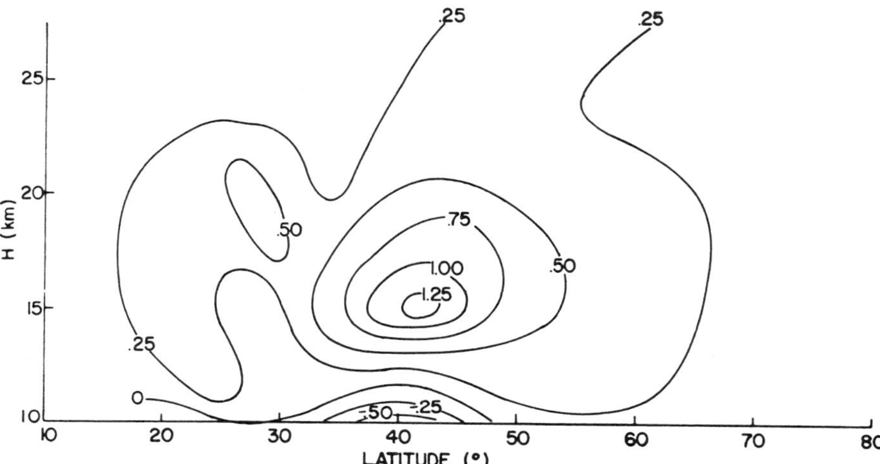

Fig. 3.14. Mean meridional temperature gradient (degrees Celsius per degree latitude) for July from radiosonde data obtained at stations near the 80° W meridian in the Northern Hemisphere during 1959–1963.

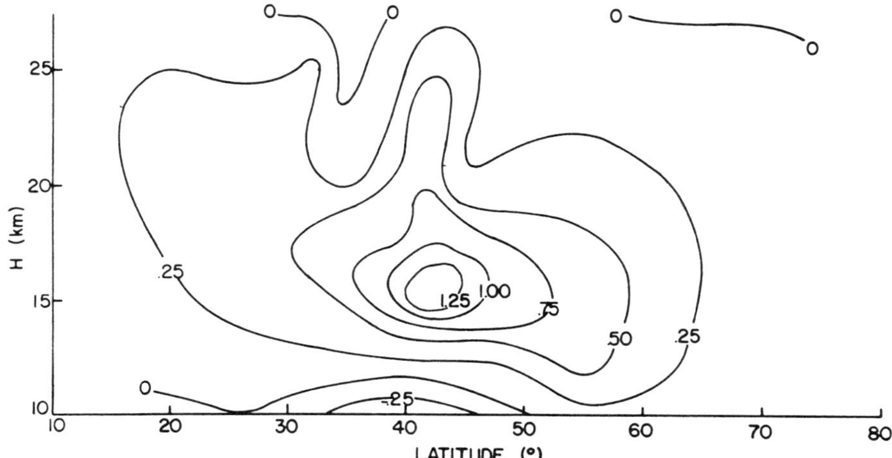

Fig. 3.15. Mean meridional temperature gradient (degrees Celsius per degree latitude) for August from radiosonde data obtained at stations near the 80° W meridian in the Northern Hemisphere during 1959–1963.

temperature gradient in the 40- to 45-degree latitude region appears in the May mean data and the gradient is well established in June although it does not reach its peak value of $+1.25°$ until July. There is a general consolidation of the summer thermal structure in the lower stratosphere throughout the period from May to July, but the shift from one circu-

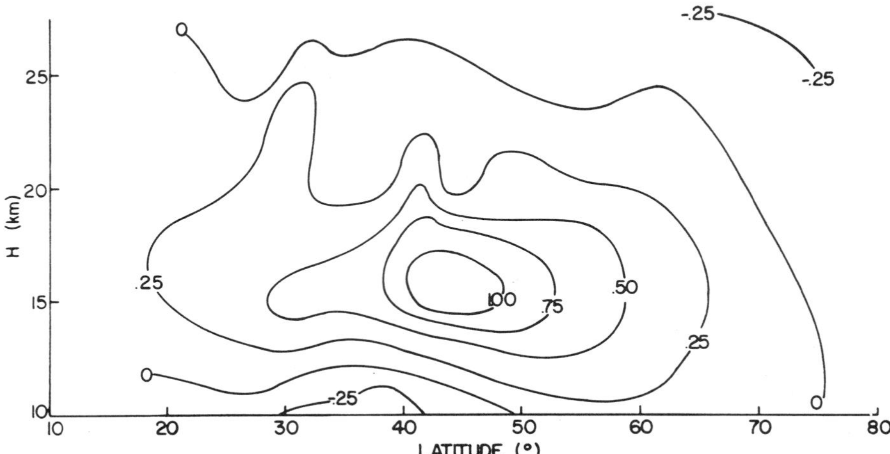

Fig. 3.16. Mean meridional temperature gradient (degrees Celsius per degree latitude) for September from radiosonde data obtained at stations near the 80° W meridian in the Northern Hemisphere during 1959–1963.

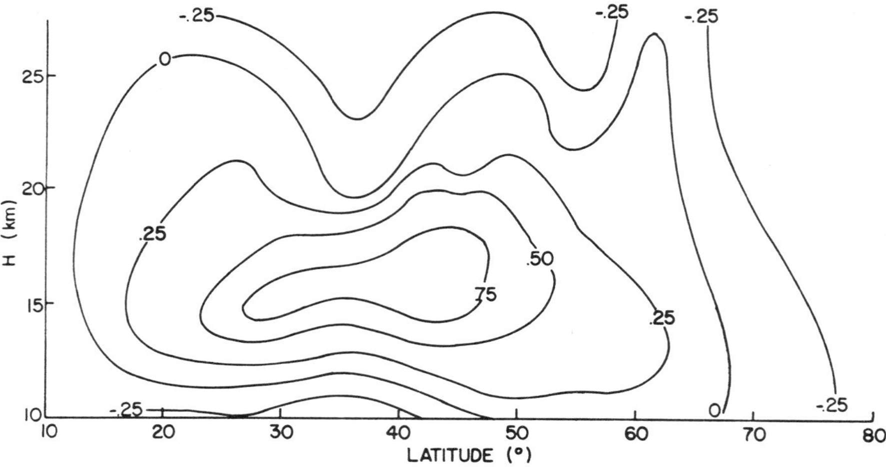

Fig. 3.17. Mean meridional temperature gradient (degrees Celsius per degree latitude) for October from radiosonde data obtained at stations near the 80° W meridian in the Northern Hemisphere during 1959–1963.

lation system to the other would appear to occur very rapidly between April and June.

On the other hand, the change from a summer to a winter meridional temperature structure occurs in a progressive fashion from September through January. The equatorward extension of meridional temperature gradient contour lines has begun in the September mean data and is

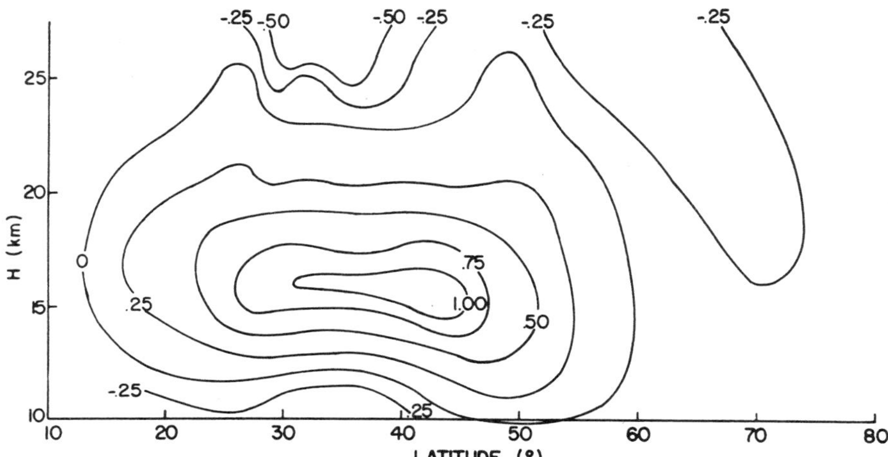

Fig. 3.18. Mean meridional temperature gradient (degrees Celsius per degree latitude) for November from radiosonde data obtained at stations near the 80° W meridian in the Northern Hemisphere during 1959–1963.

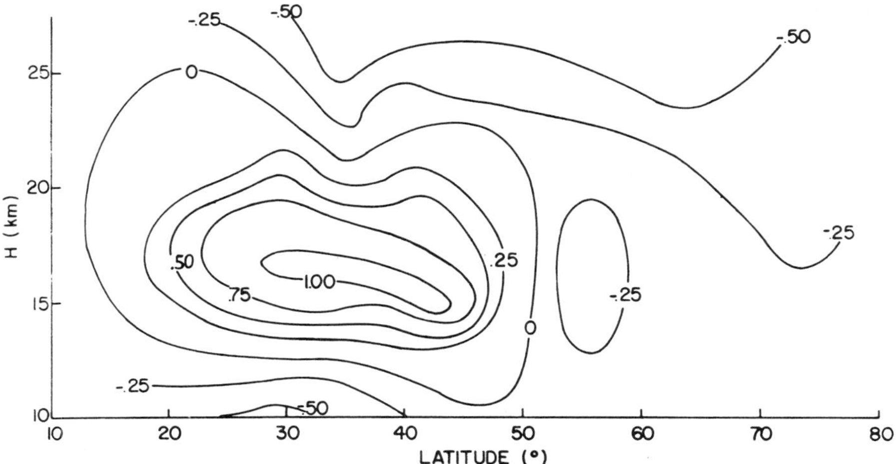

Fig. 3.19. Mean meridional temperature gradient (degrees Celsius per degree latitude) for December from radiosonde data obtained at stations near the 80° W meridian in the Northern Hemisphere during 1959–1963.

progressively strengthened during the fall and early winter months, although the extent of strong meridional temperature gradients is a maximum during this period ranging from lower midlatitudes to upper midlatitudes.

The equatorward shift of strong positive meridional temperature gradients in the lower stratosphere is accompanied by the appearance of weak negative meridional temperature gradients in the polar regions. These regions of negative gradient achieve their maximum in January when the —0.25 contour dominates the higher latitudes throughout the lower stratosphere. The contour lines during this season are near vertical in the 45- to 60-degree latitude region and therefore indicate rather uniform conditions in the vertical.

The summer season in the lower stratosphere is dramatically different in that the positive meridional temperature gradients extend well into high latitudes to dominate the structure in that region. To illustrate these points, one may note that the zero contour line is found as far equatorward as 50 degrees latitude in January while during the summer season the +0.50°C per degree latitude contour extends poleward almost to the 60-degree latitude line.

Relatively little annual change is observed in the meridional temperature gradient in equatorial and midlatitude zones. The equatorward shift of peak gradients which appear during the winter season serves to pack the contour lines in lower latitudes, but does not result in an extension of the strong meridional temperature gradient zone into equatorial regions. For example, the meridional temperature gradient at 20 degrees latitude remains at approximately +0.30 in the central lower stratosphere throughout the year. The shift of peak meridional temperature gradients during the summer season results in a flattening of the gradient throughout the lower stratosphere at lower latitudes during the summer season.

During the winter season, the zero contour line falls near the location of the tropopause, at least in lower latitudes. One observes this zero contour consistently during the winter months to appear at around the 12- to 13-km altitude level. During the summer months, it drops significantly into the 10- to 11-km altitude range at lower latitudes. At middle latitudes between 30 and 45 degrees, there is consistently an upward bulge of the tropospheric thermal structure regime to higher altitudes. This is particularly pronounced during the summer months with upward positions of contour lines at least a kilometer higher at 40 degrees latitude than at their lower-latitude counterparts.

Negative meridional temperature gradients appear in the lower stratosphere at high latitudes only during the winter months of December, January, and February. The downward and poleward extension of the characteristic lower stratospheric meridional temperature structure is first obvious in the March data and becomes quite pronounced throughout the summer season with the zero contour falling well below 10-km altitude at latitudes above 45 degrees.

Much detailed structure is to be observed in the upper boundary of the lower stratosphere, which separates it from the upper stratospheric regions. This boundary, which has been defined as the stratonull, shows variability with latitude even in these mean monthly data. One notes a comparatively simple structure in the January data except for a strong upward protuberance of the lower stratospheric meridional temperature regime at around 32 degrees latitude until, at high latitudes, the stratonull surface rises sharply as the characteristic lower stratospheric thermal regime extends into upper stratospheric altitudes over the polar regions. A consolidation of this positive meridional temperature gradient back into the lower stratosphere is evidenced in the February data with the development of a well-defined arc of +0.50°C per degree latitude contour in the 45- to 50-degree latitude band. The perturbation which was observed at 32 degrees in the stratonull surface in January is also still apparent, although somewhat weaker, in the February data. Inspection of the March chart shows strong deviations in the stratonull level both at the 32- and 43-degree latitude locations. In this case, the upper stratospheric invasions of lower stratospheric territory appear to be of nearly equal amplitude as the upward extensions of lower stratospheric meridional temperature structure. In April, when the maximum positive meridional temperature gradient in the lower stratosphere has achieved its furthest equatorward extension, the zone of the interaction between the upper and lower stratosphere has also achieved its maximum equatorward displacement with a peak invasion of lower stratospheric structure into the upper stratosphere occurring at about 32 degrees latitude with a strong, well-developed surge of upper stratospheric structure down into the lower stratosphere appearing at about 35 degrees latitude. These strong latitudinal variations in temperature structure in these mean data are indicative of strong interactions between these two atmospheric layers in the 30- to 35-degree latitude band as the winter season dissipates and the reversal to summer circulation is initiated.

Perturbations in the contour lines which indicate the separation between the upper and lower stratosphere are maintained even during the summer season when most of the interaction zone is relatively calm, although they are definitely weaker than in the late winter case when significantly weaker perturbations of the contour lines in the 30- to 35-degree latitude zone in May and June are noted, and even more obvious in August when the peak gradients in the 40- to 45-degree latitude band begin to show an upward extension.

The summer meridional temperature gradient would seem to be characterized by an axis of maximum intensity which slopes downward in the poleward direction, ranging from the 17-km altitude of lower midlatitudes, characteristic of the winter, to 15 km and lower in the upper midlatitudes. On the other hand, the winter season would appear to be more horizontally stratified relative to the intensity of the contour lines in middle and lower latitudes with an upward tilt of this axis in the higher latitudes, particularly during the January and February period.

The charts from January through April present every evidence of an equatorward-moving disturbance affecting the interaction zone between the lower and upper stratosphere. This may be misleading in that these are mean data and the disturbances which they depict there may be actual breaks in the stratonull surface. This would be analogous to the breaks which are found in the tropopause separating the troposphere and the lower stratosphere.

The stratonull level shows its mean altitude of 25 km in lower latitudes, dipping to as low as 23 km at about 35 degrees latitude and then rising well above 30 km in the upper and polar latitudes of the stratosphere. It is clear, therefore, that through May there is a gradual equatorward shift of the center of maximum meridional temperature structure as the winter season decays and the summer season begins to develop. The peak intensity has drifted from the mid-30's into the 20-degree latitude zone with a general lowering of the center of activity from about 17½- to about 16½-km latitude. The middle and upper latitudes of the lower stratosphere during the month of May exhibit in the mean a rather weak meridional temperature structure. The mean chart of the meridional structure for the month of June evidences a decidedly different structure than had been previously presented. The maximum positive meridional temperature gradient of over 1°C per degree latitude occurs in the 40- to 45-degree latitude zone. Altitude ranges are from about 14 to 17 km for this particular contour line. The +0.5° meridional temperature gradient contour extends to almost 60 degrees latitude where in previous months during the year it had not extended poleward above 50 degrees latitude. Therefore, there is a decided poleward shift of the maximum meridional temperature gradient which characterizes the summer lower stratosphere.

The high-latitude and polar regions of the lower stratosphere are characterized by relatively flat contours of positive meridional temperature gradient during the summer season. A perturbation still exists in the 30-degree latitude zone with upward extensions of the typical lower stratospheric meridional temperature gradient driving the stratonull surface upward from its general 25- or 26-km altitude region in a local zone. The stratonull level appears to be at higher altitudes in upper and polar latitudes during this season. An upward extension of the tropospheric regime is also apparent in the 30- to 40-degree latitude zone.

The July data (Fig. 3.14) further illustrate this northward shift of the center of stratospheric positive meridional temperature gradient. In the July case we have a maximum gradient of over +1.25°C per degree latitude centered at about 42 degrees latitude and an altitude of 15 km. The latitude of the +0.5°C per degree latitude contour has shrunk to the 33- to 54-degree latitude range, although a small segment of the 0.5° contour remains at higher altitudes in the 25- to 30-degree latitude belt.

The July data show a further weakening of the meridional temperature gradient in the lower stratosphere and tropical regions with a continued presence of a moderate positive meridional temperature gradient well into the high-latitude zone. The upward extension of the tropospheric regime is again evidenced in the 40-degree latitude belt. The stratonull is not clearly indicated in these data, but would appear to slope upward at higher latitudes as in the June case.

The August data (Fig. 3.15) show a strengthening of the positive meridional temperature gradient of the middle lower stratosphere in the 40- to 45-degree latitude region. Values greater than +1.25°C per degree latitude occur between 15- and 16-km altitude at above 42½ degrees latitude and an upward extension of this strong meridional temperature gradient is evidenced all the way to 27-km altitude over this 40- to 45-degree latitude zone. Moderate positive meridional temperature gradients extend well into high-latitude regions with general tendencies of downsloping toward the poles, while further weakening of meridional temperature gradients in the subtropical region is indicated during August.

Apparently the stratonull surface is quite high during this season, dipping downward to approximately 23-km altitude at about 35 degrees latitude, while the tropospheric intrusion into lower stratospheric regions is most pronounced between 35 and 45 degrees latitude. August marks the maximum depression of the zero meridional temperature gradient line separating the troposphere and stratosphere at low latitudes, touching the 10-km level at about 25 degrees latitude. The September meridional cross section of temperature gradient (Fig. 3.16) exemplifies a well-defined stratonull layer averaging around 25-km altitude extending from equatorial to polar regions. The lower separation of the stratosphere and the troposphere is similarly well defined by a zero meridional temperature gradient that dips to about 11-km altitude in the 20- to 30-degree latitude band, rises to 12-km altitude around 40 degrees latitude, and then passes below the 10-km level near 50 degrees. While the upper portion of the lower stratosphere is fairly well defined at the stratonull level, considerable detail structure is in evidence in the +0.25°C per degree latitude contour line with a peak occurring at about 42 degrees latitude and a similar peak occurring at about 32 degrees latitude. The strongest point in the temperature gradient occurs near 45 degrees latitude at about 16-km altitude.

The absence of strong meridional temperature gradients in the October data (Fig. 3.17) denotes some weakness in the temperature structure, but a strengthening of the contour lines, particularly in their development toward lower latitudes, is indicated. A withdrawal of the +0.5° contour away from the 60-degree latitude back into its winter position near 50 degrees is indicated. The development of the most intense gradients at low latitudes appears to occur around 15-km altitude during the month of October. The stratonull level is well defined but has rather large amplitudes of height variations even in these mean data. Maximum intrusions of upper stratospheric air to lower altitudes occur in the 35- to 40-degree latitude band with the zero meridional temperature gradient contour extending as low as 20-km altitude at that point. A secondary dip of the zero meridional temperature gradient contour to about 22-km altitude is noted at about 55 degrees latitude. The lower boundary of the positive meridional temperature gradient region which characterized the lower stratosphere has lifted some from its summer minimum to a mean altitude of almost 12 km during the month of October.

The mean cross-sectional data for November (Fig. 3.18) show a well-developed structure encompassing most of the lower stratosphere. Almost symmetrical contours ranging from a maximum strength of slightly over +1.00°C per degree latitude appear at around 16-km altitude and extend from 31 to 46 degrees latitude. The +0.5° contour line has retreated to about 51 degrees latitude and extends well into subtropical regions. The zero contour has risen to approximately 12 km in middle and low latitudes and extends down only to the 10-km level in high latitudes. The stratonull boundary is relatively clear-cut in the lower 20-km altitude range, with upward extensions occurring at around 26 and 49 degrees latitude.

The most notable change appearing in the December data (Fig. 3.19) is the further drift of the maximum temperature gradient toward low latitudes. A negative meridional temperature gradient is in evidence in the lower stratosphere between 50 and 60 degrees latitude while the center of positive meridional temperature gradient in general is from the 50-degree latitude line equatorward. Values of over +1°C per degree latitude are indicated, ranging from an altitude of about 15 km at 43 degrees latitude upward at lower latitudes to about 17-km altitude at 28 degrees latitude. The stratonull layer has dipped to lower altitudes, ranging from near 25 km in subtropical regions downward to 21 km at about 35 degrees latitude and to the lower 20-km region at 50 degrees latitude.

In summary then, we find the winter season is marked by the development of a strong positive meridional temperature gradient in lower latitudes, centering at 17½-km altitude at about 25 degrees latitude. Contours of temperature gradient with latitude are almost vertical in the high latitudes and polar regions, with negative gradients in evidence at the higher latitudes. The zero meridional temperature gradient contour is in general in the 50-degree latitude belt.

3.5 Circulation

In midlatitudes, the circulation of the lower stratosphere is dominated by the circulation of the troposphere. Maximum speeds of the tropospheric circulation systems are to be found at or slightly below the tropopause, and the midlatitude lower stratosphere is the scene of an easterly thermal wind which produces the wind minimum or inflection point at its upper surface and marks the level of the stratonull surface. The thermal wind equation may be expressed as

$$\frac{\partial \mathbf{V}_T}{\partial h} \times \mathbf{k} = \frac{g}{2\omega \sin \varphi T} \nabla T, \tag{3.1}$$

where $\partial \mathbf{V}_T/\partial h$ is the geostrophic wind gradient with height, T is the temperature, and ∇T is the temperature gradient in the horizontal, and states that the fact of an easterly zonal wind gradient with height must be ascribed to a positive meridional temperature gradient. Such a temperature gradient is illustrated for this region by the data for Swan Island (17° 24′ N) in the winter season in Fig. 3.6, and by the temperature differences between Swan Island and Coral Harbour (64° 11′ N) in the summer season in Fig. 3.7.

The detailed data on thermal gradient structure of the lower stratospheric region presented in Figs. 3.8 through 3.19 illustrate the mode of annual variation of this parameter. A positive meridional temperature gradient is thus characteristic of the entire summer hemisphere in the lower stratosphere, but is limited to lower and middle latitudes during the winter season. Above about 60 degrees latitude lack of insolation in the winter night results in a negative meridional temperature gradient in the lower stratosphere and effectively eliminates the stratonull from that region. In general, the positive meridional temperature gradient of lower and middle latitudes results from an over-all temperature difference

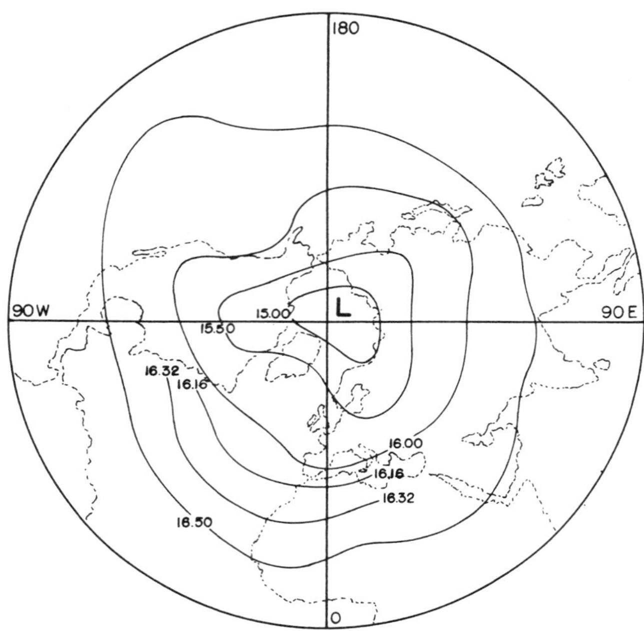

Fig. 3.20. One hundred-millibar chart for 12 January 1963 over the Northern Hemisphere. Contours are heights in kilometers.

of approximately 30° from the —80°C of the tropical tropopause to the nominal midlatitude —50°C temperatures of the central lower stratosphere.

Numerous treatments of the lower stratospheric circulation systems are available in the literature, so we will confine our attention here to a few cases of special interest. An intensive international observational effort in the lower stratosphere was made as a part of the International Geophysical Year program, and analyses of these data by Teweles *et al.* (1960) and others have illustrated the general and many of the detailed features of these circulation systems. Routine observation of the lower stratosphere has proved feasible and has continued. A program for the assimilation of these data into hemispheric analysis of the circulation structure of the upper reaches of balloon radiosonde observations has been conducted by the Institut für Meteorologie und Geophysik der Freien Universität Berlin under direction of Richard Scherhag and the sponsorship of the United States Army Research Office. Since 1958, daily charts have been prepared and published as volumes of the Meteorologische Abhandlungen of the University, along with certain summary analyses of these data.

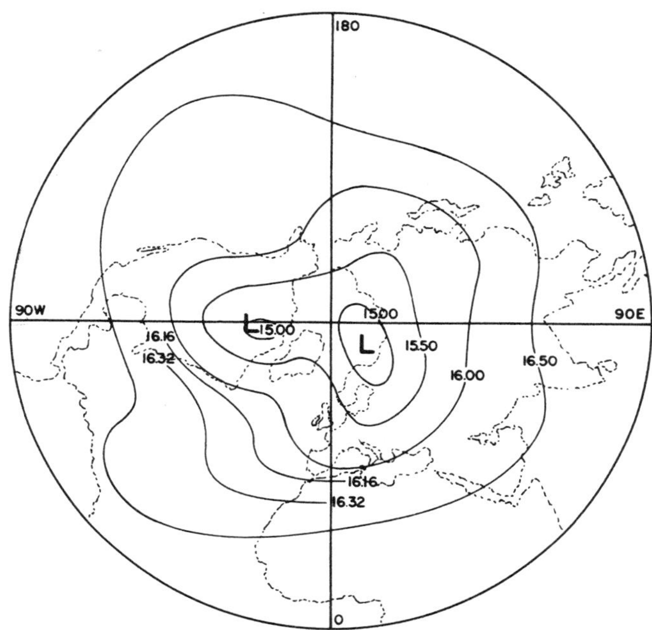

Fig. 3.21. One hundred-millibar chart for 14 January 1963 over the Northern Hemisphere. Contours are heights in kilometers.

The 100-mb constant pressure surface is chosen for our purposes since it lies near the level of the tropical tropopause, and thus represents the most intense portion of the lower stratospheric circulation system. A series of these maps is presented in abbreviated form in Figs. 3.20 through 3.30, and may be studied in Band XL, Heft 1, of the Meteorologische Abhandlungen. These particular charts were selected to depict progress of lower stratospheric circulation changes (or lack thereof) during a period of intensive change in the stratospheric circulation which will be described in some detail later. The time period covered here is considered to be a rather typical example of the type of phenomena discovered by Richard Scherhag in the winter of 1951–1952 in high-altitude balloon soundings and labeled "sudden warmings." The large temperature changes which he found are characteristic of higher levels, in the base of the stratospheric circulation system, but must surely be reflected in the lower stratospheric region as a response to a driving mechanism.

The Northern Hemisphere 100-mb chart for 12 January 1963 presented in Fig. 3.20 is generally representative of the topography of this pressure surface during the winter months. A deep cold low extending

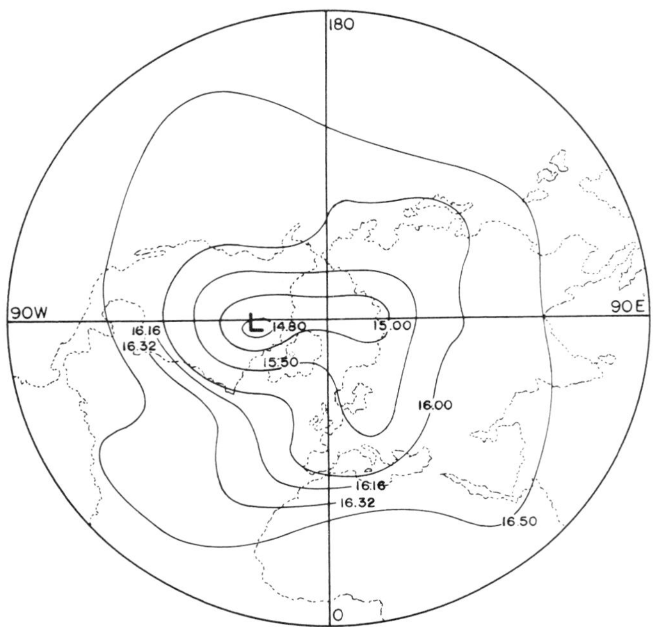

Fig. 3.22. One hundred-millibar chart for 16 January 1963 over the Northern Hemisphere. Contours are heights in kilometers.

to below 15-km altitude centered near the pole exhibits central temperatures of approximately —75°C. A lack of symmetry in the height contours is characteristic, with extensions to lower latitudes apparent over the continental land masses of North America and Eurasia. The temperature pattern is essentially symmetrical with the height contours poleward from 60 degrees latitude but shows marked variations in middle latitudes. A warm center with temperatures above —55°C is located in the southern extreme of the trough over the United States as is indicated by the strong curvature of the 16-km contour, and strong zonal winds are the rule over the eastern portions of North America.

A second warm center in elongated form extends from Japan toward Alaska, with warmest temperatures above —50°C, with a strong westerly circulation over southern Japan along the eastern edge of the warm core and in the center of the trough which extends into that region. The large expanse of middle latitude and tropical regions is characterized by a rather flat height contour gradient with maximum heights in the tropics, and an almost symmetric distribution of temperature contours, with temperatures falling from the warm ridge of —50°C in upper midlatitudes to the almost uniform cold —80°C of the tropical tropopause.

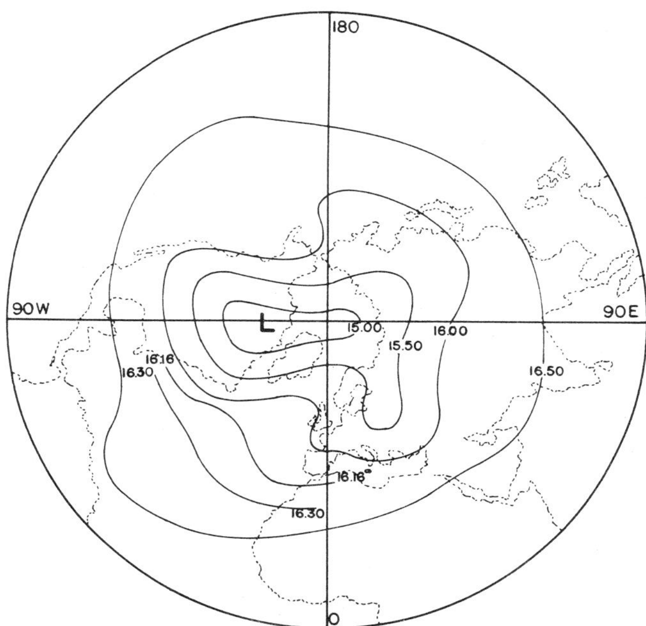

Fig. 3.23. One hundred-millibar chart for 18 January 1963 over the Northern Hemisphere. Contours are heights in kilometers.

Winds are light in the latter regions, except where the polar low–associated troughs dip equatorward over continental areas.

A considerable change in the height contour configuration is obvious from the 100-mb chart for 14 January (Fig. 3.21). The polar low has filled slightly and split into two centers, the stronger of which has drifted toward Eurasia and the other toward the Hudson Bay area. The trough over Japan has weakened appreciably and the warm cell has intensified, with temperatures warmer than −45°C. The warm center over the United States has begun to move eastward and is centered over the Great Lakes area on the 14th. A marked poleward intrusion of the height contours is noted over Greenland as the low centers separate. On 16 January (Fig. 3.22) a new warm center of approximately −55° is in evidence in the Atlantic and the warm center from central North America has begun to move into the North Atlantic. The Hudson Bay low center has intensified to dominate the polar region, and strong winds are the rule over North America. The eastern hemispheric warm cell has developed to warmer than −40°, has remained relatively stationary over Japan, and has spread to dominate the midlatitude Pacific and eastern Asia.

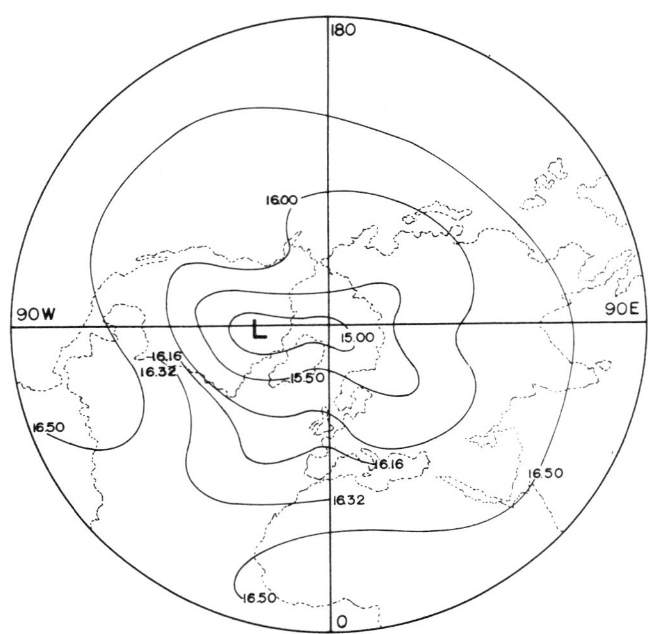

Fig. 3.24. One hundred-millibar chart for 20 January 1963 over the Northern Hemisphere. Contours are heights in kilometers.

Migration and consolidation of the Atlantic warm centers is accomplished by 18 January (Fig. 3.23) to produce an eastern Atlantic warm cell of —50° at about the 50-degree latitude point. There is evidence of advection of polar temperatures of —70°C into western portions of North America, and a shift of the warm cell poleward and eastward toward the Kamchatka Peninsula.

The Middle Atlantic warm cell strengthens until by 20 January (Fig. 3.24) a gross change in the height contour distribution is effected. Cyclonic curvature in the North Atlantic and anticyclonic curvature in the Western Atlantic result in a marked divergence about the warm cell. This is in contrast to the Eastern Hemisphere warm center, where the isotherms and isobars are roughly concurrent about a trough which extends equatorward toward Japan. Temperatures at the center, near 50 degrees latitude, are —40°C. The principal low center remains in the Hudson Bay area with the most intense circulation in the eastern United States.

By 22 January (Fig. 3.25) the warm center north of Japan has retreated southwestward over the Asiatic continent and the Atlantic warm

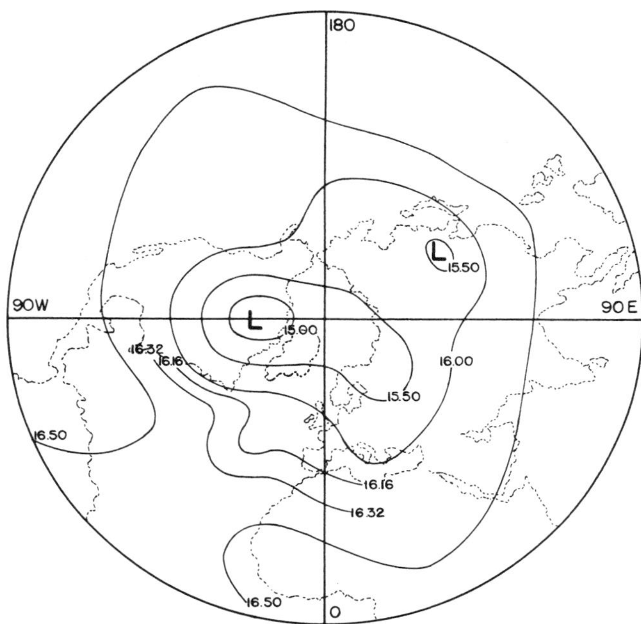

Fig. 3.25. One hundred-millibar chart for 22 January 1963 over the Northern Hemisphere. Contours are heights in kilometers.

cell has intensified to —45°C. A well-developed meridional flow along the eastern margins of North America and the North Atlantic is a most distinctive feature of this chart, and the remains of the initially roughly circumpolar low center have largely destroyed strong warm centers over the North Atlantic and Northwest Pacific. The Atlantic warm sector has spread upstream across North America during the 23rd (Fig. 3.26) and 24th (Fig. 3.27), as the warm cell slips westward from its Atlantic trough position into the lee of the Hudson Bay low center. There is a distinct poleward motion of the warm cell on the west side of the ridge, with maximum northward intrusion of the warm isotherms occurring in the Greenland area on 25 January (Fig. 3.28). Along with the divergence of height contour lines noted earlier in the South Atlantic, there is also an unusual equatorward extension of isotherms with almost the same contour pattern.

Development of this Atlantic wave was first evident on 14 January (Fig. 3.21) when, with the splitting of the polar low center, the contour lines pushed northward across Greenland. The axis of this ridge extended from 80 degrees latitude and 90 degrees west longitude into the North

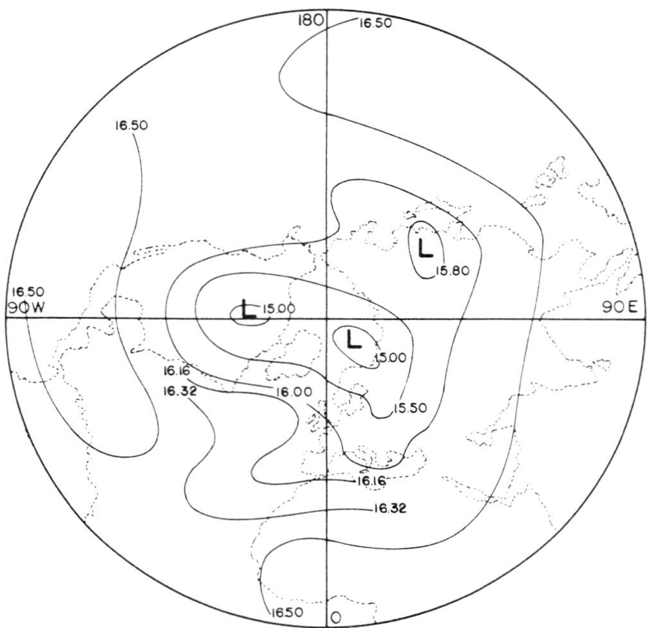

Fig. 3.26. One hundred-millibar chart for 23 January 1963 over the Northern Hemisphere. Coutours are heights in kilometers.

Atlantic and curved southwestward toward Puerto Rico. The trough on its lee had an axis which started in the Eastern Hemisphere low center and curved southwestward across Europe toward Trinidad. As the warm center developed in the North Atlantic the southern margins of the trough oscillated until on the 22nd and 23rd (Figs. 3.25 and 3.26) a strong wave was achieved. Temperatures in equatorial regions at the 100-mb level over the South Atlantic experienced a "warm wave" as temperatures of as high as —70° displaced the normal —80° tropical tropopause and the —60° isotherm was advected some 20 degrees south of its usual position.

A slow eastward drift of this trough was evidenced as it pushed equatorward, with a maximum intrusion centered at about 40 degrees west longitude and 10 degrees north latitude on 24 January (Fig. 3.27). The equatorward extension of this trough over the Atlantic is a remarkably stable feature of the 100-mb pressure surface structures for the remainder of January. Another very important feature of this series of 100-mb charts is the strong development of tropical characteristics in the South Pacific region of the Eastern Hemisphere. This first became apparent with tightening of the height contour and isothermal gradient

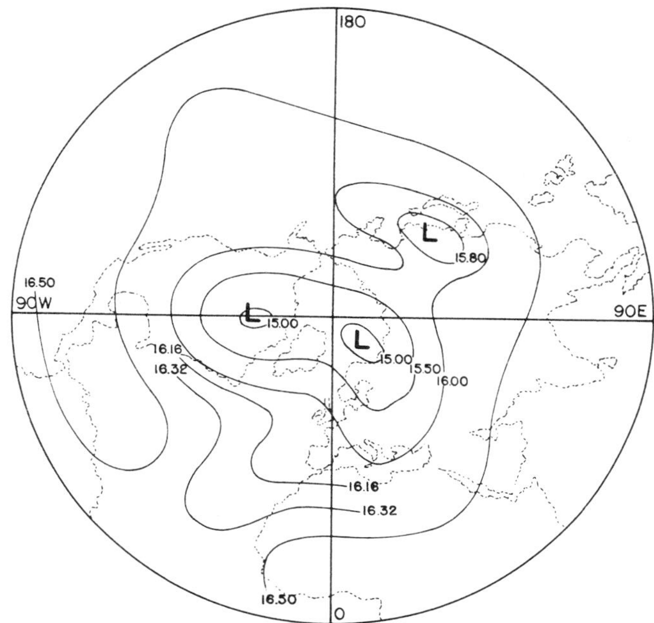

Fig. 3.27. One hundred-millibar chart for 24 January 1963 over the Northern Hemisphere. Contours are heights in kilometers.

off the coast of Japan and the appearance of a weak high center in subtropical latitudes. The —80° isotherm pushed above 25 degrees latitude over much of that quadrant of the hemisphere, well above the normal position in the winter hemisphere, and —85° temperatures are noted above 20 degrees latitude during the latter portion of the series (Figs. 3.28–3.30).

This shifting of the center of the winter polar vortex into the North American region appeared to provide the impetus for development of the Atlantic wave and the resulting incorporation of tropical regions in the polar vortex. This event occurred about 19 January and was an important feature of the 100-mb chart until 29 January (Fig. 3.30). It is matched with the occurrence of the "sudden warming" of the upper stratosphere then, since that event showed its first signs on 19 January. One can make a case for the idea that initiation of the event at 100 mb was on 14 January when the first obvious signs of development of the Atlantic wave were noted. Before this information could be considered significant a knowledge of the probability of the wave increasing in intensity to the required degree would have to be obtained. The possibility obviously

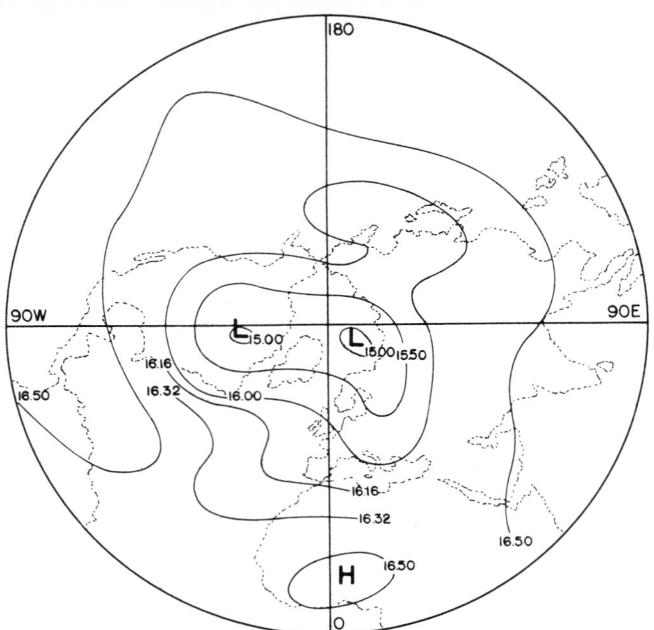

Fig. 3.28. One hundred-millibar chart for 25 January 1963 over the Northern Hemisphere. Contours are heights in kilometers.

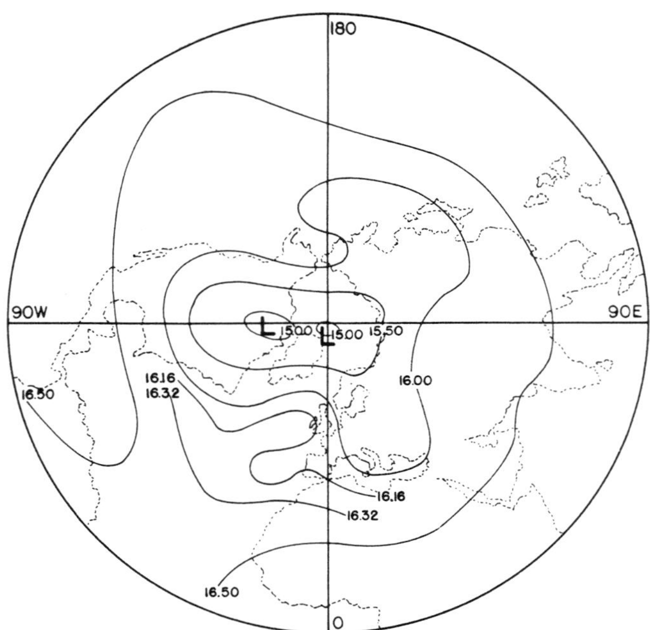

Fig. 3.29. One hundred-millibar chart for 27 January, 1963 over the Northern Hemisphere. Contours are heights in kilometers.

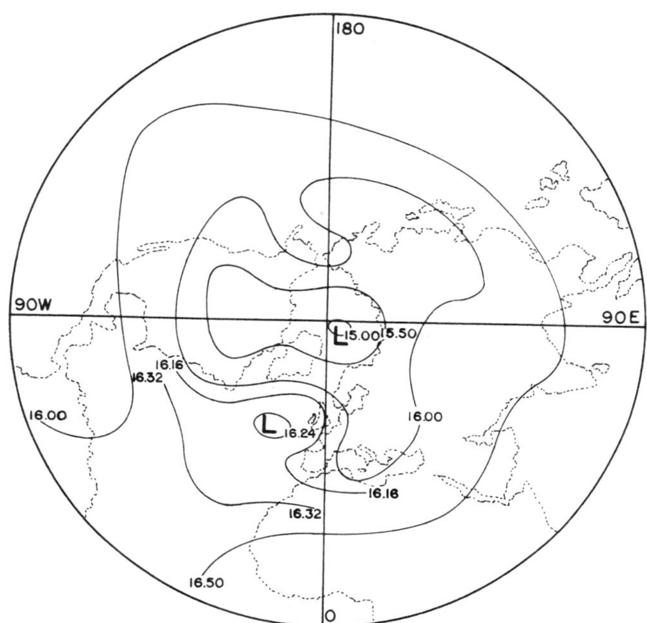

Fig. 3.30. One hundred-millibar chart for 29 January 1963 over the Northern Hemisphere. Contours are heights in kilometers.

exists that under different circumstances the genesis of the wave could have been in the Aleutian area, or possibly in central Siberia.

The "sudden warming" events are a very special class of upper stratospheric circulation changes which are observed only occasionally. There is evidence that they are members of a much larger class of events, the chief differences in which are the scales of phenomena and location of their occurrence. Thus, one gross event might produce a certain effect at a particular location while essentially the same apparent results might be produced by a large number of smaller events. It is clear from the previous discussion that the character of phenomena on the 100-mb surface will vary with location of observation points, so the situation is complex and there is a basic requirement for adequate amounts of observational data.

As an example of the circulation changes which are to be observed without a unique event of the sudden warming type, selected 100-mb charts for December and January of the winter of 1963–1964 are presented in Figs. 3.31–3.38. These maps, which are presented for 5-day intervals from 15 December to 19 January, indicate a persistent cold air mass surrounding the pole throughout the period. Inspection of the

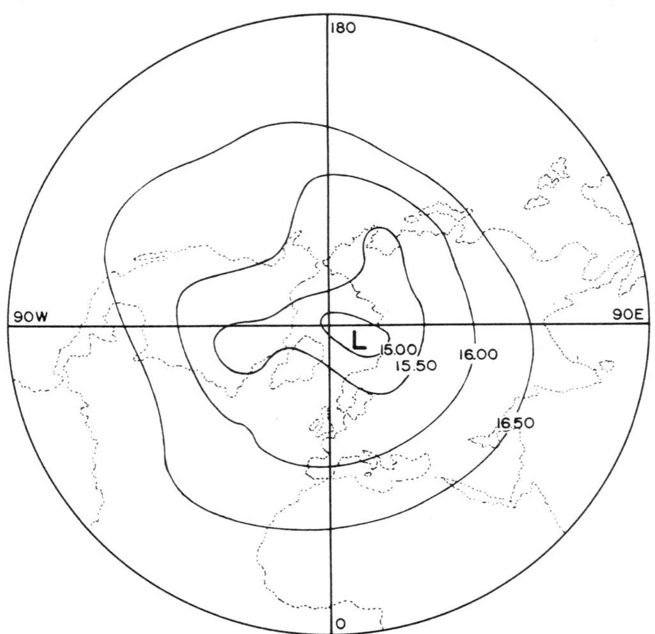

Fig. 3.31. One hundred-millibar chart for 15 December 1963 over the Northern Hemisphere. Contours are heights in kilometers.

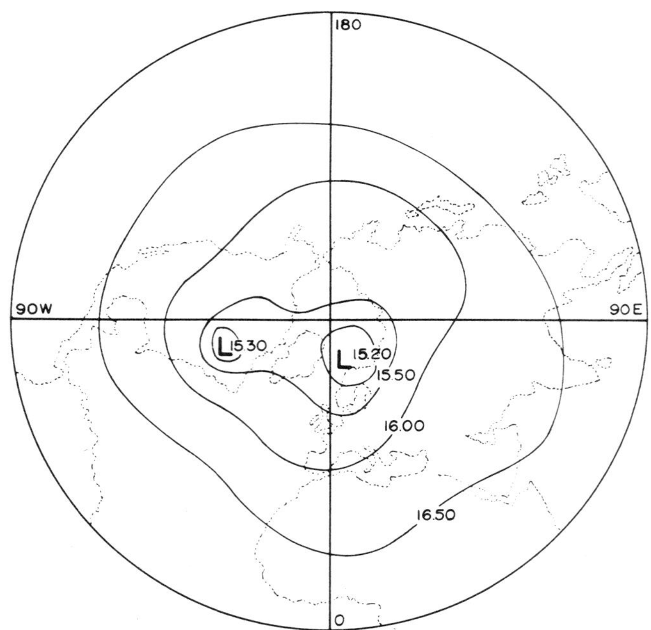

Fig. 3.32. One hundred-millibar chart for 20 December 1963 over the Northern Hemisphere. Contours are heights in kilometers.

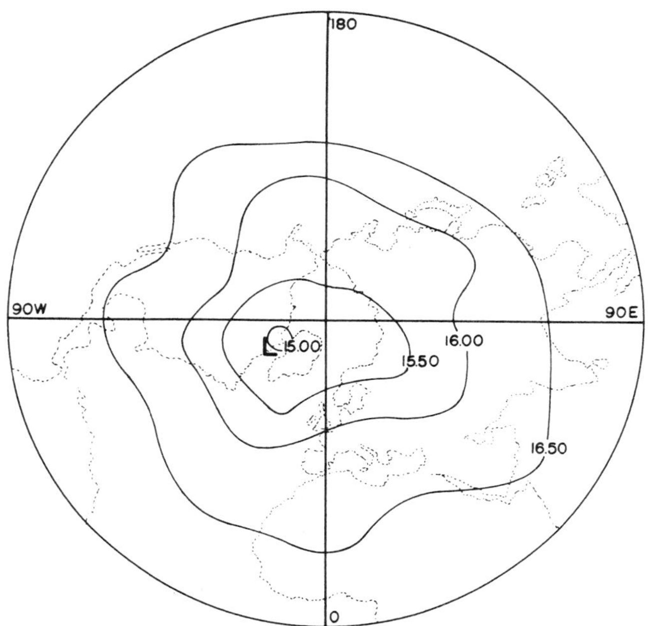

Fig 3.33. One hundred-millibar chart for 25 December 1963 over the Northern Hemisphere. Contours are heights in kilometers.

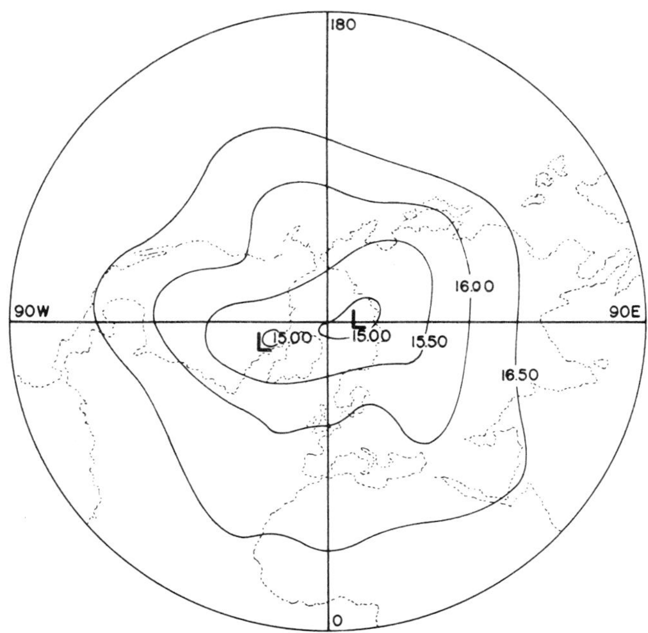

Fig. 3.34. One hundred-millibar chart for 30 December 1963 over the Northern Hemisphere. Contours are heights in kilometers.

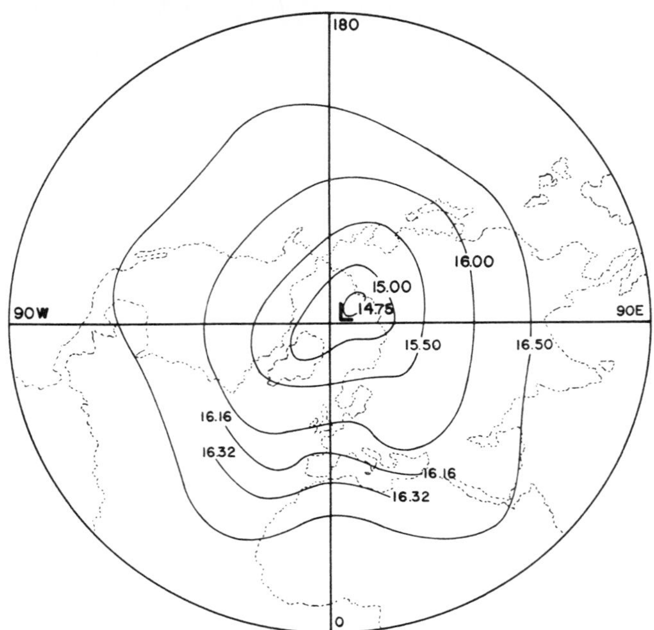

Fig. 3.35. One hundred-millibar chart for 4 January 1964 over the Northern Hemisphere. Contours are heights in kilometers.

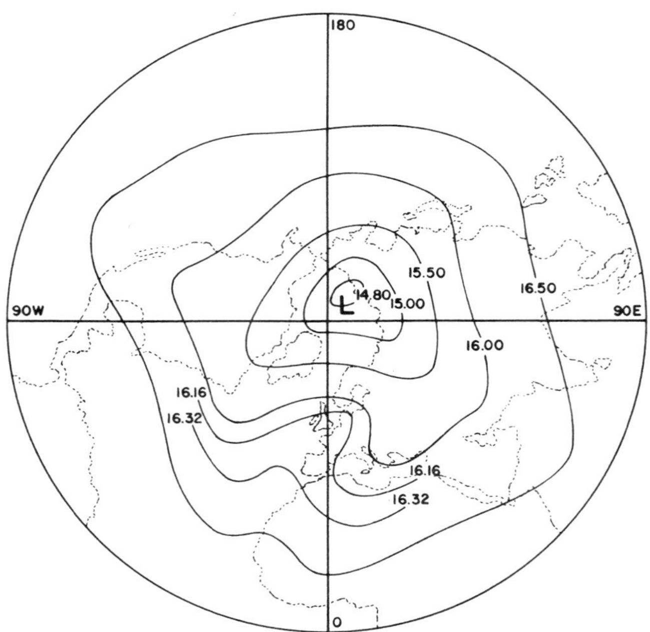

Fig. 3.36. One hundred-millibar chart for 9 January 1964 over the Northern Hemisphere. Contours are heights in kilometers.

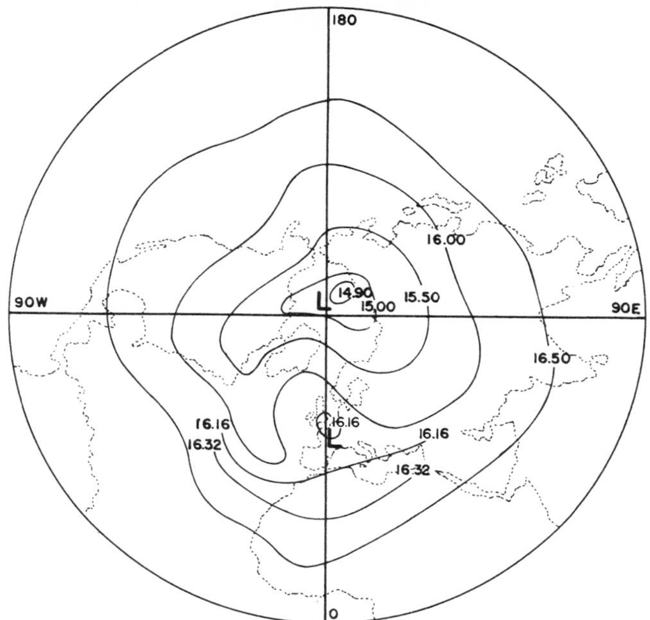

Fig. 3.37. One hundred-millibar chart for 14 January 1964 over the Northern Hemisphere. Contours are heights in kilometers.

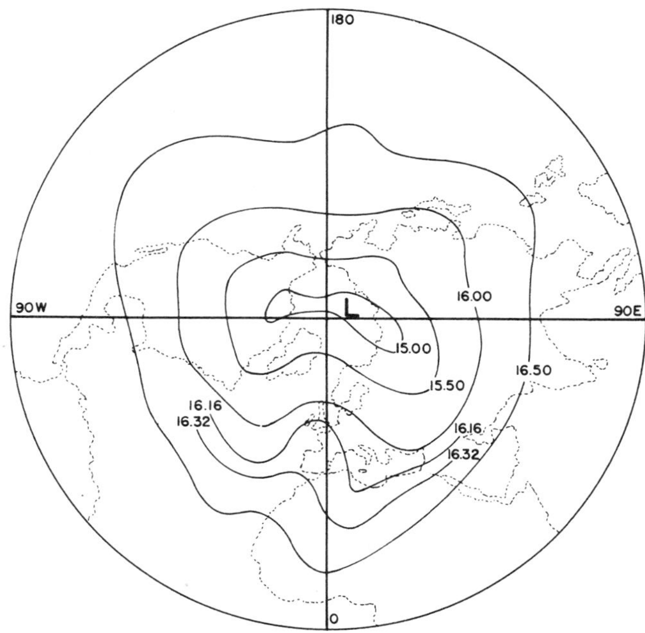

Fig. 3.38. One hundred-millibar chart for 19 January 1964 over the Northern Hemisphere. Contours are heights in kilometers.

height contours in the central portion of the low-pressure system for the January 1963 case (Figs. 3.20–3.30) and the December 1963 case (Figs. 3.31–3.38) shows that the latter situation is significantly weaker, with a minimum height of some 14.96 km on 15 December 1963 (Fig. 3.31) compared with 14.72 km on 12 January 1963 (Fig. 3.20) before initiation of the sudden warming of that winter.

A weak trough is located in the Hudson Bay area on 15 December, and a wave is evident in the Atlantic. This is a situation that is quite similar to the development which was observed at 100-mb level in January 1963, but inspection of subsequent maps shows that neither the low center nor the wave follows the development pattern which occurred in the earlier case. The low intensified somewhat and started a slow eastward movement, but on 25 December it began to move northward and a consolidation of the polar low was initiated. The eastward movement of the trough caused the major axis to be oriented eastward from its usual position over the North American continent to lie across Greenland. Maximum displacement is noted on about 4 January 1964, after which it began a slow westward upstream rotation as the low center intensified.

Minimum heights in the pressure region are noted on 4 January (Fig. 3.35), with the lowest value near 14.72 km at a location of 75–80 degrees latitude in the direction of Japan from the pole. This is in marked contrast to the situation in mid-January 1963 when the low-pressure center, prior to the circulation system breakdown, was elongated across the pole toward the Hudson Bay area with a minimum height of less than 14.80 km over the Hudson Bay on 16 January (Fig. 3.22). In the January 1963 case the low-pressure center remained in the North American region throughout the warming period. During the winter of 1963–1964 at about the turn of the year the low-pressure system intensified in the eastern Asiatic region as mentioned above, and the center of gravity stayed in the quadrant the remainder of the principal winter season. While the trough remains a feature of the 100 mb contours, it is strongly curved eastward into the central Atlantic, and the wave does not intensify as it had during the warming case.

Encroachment of tropical characteristics into the winter hemisphere in the western Pacific is a feature of each of the situations illustrated here, with —85°C a persistent feature of the 15-degree latitude of that region. The warm ridge which is characteristic of the winter upper mid-latitudes has a strong development in the Japan area in both cases. In January 1963 it showed some preference for a continental location with temperatures of —40°C. In December of the same year and January 1964 this warm sector is about 5° cooler, and shows a preference for a marine location in the Northern Pacific. The —45° isotherm is found in the Aleutian region throughout the period and invades the Alaskan coast on 4 and 14 January 1964 (Figs. 3.35 and 3.37).

3.6 Water Vapor

It has been generally assumed that the stratosphere is a dry layer of the atmosphere. This assumption is in general agreement with the theory that the stratosphere is a quiescent, uninteresting region of our environment. Lack of visual evidence of condensation in these high levels has been cited as proof of a dry regime, but this source of data is questionable since the maximum precipitable water which could be present in this rarefied air is likely to be below the threshold of visual observation from the ground except under ideal conditions. At present, there is no firm evidence that the moisture structure of stratospheric regions is significantly more variable than that observed in the troposphere. Actually, clouds (nacreous–mother-of-pearl) have been observed in upper portions of the lower stratosphere, apparently in relation to strong wave motions generated by orographic effects in the troposphere. These observations indicate saturation and thus provide evidence of a wide range in relative humidity from the few percent measured immediately above the tropopause to the saturation of these cloud regions.

A principal reason for the assumption of a dry stratosphere concerns the presence of the so-called "cold trap" of the tropopause. If one considers that all stratospheric moisture has its origin in vertical transport by convection phenomena, he can amply justify extremely low values of water vapor mixing ratio, particularly in the equatorial stratosphere where the tropopause is coldest and offers a maximum barrier to upward convection of the tropospheric moisture. Investigators have generally fallen back on assumptions that the water vapor mixing ratio remained constant with height, which is tantamount to assuming a dry-air exponential decrease with height. The point at which this assumed distribution was anchored has risen in altitude through the years as observational techniques improved, but has invariably assigned negligible importance to the role of water vapor in stratospheric physical processes.

A highly important though neglected process for transfer of water vapor is the diffusion of water molecules over the intervening atmosphere between the principal source at the ocean's surface and the upper stratosphere. The "cold trap" loses much of its significance as a barrier for the transfer of water vapor, being reduced to the function of a leaky valve which can have an influence on the time constant of the system but exercises no control on the level to which the reservoir is filled.

Clearly the long-term equilibrium, neglecting advective transport, will result in a distribution of water vapor in the lower stratosphere which has a maximum in equatorial regions and a minimum in polar regions. This picture of a moist equatorial stratosphere resulting from its juxtaposition to the gross source of water vapor which is characteristic of the low-latitude troposphere will undoubtedly be modified by transport mechanisms, both in the troposphere and through the tropopause interface. These factors should simply introduce complexity into the lower stratospheric moisture distribution and reduce the extreme range of the water vapor concentration to be observed in the lower troposphere.

Under static equilibrium conditions, diffusive separation of the atmospheric gases would occur as a result of the heterogeneous composition of air in accord with the density distributions prescribed by Eq. (1.2). Relative concentration of water vapor would increase with altitude as a result of its smaller molecular weight. In the troposphere this condition is not realized because the environment is far fram static. Vertical motions induced by turbulence result in strong mixing between the surface and tropopause. This very efficient transport mechanism is hampered by the precipitation process which keeps the relative humidity below saturation. As has been pointed out by Junge (1963), this situation changes at the tropopause above which the water vapor concentration continues to fall with altitude for several kilometers at rates far in excess of that expected from diffusive considerations.

It is interesting to note that it is in this altitude region that a sharp negative lapse rate of ozone is observed (Craig, 1965). This anomaly in water vapor concentration coincident with the sharp destruction of ozone diffusing downward from above suggests the presence of a chemical reaction involving these constituents. A probable reaction would appear to be

$$H_2O + O_3 \rightarrow H_2O_2 + O_2 \tag{3.2}$$

in a region where the number concentrations of these elements are equal to an order-of-magnitude approximation. Under such a situation the negative lapse rate of water-vapor mixing ratio would become slightly positive in this thin layer due to the heavier molecular weight (approximately 34) of hydrogen peroxide (H_2O_2). This volatile substance would at the same time preserve the water vapor and reduce the frost point below that expected for the ambient water concentration. Dissociation of the hydrogen peroxide in sunlight through the reaction

$$H_2O_2 + h\nu \rightarrow H_2O + O \tag{3.3}$$

could then restore the water vapor molecules from the level of the stratonull upward to the usual negative mixing ratio lapse rate.

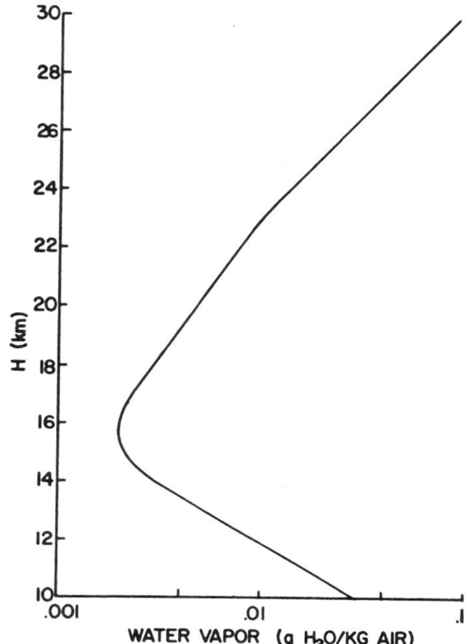

Fig. 3.39. Illustration of the water vapor mixing ratios that have been observed in the stratosphere by Barclay et al. (1960), Mastenbrook and Dinger (1960), and Murcray et al. (1960a).

These factors point toward diffusion control of the water vapor equilibrium of the equatorial lower stratosphere. Known advective processes in this region appear to be negligible in general compared with the efficiency of the diffusion process. That is very likely not true of the middle and high latitudes in the lower stratosphere. Danielson (1964) has shown that active intrusions of stratospheric air into the troposphere occur in middle and upper latitudes as a part of the circulation systems associated with cyclonic vortex systems. This outflow of lower stratospheric mass will materially alter the water vapor content of the lower stratosphere, and, as is indicated in the above discussion, a lowering of the mixing ratio can be expected to result from advective processes in general.

Measurements of lower stratospheric water vapor content have been conducted in midlatitudes with results as indicated in Fig. 3.39. Mixing ratio values of the order of 0.002 to 0.004 gm per kilogram of air are reported at 16-km altitude. The mixing ratio has values of 0.01 to 0.02 gm per kilogram of air at the top of the lower stratosphere according to these measurements and increases with height in the lower portion of the

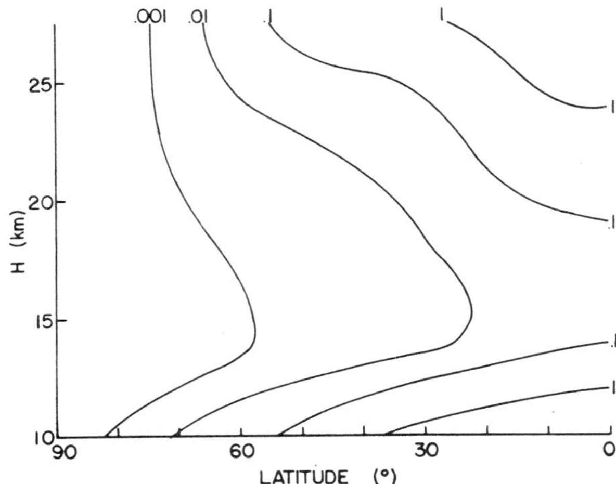

Fig. 3.40. Model water vapor meridional distribution postulated from available measurements and new data on stratospheric circulation. Contour units are grams per kilogram.

upper stratosphere. A model water vapor distribution of the lower stratosphere along a meridian is then postulated to have the form presented in Fig. 3.40. Here we assume values of 0.01 gm per kilogram at the 16-km altitude base of the stratosphere in equatorial regions and 0.001 gm per kilogram of the 11-km base of the stratosphere in polar regions.

The stratonull surface defines the upper boundary of the lower stratosphere. It varies in altitude, with a mean height of approximately 24 km in middle and low latitudes. The stratonull surface slopes upward to near 30 km in polar regions and is thus warmer at high latitudes. Assuming a constant relative humidity then prescribes a water vapor distribution which has greatest values in the upper warmer portions of the lower stratosphere, with maximum values in the extreme upper portion of the polar stratosphere.

The above estimate of the general distribution of water vapor in the lower stratosphere is based on a relatively static convective and advective structure. Clearly that is not a reasonable assumption, since pronounced circulation patterns are known to exist, even to the extent of mass interchange across the tropopause. This interlocking of circulation systems between our rather arbitrary layered atmospheres precludes the resolution of these complications until a more comprehensive picture is available. The cross section presented in Fig. 3.40 is a first approximation to the actual distribution which must suffice until additional data permit evaluation of the more detailed structure that undoubtedly exists.

3.7 Particulate Matter

While the lower stratosphere is marked by an absence of internal sources of particulate contamination, it is the scene of a flux of particulate matter through sedimentation of meteoroid material from the higher atmosphere and convective injection across the lower boundary from the troposphere. The latter process has a far greater potential owing to the gross contamination which exists in certain tropospheric regions but is limited in efficiency because of the considerable stability of the lower stratospheric temperature structure. Influx from above is extremely efficient, although the total input from this region cannot be high because of the relatively weak source. In addition to these steady state fluxes of particulate matter into the lower stratosphere, there are occasional point injections precipitated by volcanic eruptions, large meteor entries, and jet and rocket motor exhausts which may significantly enhance the particulate population for a restricted space and time interval. None of the above sources of lower stratospheric contamination can be expected to provide a uniform input, so it must be assumed that there will be a variable particulate concentration.

Observation of micron-size particles in the 10- to 25-km altitude atmospheric region is a comparatively difficult task. Techniques range from aspirated sampling on a balloon platform to impact collection through ports on high-speed aircraft. The former technique suffers from possible contamination of the sample by the large platform system, while the latter is subject to aerodynamic and collection errors. In general, the balloon-borne systems should produce measurements which are in excess of the actual counts, and the aircraft systems should produce an underevaluation of the actual concentration. Both techniques can be expected to suffer difficulties in evaluation of the small-particle concentration, and thus a bias toward the larger particles is probably contained in all of the observational data. Chagnon and Junge (1961) have performed extensive analyses of these data and concluded that a nominal lower stratospheric contamination of less than one particle per cubic centimeter exists. A summary of the available data is presented in Fig. 3.41. An interesting feature of these data is the weak peak in concentration at approximately 20-km altitude. Chagnon and Junge have hypothesized the particles to be a local condensate of atmospheric sulfate gases. These particles were in the 10^{-5}- to 10^{-4}-cm radius range.

Influx of particulate material through the upper boundary of the lower stratosphere is almost certainly governed by sedimentation of meteoroid debris collected by the higher atmosphere through interaction of solar meteoroid material and the earth's gravitational field. Consideration of the capture process leads to the expectation of higher concentrations of

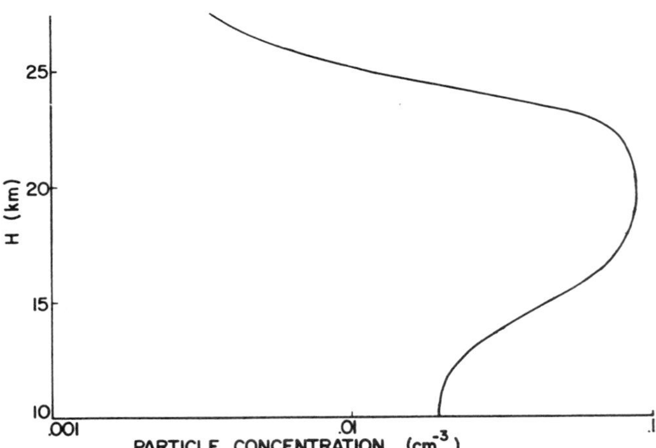

Fig. 3.41. Mean vertical particulate structure in the stratosphere obtained from five soundings summarized averaging the data presented by Chagnon and Junge (1961, Figure 5). (Courtesy Journal of Meteorology)

the flux at higher latitudes, with a maximum in polar regions, although not necessarily at the poles. Observation of sunlight scattering intensity and polarization by noctilucent cloud particles (Witt, 1960) has established that these particles are largely in the 10^{-6} to 10^{-5}-cm size range. Assuming a density of the material which composes the particles of 2 gm/cm^3, Junge *et al.* (1961) have calculated the still air fall rates of such particles in the lower stratosphere to be that illustrated in Fig. 3.42. Fall rates of all the lower stratospheric particles received from above are less than 1 cm/sec, and a majority of these particles have vertical fall velocities of hundredths and thousandths of a centimeter per second. These fall rates are sufficiently small that gross modification of the vertical distribution by vertical winds is to be expected, both on local and on general scales.

A decrease in fall rate of approximately one order of magnitude is noted (Fig. 3.42) over the altitude range from the upper to the lower boundary of the lower stratosphere. This distribution of fall velocities is the result of the normal positive density lapse rate, and has the predictable result of an increasing particulate concentration with decreasing height. The concentration will be increased owing to this factor alone by almost an order of magnitude at the base of the lower stratosphere for a given input at the top. It should be remembered that the number density may be depleted by any coagulation process, and that the importance of such processes will be enhanced by increased concentration. Residence times in an environment are important in evaluating the

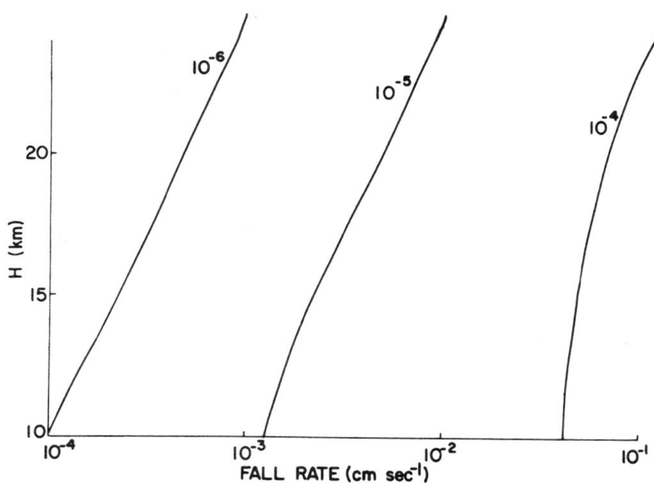

Fig. 3.42. Fall rates of spherical particles of 2-gm/cm³ density in the lower stratosphere, according to Junge et al. (1961). (Courtesy Journal of Meteorology)

efficiency of growth processes such as condensation and evaporation or the development of an equilibrium charge condition in a field of ionized molecules. Reasonable estimates of the residence times during which these particles populate a 1-km layer at the top of the lower stratosphere (approximately 25-km altitude) are almost 1 month for the larger particles (0.3-micron radius) and of the order of 3 years for the smaller particles (0.01-micron radius). Similar estimates for the same particles in the lowest kilometer layer of the lower stratosphere are in excess of 6 and 30 years, respectively. Such residence times are obviously quite long relative to the time constants of most modifying processes, so it must be assumed that the particulate population which is of extraterrestrial origin will be significantly modified in size distribution and number concentration.

Similar reasoning applies to all small particles located in the lower stratosphere, whether introduced from above or below. The latter class of particles may be injected forcefully by volcanic eruptions or atomic explosions or wafted into the lower stratosphere by turbulent mixing of the nuclei residue of cloud droplets which have evaporated after rising to the upper troposphere by convective transport. These particles are known to fall into the 0.1-micron radius size range, and are thus probably larger than most of the particulate material which arrives at the base of the lower stratosphere through sedimentation. Junge *et al.* (1961) have considered possible sinks for such particles in the form of coagulation and predicts a diminution of the number density in the lower portion of this atmospheric region if only the settling particles are considered. The situation is complicated by the tropospheric source of small particles rising through the tropopause. It is probable that the combined result of these factors will generally be a varying positive lapse rate of particulate concentration in the lower stratosphere.

As a matter of fact, Junge *et al.* (1961) have on several occasions observed an enhancement in particulate concentration in a layer centered about the 20-km level. Their measurements show a tropopause concentration of the order of 0.04 particle per cubic centimeter over the mid-latitude United States in both summer and winter seasons, increasing to almost 0.1 particle per cubic centimeter at an altitude of 20 km, and decreasing sharply to less than 0.01 particle per cubic centimeter at the stratonull altitude. The authors have classed the observed size distributions into three ranges, attributing a tropospheric origin to those particles in the sub-0.1-micron (10^{-5} cm) radius range and considering those in the 0.1- to 1-micron radius range as possibly of internal stratospheric origin, and the larger particles as extraterrestrial in origin.

This breakdown is based on the postulate that the particles in the 20-km layer are formed by coagulation growth of ionized sulfate molecules which are normally present in the lower troposphere and are thus assumed to be present in the lower stratosphere. Chemical analysis of collected particulate material has indicated the presence of sulfate; however, the evidence for the *in situ* growth of the particles is considerably less firm. It appears at this time to be rather likely that there is an influx from above of a significant number of particles of meteoric origin in the 0.01 to 0.1 size range. It has been suggested (Junge *et al.*, 1961) that these particles would be lost through coagulation into larger particles by contact as a smaller particle takes part in Brownian motion of the air molecules. While such a trend undoubtedly is initiated, present information makes it appear unlikely that the process would transmute all of these particles into the greater than 1-micron radius range. The particulate size distributions available to date are thus probably composites of those two principal sources of such material abetted by coagulation and condensation processes.

If this is so, the observed peak in particulate concentration at about 20-km altitude is probably the result of increasing particle concentration due to the reduced fall velocity field as they enter the more dense lower stratosphere. Below the 20-km peak, coagulation probably results in a reduction in over-all concentration of these particles, and the general increasing trend with decreasing altitude which is characteristic of the troposphere establishes the observed positive lapse rate profile in the extreme lower stratosphere.

There are many problems associated with the sampling of small particulate material in the lower stratosphere. Generally the data are too meager to depict the detail structure of lower stratospheric particulate distribution, although certain of the more sensitive sampling systems (Junge *et al.*, 1961) show a wealth of small-scale structure in the vertical profiles. The horizontal extent of these layered structures is largely unknown. Such knowledge would provide an indicator relative to the source of the particles as well as depicting the structure of circulation and turbulence in this region.

3.8 Ozone in the Lower Stratosphere

Early in the 1900's came the realization that absorption of solar ultraviolet energy by an atmospheric constituent was the cause of the warm temperatures observed immediately above the tropopause. Identification of ozone as the absorbing element was quickly accomplished by spectrometric techniques, and the determination of the total atmospheric ozone concentration through solar absorption measurements was initiated. These data provided the knowledge that ozone concentration was greater at higher latitudes and thus led to considerable interest in the nature of reaction rate coefficients and advective transports which could result in the observed resultant distributions. These data could not be accurately determined without detailed information on the vertical profiles of ozone concentration, so this was the focal point of the second phase of exploration of lower stratospheric chemical composition.

Analysis of the spectral absorption of solar radiation as a function of solar elevation angle led to the discovery of an anomaly termed the "*Umkehr* effect" which provided a technique for determination of the vertical profile of atmospheric ozone. The system was easy to employ, although restricted to fair weather, but did not provide great resolution or accuracy in the profile determination. It was applied extensively in network-type studies during the first half of the Twentieth Century so that our knowledge of the global ozone structure in the very general sense was comparatively broad before direct measurement of the detail structure was attempted. A most important result of these *Umkehr* observations was the discovery that the maximum concentration of ozone occurred immediately above the tropopause and that the concentration decreased in the upper portions of the stratosphere where the heat input had its maximum effect. These data were adequate for the theorists to expound on relatively satisfactory processes for the generation of ozone and its subsequent destruction to account for the general features of the observed distribution.

Photochemical reactions which may occur in the solar radiant flux of the lower stratosphere to produce ozone generally involve an intermediate step in which atomic oxygen is formed and destroyed. A major sink for atomic oxygen is the combination with molecular oxygen to form the ozone molecule O_3. Ultraviolet radiation in the shorter-wavelength bands is required to produce atomic oxygen with facility, and since these radiations are absorbed in the upper portions of the stratosphere, the presence of atomic oxygen or its active participation in the generation of ozone in the lower stratosphere is generally negligible. Ozone molecules are very efficient in absorbing ultraviolet energy in the 2000–3000 angstrom wavelength band. The solar flux in this wavelength region amounts to approximately 0.02 calorie per square centimeter per minute, or roughly 1% of the solar constant. Most of the energy thus obtained by the ozone molecules is exchanged through kinetic collisions into the thermal reservoir of internal energy of the air as a result of the considerable density of neutral particles available for such interaction in the lower stratosphere. The relatively pure air of the lower stratosphere provides little opportunity for oxidation reduction of ozone, with the resulting lifetime of ozone molecules falling in the range of months and years. In fact, the strongest agent for destruction of ozone after it has attained the protected environment of the lower stratosphere appears to be diffusive and turbulent mixing to the gross sink characterized by the troposphere.

The processes which form and destroy ozone through the catalytic activity of ultraviolet radiation can be expected to have their maximum intensity at the subsolar point and to become less intense at a given altitude level as the more oblique intercept of the solar flux causes the absorption levels to rise in altitude and the received energy to be distributed through a greater volume. This is a matter of paramount importance in the establishment of thermal equilibrium in the upper stratospheric region, but is of very little importance to the lower stratosphere except in that the debris of this furnace activity serves to supply the lower stratosphere with an adequate supply of ozone molecules for the absorption of available solar ultraviolet energy. The tropopause is found at relatively high altitudes (approximately 17 km) in equatorial regions, and the turbulent mixing of the troposphere appears to be adequate to confine the ozone storage region to above that level. The ozone molecules which are formed in great quantity at the 50-km altitude level settle gradually owing to their greater molecular weight (48 versus 29) to produce a negative gradient with height as they penetrate the zone where their destruction is progressively less likely. Since this process is cut off at a high level in the equatorial region and a new, more efficient dilution and destruction process dominates the scene, we find the minimum concentration of total ozone in the tropical regions.

The tropopause is found at low altitudes in polar areas. Thus the polar lower stratosphere forms an especially efficient reservoir for the collection and protection of ozone molecules. Newly generated ozone molecules formed in the polar stratopause region settle into this deep storage unit. It has been postulated that advective motions transport lower latitude ozone into the polar regions, although the exact circulation which effects this transport is not clear. The lower stratosphere has its greatest depth in polar regions during the winter period. Ozone concentration increases with decreasing height until this process is stopped at the tropopause by the turbulent mixing of the troposphere. Thus it is to be expected that maximum values of concentration will be observed near the base of the polar lower stratosphere and that the summation of all ozone in a vertical column will yield maximum values in that region. There appear, therefore, to be no significant inconsistencies between the theoretically expected distribution of atmospheric ozone and the general information which has become available thus far.

In addition to the obvious advantages of having ozone in the stratosphere to shield us from the sun's ultraviolet flux and the fortuitous production of an acoustic wave guide in the lower atmosphere by the ozonosphere heating, and aside from the inherent difficulties in operating in an environment contaminated by this toxic element, a major expected virtue of the ozone layer is its usefulness as a tracer of atmospheric motions. Analysis techniques using natural ozone as the tracer are most likely to be productive in providing data on the efficiency of the tropopause as a barrier to interchange of mass between the troposphere and stratosphere. The considerable stability evidenced by this element in its native environment indicates that it would function well, although the deleterious effects of mixing and oxidation reactions with particulate material in tropospheric air may represent strong limitations on the conservation properties with time or distance away from the lower stratosphere.

Regener (1960) has perfected a balloon-borne ozone measuring instrument which has the sensitivity and simplicity which are required for synoptic observation of lower stratospheric ozone. The instrument design is based on a chemiluminescent reaction of a prepared sampling surface which responds to the natural ozone molecules in a 100-cm^3 air sample as it is aspirated into and out of the system. Design of the sampling cavity is such that all of the ambient ozone is reduced while in the field of view of a photomultiplier tube, the output of which is then telemetered back to the launch site. A data point is obtained from a

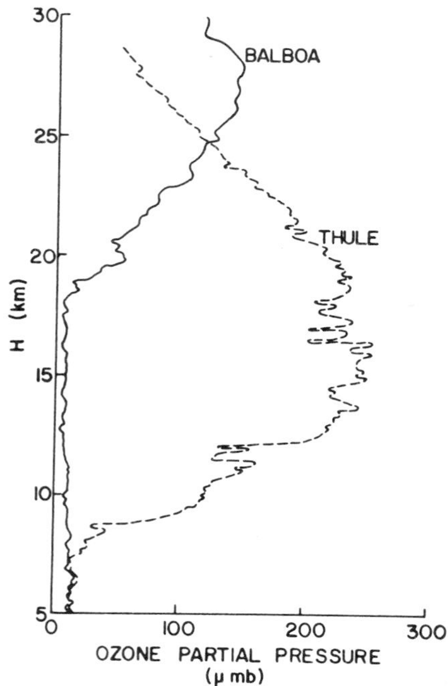

Fig. 3.43. Vertical profiles of ozone partial pressure representative of low- and high-latitude locations obtained by the U. S. Air Force ozonesonde network on 20 March 1963 using a Regener sonde. (Courtesy U. S. Air Force Cambridge Research Laboratories)

15-sec sample every 30 sec for three consecutive time intervals, after which there is a 30-second delay for calibration purposes. With an ascent rate of 300 meters per second this system provides for a sensitivity to variations in the vertical profile of 150 meters through most of the sounding and a maximum value of 300 meters. One of the most interesting points of information illustrated in these data is the large amount of detailed structure which makes up the ozone profiles in the lower stratosphere.

Examples of the type of data obtained from this measuring system are presented in Fig. 3.43. The high weak ozone layer which is typical of the tropical regions is illustrated by the sounding at Balboa, Canal Zone (latitude 9.0° N, longitude 79.6° W) obtained on 20 March 1963. A sounding on the same day at Thule, Greenland (latitude 76.5° N, longitude 68.8° W) produced the data represented by the dashed curve of Fig. 3.43. These ozone profiles demonstrate the gross latitudinal variations which result from the formation and destruction factors dis-

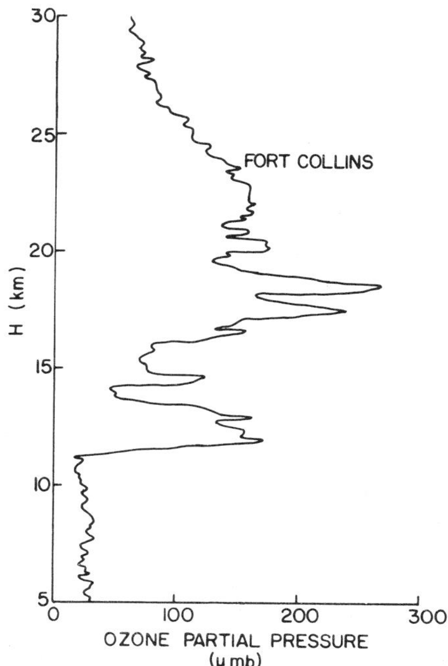

Fig. 3.44. Ozone partial pressure profile for 20 March 1963 at Fort Collins, Colorado obtained by the U. S. Air Force ozonesonde network using the Regener sonde. (Courtesy U. S. Air Force Cambridge Research Laboratories)

cussed above and, in addition, point to a significant difference in the detail structure to be observed in these different regimes. The comparatively smooth curve for tropical regions is probably a result of the higher altitudes at which the ozone occurs there, while the small-scale variations of as much as 50 μmb in the 16- to 17-km altitude range of the polar data may well be characteristic of the scale of variability in that low-altitude range. This concept is borne out by the characteristics of the upper portions of the polar data, where the variability is of the same order as, or even more conservative than, that observed in the tropical data. These profiles are rather characteristic of the type of data that have thus far been observed in these regions with the Regener sonde during the spring season when the ozone concentration is at its maximum.

On occasion, the ozone profiles obtained with the Regener sonde in middle latitudes exhibit significantly different characteristics from these more conservative equatorial and polar data. Such a case is illustrated in Fig. 3.44 in which the profile obtained at Fort Collins, Colorado

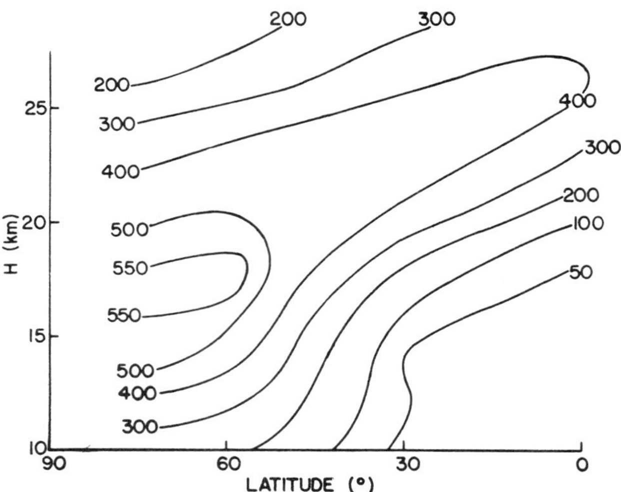

Fig. 3.45. Mean meridional contours of ozone concentration over North America for January and February 1963. Contours are in micrograms per cubic meter.

(latitude 40.6° N, longitude 105.1° W) on the same 20 March 1963 date of the above data is presented. The profile is indeed intermediate between the limiting cases, but a variability in vertical ozone concentration different from either the equatorial or polar case is evident. These data can be interpreted to represent latitudinal extensions of typically tropical and polar regimes in layered structures across this middle latitude station. These data are considered evidence of the utility of ozone as a tracer material for observation of advection in this region of the atmosphere. Since there are other ways in which this type of vertical structure can be generated, such as local circulations which involve subsidence and convection, care must be exercised in interpretation of such data until the relative importance of the various possible contributors to the observed structure can be evaluated. It is clear from these data and the considerable number of similar profiles which have been accumulated that observation of ozone distribution in space and time is a very informative technique for furthering our understanding of the mechanics of lower stratospheric circulation.

The U. S. Air Force Cambridge Research Laboratories have supported the application of the Regener sonde to systematic weekly observations of ozone profiles at eleven locations in the North American region on a 1200Z Wednesday schedule. This network provides data on a segment of the globe bounded by 9 and 77 degrees north latitude and 60 and 148 degrees west longitude. Network observations were begun in January 1963 and volumes of the data are published at intervals (Hering

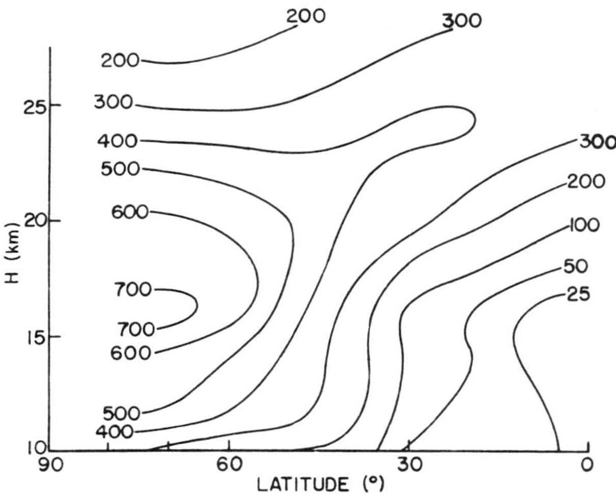

Fig. 3.46. Mean meridional contours of ozone concentration over North America for March and April 1963. Contours are in micrograms per cubic meter.

et al., 1964) under the title "Ozonosonde Observations Over North America." Summaries of the network data for bimonthly periods are included in the above publications, and these data for the first eight months of 1963 were used to plot the meridional cross sections presented in Figs. 3.45–3.48. Contours of ozone concentration in micrograms per cubic meter were prepared from the published values at 10-degree latitude intervals to illustrate the meridional distribution and its variability through the peak of maximum atmospheric ozone concentration.

Typical midwinter data are illustrated in Fig. 3.45. Peak values of ozone concentration are found in the 17-km altitude region of high latitudes, with values reaching above 550 μg per cubic meter. Concentrations above 400 are found at all latitudes, with a pronounced increase in altitude of approximately 10 km from polar to equatorial regions. The very steep slope of the contours in lower midlatitudes is evidence of the subtropical extent of the high tropical tropopause, although the rather abrupt latitudinal variations probably represent appearance of midlatitude mixing into upper tropospheric altitudes rather than such dramatic modification of the tropopause structure. It is probably in this zone that the most informative analysis of atmospheric circulation processes can be obtained from the ozone network data. The rather smooth decrease in ozone concentration with increasing latitude just above the stratonull level correlates well with the expected diminution of solar ultraviolet input to the stratopause level with increasing latitude.

A rather pronounced change is to be observed in the mean data for

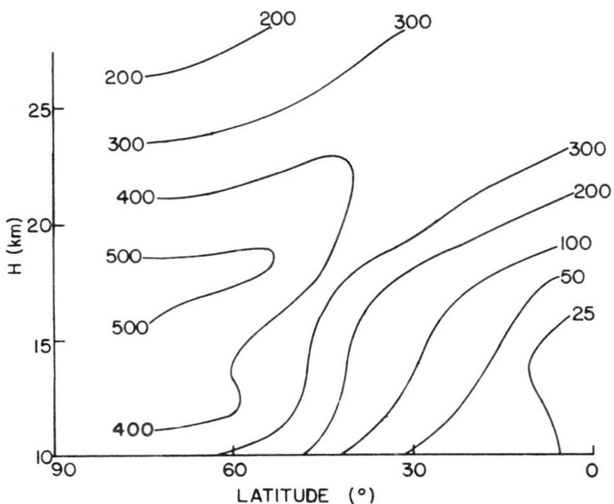

Fig. 3.47. Mean meridional contours of ozone concentration over North America for May and June 1963. Contours are in micrograms per cubic meter.

March and April of 1963. A maximum value of the ozone concentration of well over 700 micrograms per cubic meter appears at near 16-km altitude in the Arctic regions, and at the same time there is a withdrawal of the 400 contour away from equatorial regions. The high concentrations which are characteristic of high-latitude lower-stratospheric regions have pushed equatorward, with the 100 contour appearing near vertical in the 10- to 15-km altitude range just above 30 degrees latitude. A principal characteristic of the change which occurs during this period concerns the slight lowering of the main center of ozone concentration at high latitudes and the expansion of this region into the lower portions of the lower stratosphere and across the tropopause into the troposphere in middle and lower latitudes. It is indeed interesting, although perhaps coincidental, that the axis of the zone of maximum ozone concentration has essentially the same geometry as the circulation disturbances which play a major role in controlling the heat balance of the stratosphere and the troposphere. These events, termed "sudden warmings" when they are intense, appear to be initiated in the upper stratosphere in equatorial regions and proceed toward high latitudes and descend so that they are observable by 30-km peak altitude balloon observation systems from midlatitudes poleward.

Data for May and June of 1963 show a weakening of gradients in the lower stratospheric ozone distribution. Peak concentrations have fallen to near 500 μg per cubic meter and the level of this maximum

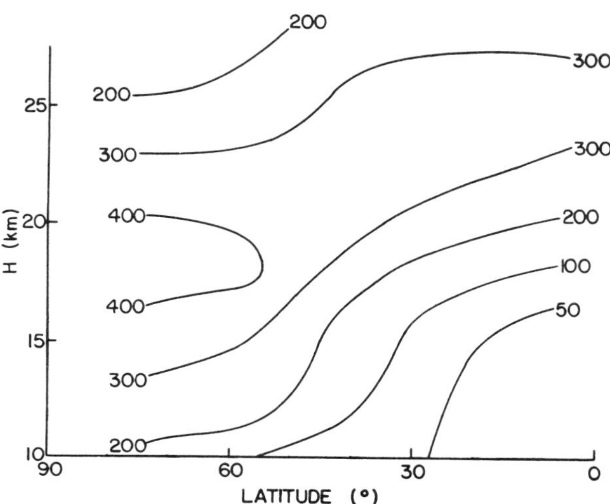

Fig. 3.48. Mean meridional contours of ozone concentration over North America for July and August 1963. Contours are in micrograms per cubic meter.

has risen to above 17 km. The 400 contour has completed withdrawal from equatorial regions to 40 degrees latitude and above. The surge of ozone into lower latitudes in the 10- to 15-km altitude range has begun to withdraw. It is interesting to note that ozone concentrations are higher in the tropical upper troposphere in the summer data for July and August 1963 presented in Fig. 3.48, although the gradient of ozone concentration in middle latitudes is significantly weaker. The high-latitude zone of maximum concentration has lifted to above 18-km altitude.

Now the maximum temperature gradients observed in the lower stratosphere during the winter season are found to be located in the 30-degree latitude zone at an altitude of approximately 17 km during the month of April. This corresponds roughly to the location where the contour lines are forced near vertical as the equatorward push of ozone concentration leads to maximum gradients. Maximum summer temperature gradients were found above 40 degrees latitude at an altitude of about 15 km, which again appears as the location of maximum slope of the contours of equal ozone concentration. It is not at all clear from these mean data as to whether these vertically oriented surfaces of equal concentration are the drivers of the lower stratospheric circulation because of their ability to heat the atmosphere at that point through excessive absorption of ultraviolet radiation, or whether that concentration is a by-product of advective transport in a circulation that has its genesis in other regions. The development of the companion effects so late in the winter season and the immediate breakup of the winter circulation would indicate that this particular phenomenon is the result of processes that are precipitated by the maturing of the winter circulation.

REFERENCES

Angell, J. K., and J. Korshover. (1962). The biennial wind and temperature oscillations of the equatorial stratosphere and their possible extension to higher altitudes. *Monthly Weather Rev.* **90,** 127–132.

Arnold, A. (1952). Turbulence in the stratosphere. *Bull. Am. Meteorol. Soc.* **33,** 77.

aufm Kampe, H. J. (1953). Neuere Stratosphärendaten. *Meteorol. Rundschau* **6,** 99–100.

Austin, J. A., and L. Krawitz. (1956). 50 millibar patterns and their relationship to tropospheric changes. *J. Meteorol.* **13,** 152–159.

Badgley, F. I. (1957). Response of radiosonde thermistors. *Rev. Sci. Instr.* **28,** 1079–1084.

Bannon, J. K. (1958). Stratospheric temperatures over the Antarctic. *Quart. J. Roy. Meteorol. Soc.* **84,** 434–436.

Bannon, J. K., and A. Gilchrist. (1956). A statistical study of the variation of temperature in the troposphere and lower stratosphere. *Quart. J. Roy. Meteorol. Soc.* **82,** 58–74.

Barclay, F. R., M. J. W. Elliott, P. Goldsmith, and J. V. Jelley. (1960). A direct measurement of the humidity in the stratosphere using a cooled-vapour trap. *Quart. J. Roy. Meteorol. Soc.* **86,** 259–264.

Barrett, E. W., L. R. Herndon, Jr., and H. J. Carter. (1949). A preliminary note on the measurement of water vapor content in the middle stratosphere. *J. Meteorol.* **6,** 367–368.

Belmont, A. D. (1957). Apparent diurnal variations of Arctic stratospheric temperatures. *Trans. Am. Geophys. Union* **38,** 462–468.

Belmont, A. D. (1962). The reversal of stratospheric winds over North America during 1957, 1958, and 1959. *Beitr. Physik Atmosphäre* **35,** 121–135.

Belmont, A. D. (1963). An atlas of prevailing monthly zonal winds in the stratosphere over the Northern Hemisphere. *Proc. Intern. Symp. Stratospheric Mesospheric Circulation* pp. 537–640. Inst. Meteorol. Geophys., Freien Univ. Berlin.

Belmont, A. D., and D. G. Dartt. (1964). Double quasi-biennial cycles in observed winds in the tropical stratosphere. *J. Atmospheric Sci.* **21,** 354–360.

Berkofsky, L. (1960). Internal gravity-vorticity lee waves over mountains. *J. Geophys. Res.* **65,** 3685–3691.

Berkofsky, L., and R. Shapiro. (1964). A dynamical model for investigating the effects of periodic heat sources on the equatorial stratosphere. U. S. Air Force, Cambridge Research Laboratories, L. G. Hanscom Field, Massachusetts, Environmental Research Papers, No. 57.

Bickert, A., K. Labitzke-Behr, K. Petzoldt, R. Scherhag, and G. Warnecke. (1961–1962). Tägliche Höhenkarten der 50 mbar-Flache für das Jahr 1960. Teil I-IV. *Meteorol. Abhandl. Inst. Meteorol. Geophys. Freien Univ. Berlin* **15,** 100 pp.

Bleichrodt, J. F., J. Blok, and R. H. Dekker. (1961). On the spring maximum of radioactive fallout from nuclear test explosions. *J. Geophys. Res.* **66,** 135–141.

Borden, T. R., Jr., W. S. Hering, H. S. Muench, and G. Ohring. (1960). Contributions to stratospheric meteorology, Volume II. U. S. Air Force, Cambridge Research Laboratories, Bedford, Massachusetts, 206 pp. Geophysics Research Directorate, G. R. D. Notes No. I, (OTS order No. PB 154898).

Bork, I., K. Labitzke-Behr, Z. Petkosek, K. Petzoldt, R. Scherhag, and G. Warnecke. (1961a). 100 mbar-Karten der Berliner Wetter Karte, herausgeg. v., Institut für Meteorologie und Geophysik der Freien Universität Berlin, Jahrgang 9, 1960 und Jahrgang 10, 1961.

Bork, I., K. Labitzke-Behr, Z. Petkosek, K. Petzoldt, R. Scherhag, and G. Warnecke. (1961b). Ergebnisse der Aufstiege d. Radiosendenstation Berlin-Tempelhof d. Inst. f. Meteorologie u. Geophysik d. Freien Universität Berlin, Bd. XI, Heft 3.

Bork, I., K. Labitzke-Behr, Z. Petkosek, K. Petzoldt, R. Scherhag, and G. Warnecke. (1962). Preliminary daily Northern Hemisphere 10-millibar synoptic weather maps of the year 1961, Part I-IV. *Meteorol. Abhandl. Inst. Meteorol. Geophys. Freien Univ. Berlin* **20,** 100 pp.

Bork, I., K. Labitzke-Behr, Z. Petkosek, and R. Scherhag. (1962b). Preliminary daily Northern Hemisphere 30-millibar synoptic weather maps of the year 1962, Part I: January–March. *Meteorol. Abhandl. Inst. Meteorol. Geophys. Freien Univ. Berlin* **25,** 100 pp.

Bork, I., B. Kriester, K. Labitzke-Behr, Z. Petkosek, R. Scherhag, K. Sieland, R. Stuhrmann, and G. Warnecke. (1962c). Tägliche Höhenkarten der 100 mbar-Fläche sowie monatliche Mittelkarten für das Jahr 1962. *Meteorol. Abhandl. Inst. Meteorol. Geophys. Freien Univ. Berlin.* pp. 23, 110 pp.

Borovikov, A. M., G. I. Golyshev, and G. A. Kokin. (1963). Some characteristics of the atmosphere of the Southern Hemisphere. *In* "First International Symposium on Rocket and Satellite Meteorology" (H. Wexler and J. E. Caskey, Jr., eds.), pp. 164–172. Wiley, New York.

Boville, B. W. (1960). The Aleutian stratospheric anticyclone. *J. Meteorol.* **17,** 329–336.

Boville, B. W. (1961). A dynamical study of the 1958–59 stratospheric polar vortex. McGill University, Montreal, Canada, Sci. Report No. 9, Contract AF19 (604)-3865, 134 pp.

Boville, B. W. (1963). What are the causes of the Aleutian anticyclone? *Proc. Intern. Symp. Stratospheric Mesospheric Circulation,* pp. 107–121. Inst. Meteorol. Geophys. Freien Univ. Berlin.

Boville, B. W., C. V. Wilson, and F. K. Hare. (1961). Baroclinic waves of the polar-night vortex. *J. Meteorol.* **18,** 567–580.

Brasefield, C. J. (1950). Winds and temperatures in the lower stratosphere. *J. Meteorol.* **7,** 66–69.

Brasefield, C. J. (1954). Measurement of atmospheric humidity up to 35 km. *J. Meteorol.* **11,** 412–416.

Brewer, A. W. (1949). Evidence for a world circulation provided by measurements of helium and water vapor distribution in the stratosphere. *Quart. J. Roy. Meteorol. Soc.* **75,** 351–363.

Brewer, A. W., B. M. Cwilong, and G. M. B. Dobson. (1948). Measurement of absolute humidity in extremely dry air. *Proc. Phys. Soc.* (*London*) **A60,** 52–70.

Bull, G. A., and D. G. James. (1956). Dust in the stratosphere over western Britain on April 3 and 4, 1956. *Meteorol. Mag.* **85,** 293.

Callendar, G. S. (1958). On the amount of carbon dioxide in the atmosphere. *Tellus* **10,** 243–248.

Chagnon, C. W., and C. E. Junge. (1961). The vertical distribution of sub-micron particles in the stratosphere. *J. Meteorol.* **18,** 746–752.

Chapman, S. (1930). A theory of upper atmospheric ozone. *Mem. Roy. Meteorol. Soc.* **3,** 103–125.

Chiu, W. C. (1958). The observed mean monthly wind fields in the lower stratosphere and upper troposphere over North America. *J. Meteorol.* **15,** 9–16.

Chiu, W. C. (1960). Studies of the properties of the wind and temperature field of the upper troposphere and lower stratosphere. New York University, Final Report, Contract AF19(604)-1755.

Chiu, W. C., and R. S. Greenfeld. (1959). The relative importance of different heat-exchange processes in the lower stratosphere. *J. Meteorol.* **16,** 271–280.

Chiu, W. C., S. Hellerman, and R. S. Greenfeld. (1956). Mean monthly temperature, pressure and wind fields in the lower stratosphere and upper troposphere over North America. New York University, Report under Signal Corps Contract DA-36-039-sc-64673.

Conover, W. C. (1961). An instance of a stratospheric "explosive" warming. *J. Meteorol.* **18,** 410–413.

Court, A. (1941). Tropopause disappearance during the Antarctic winter. *Bull. Am. Meteorol. Soc.* **23,** 220–238.

Court, A., A. J. Kantor, and A. E. Cole. (1962). Supplemental atmosphere. U. S. Air Force, Cambridge Research Laboratories, L. G. Hanscom Field, Massachusetts, Meteorological Development Laboratory Project 8624, AFCRL-62-899.

Craig, R. A. (1950). The observations and photochemistry of atmospheric ozone and their meteorological significance. *Meteorol. Monographs* **1,** 50.

Craig, R. A. (1951). Radiative temperature changes in the ozone layer. *Compendium Meteorol.* pp. 292–302.

Craig, R. A., and W. S. Hering. (1959). A stratospheric warming of January–February 1957. *J. Meteorol.* **16,** 91–107.

Craig, R. A., and M. A. Lateef. (1962). Vertical motion during the 1957 stratospheric warming. *J. Geophys. Res.* **67,** 1839–1854.

Craig, R. A. (1965). "The Upper Atmosphere," p. 509. Academic Press, New York.

Danielsen, E. F. (1964). Radioactivity transport from stratosphere to troposphere. *Mineral Ind. Penn. State Univ.* **33,** 1–7.

Dartt, D. G., and A. D. Belmont. (1964). Periodic features of the 50-millibar zonal winds in the tropics. *J. Geophys. Res.* **69,** 2887–2893.

Davis, P. A. (1960). The diurnal variation of temperature near the tropopause due to a diurnal variation of terrestrial radiation. New York University, Final Report, Contract AF19 (604)-1755.

Defant, F., and H. Taba. (1957). The threefold structure of the atmosphere and the characteristics of the tropopause. *Tellus* **9,** 259–274.

Dobson, G. M. B. (1951). Recent work on the stratosphere. *J. Roy. Meteorol. Soc.* **77,** 488–492.

Dobson, G. M. B. (1957). Observers handbook for the ozone spectrophotometer. *Ann. Intern. Geophys. Yr.* **5,** 46–89.

Dobson, G. M. B., and D. N. Harrison. (1926). Measurements of the amount of ozone in the earth's atmosphere and its relation to other geophysical conditions. *Proc. Roy. Soc.* **A110,** 660–693.

Dobson, G. M. B., A. W. Brewer, and B. M. Cwilong. (1946) Meteorology of the lower stratosphere. *Proc. Phys. Soc. (London)* **A185,** 144–175.

Duncan, L. D. (1960a). Theoretical performance of the Arcas. U. S. Army Electronics Research and Development Activity, White Sands Missile Range, New Mexico, Special Report 34.

Duncan, L. D. (1960b). Automatic rocket impact predictor. *IRE, Trans. Military electron* **4,** 243–245.

Dütsch, H. U. (1959). Vertical ozone distribution over Arosa. Lichtklimatisches Observatorium, Arosa, Final Report, Contract AF61-(519)-905.

Dütsch, H. U. (1963). Ozone and temperature in the stratosphere. *Proc. Intern. Symp. Stratospheric Mesospheric Circulation* pp. 271–291. Inst. Meteorol. Geophys., Freien Univ. Berlin.

Ebdon, R. A. (1960). Notes on the wind flow at 50 mb in tropical and subtropical regions in January 1957 and January 1958. *Quart. J. Roy. Meteorol. Soc.* **86,** 540–543.

Ebdon, R. A. (1961). Some notes on the stratospheric winds at Canton Island and Christmas Island. *Quart. J. Roy. Meteorol. Soc.* **87,** 322–331.

Ebdon, R. A., and R. G. Veryard. (1961). Fluctuations in equatorial stratospheric winds. *Nature* **189,** 791–793.

Elford, W. G. (1959). Winds in the upper atmosphere. *J. Atmospheric Terrest. Phys.* **15,** 132–136.

Eliassen, A., and E. Kleinschmidt. (1957). Dynamic meteorology. *In* "Handbuch der Physik" (J. Bartels, eds.), Vol. 48, pp. 1–154. Springer, Berlin.

Eliassen, A., and E. Palm. (1961). On the transfer of energy in stationary mountain waves. *Geofys. Publikasioner, Norske Videnskaps-Akad. Oslo* **22,** 1–23.

Farkas, E. (1961). Springtime temperature changes in the Antarctic stratosphere. *New Zealand J. Geol. Geophys.* **4,** 372–386.

Farquharson, J. S. (1952). Cloud in the stratosphere. *Meteorol. Mag.* **81,** 341–345.

Faust, H., and W. Attmannspacher. (1959). Cell structure of the atmosphere. Final Report, DA-91-508-EUC-387, October, AD 231 502, and AD 231 503.

Feely, H. W., and J. Spar. (1960). Tungsten-185 from nuclear bomb tests as a tracer for stratospheric meteorology. *Nature* **188,** 1062–1064.

Fesenkov, V. G. (1949). The mass of the atmospheric residue of the Silhote-Alin meteorite. *Dokl. Akad. Nauk SSSR* **66,** 231-238.

Fowle, F. E. (1928). Ozone in the Northern and Southern Hemispheres. *J. Terrest. Mag. Atmosph. Elect.* **33,** 151–157.

Frogner, E. (1962). Temperature changes on a large scale in the Arctic winter stratosphere and their probable effects on the tropospheric circulation. *Geofys. Publikasioner, Norske Videnskaps-Akad. Oslo* **23,** 1–83.

Fry, L. M., F. A. Jew, and P. K. Kuroda. (1960). On the stratospheric fallout of Strontium-90: The spring peak of 1959. *J. Geophys. Res.* **65,** 2061–2066.

Gates, D. M., D. G. Murcray, C. C. Shaw, and R. J. Herbold. (1958). Near infrared solar radiation measurements by balloon to an altitude of 100,000 feet. *J. Opt. Soc. Am.* **48,** 1010–1016.

Godson, W. L. (1960). Total ozone and the middle stratosphere over Arctic and sub-Arctic areas in winter and spring. *Quart. J. Roy. Meteorol. Soc.* **86,** 301–317.

Godson, W. L. (1962). Antarctic winter stratosphere. Arctic Meteorol. Research Group, Publications in Meteor., No. 47, McGill University, Montreal, Canada, pp. 37–40.

Godson, W. L. (1963). A comparison of middle-stratosphere behaviour in the Arctic and Antarctic, with special reference to final warmings. *Proc. Intern.*

Symp. Stratospheric Mesospheric Circulation pp. 161–207. Inst. Meteorol. Geophys., Freien Univ. Berlin.

Godson, W. L., and R. Lee. (1958). High level fields of wind and temperature over the Canadian Arctic. *Beitr. Physik Atmosphare* **31**, 40–68.

Goldie, A. H. R. (1950). The average planetary circulation in vertical meridian planes. *Cent. Proc. Roy. Meteorol. Soc.* pp. 175–180.

Gotz, F. W. (1951). Ozone in the atmosphere. *Compendium Meteorol.* pp. 275–291.

Goody, R. M. (1954). "The Physics of the Stratosphere." Cambridge Univ. Press, London and New York.

Hansen, C. F. (1957). Some characteristics of the upper atmosphere pertaining to hypervelocity flight. *Jet Propulsion* **27**, 1151–1156.

Hanson, K. J. (1958). A case study of the explosive stratospheric warming over the Antarctic, October, 1958. *In* "Antarctic Meteorology," pp. 128–137. Pergamon Press, Oxford.

Hanson, K. J. (1960). Explosive warming in the Antarctic stratosphere. *Trans. Amer. Geophys. Union* **41**, 384–388.

Hare, F. K. (1960). The disturbed circulation of the Arctic stratosphere. *J. Meteorol.* **17**, 36–52.

Hare, F. K., and B. W. Boville. (1961). Studies in Arctic and stratospheric meteorology. *Meteorologie* [4] **42**, 35.

Harris, M. F., F. G. Finger, and S. Teweles. (1962). Diurnal variation of wind, pressure, and temperature in the troposphere and stratosphere over the Azores. *J. Atmospheric Sci.* **19**, 136–149.

Haskell, N. A. (1950). Diffraction effects in the propagation of compressional waves in the atmosphere. U. S. Air Force, Cambridge Research Laboratories, Geophysical Research Paper No. 3, p. 43.

Haurwitz, B. (1938). Atmospheric ozone as a constituent of the atmosphere. *Bull. Am. Meteorol. Soc.* **19**, 417–424.

Haurwitz, B. (1946). Relations between solar activity and the lower atmosphere. *Trans. Am. Geophys. Union* **27**, 161–163.

Helliwell, N. C., J. M. Mackenzie, and M. J. Kerley. (1957). Some further observations from aircraft of frost-point and temperature up to 50,000 feet. *Quart. J. Roy. Meteorol. Soc.* **83**, 257–262.

Hering, W. S. (1964). Ozonesonde observations over North America, Volume I. U. S. Air Force, Cambridge Research Laboratories, L. G. Hanscom Field Massachusetts, Report 64-30, 528 pp.

Hess, S. (1948). Some new mean meridional cross-sections through the atmosphere. *J. Meteorol.* **5**, 293–300.

Houghton, J. T., T. S. Moss, and J. P. Chamberlain. (1958). An air-borne infrared solar spectrometer. *J. Sci. Instr.* **35**, 329–333.

Houghton, J. T., T. S. Moss, J. S. Seely, and T. D. F. Hawkins. (1957). Some observations of the infrared solar spectrum from a high-flying aircraft. *Nature* **180**, 1187–1188.

Jacobs, L. (1954). Dust cloud in the stratosphere. *Meteorol. Mag.* **83**, 115.

Jensen, C. E. (1960). Energy transformation and vertical flux processes over the Northern Hemisphere. *J. Geophys. Res.* **64**, 1145–1156.

Jenkins, K. R. (1962). Empirical comparisons of meteorological rocket wind sensors. *J. Appl. Meteorol.* **1**, 196–202.

Jones, F. E., and A. Wexler. (1960). A barium fluoride film hygrometer element. *J. Geophys. Res.* **65**, 2087–2095.

Julian, P., L. Krawith, and H. Panofsky. (1959). The relation between height pattern at 500 mb and 100 mb. *Monthly Weather Rev.* **87**, 251–260.

Julian, P. R. (1961). Remarks on "The disturbed circulation of the Arctic stratosphere." *J. Meteorol.* **18**, 119–121.

Julian, P. R. (1963). Some correlations of tropospheric and stratospheric pressures and temperatures in mid and high latitudes. *Proc. Intern. Symp. Stratospheric Mesospheric Circulation,* pp. 63–76. Inst. Meteorol. Geophys. Freien Univ. Berlin.

Junge, C. E. (1963). "Air Chemistry and Radioactivity." p. 382. Academic Press, New York.

Junge, C. E., C. W. Chagnon, and J. E. Manson. (1961). Stratospheric aerosols. *J. Meteorol.* **18**, 81–108.

Kaplan, L. D. (1960). The influence of carbon dioxide variations on the atmospheric heat balance. *Tellus* **12**, 204–208.

Kay, R. H. (1951). The apparent diurnal temperature variation in the lower stratosphere. *Quart. J. Roy. Meteorol. Soc.* **77**, 427–434.

Kay, R. H. (1954). The measurement of ozone vertical distribution by a chemical method to heights of 12 km from aircraft. *In* "Rocket Exploration of the Upper Atmosphere" (R. L. F. Boyd and M. J. Seaton, eds.), pp. 208–211. Pergamon Press, Oxford.

Kiseleva, M. S., B. S. Neporent, and V. A. Fursinkiv. (1959). Spectral determination of water vapor in the upper atmosphere. *Opt. Spectr. (USSR) (English Transl.)* **6**, 522–524.

Kitaoka, T. (1963a). A short comment on the relation between the increase in ozone content and the increase in stratospheric temperatures. *Proc. Intern. Symp. Stratospheric Mesospheric Circulation,* pp. 313–315. Inst. Meteorol. Geophys. Freien Univ. Berlin.

Kitaoka, T. (1963b). Some considerations on the stratosphere related to the cause of the Aleutian high. *Proc. Intern. Symp. Stratospheric Mesospheric Circulation* pp. 121–153. Inst. Meteorol. Geophys. Freien Univ. Berlin.

Kochanski, A. (1954). Thermal structure and vertical motion in the lower stratosphere. *Air Weather Ser. Tech. Rept.* **35**, 105–129.

Kochanski, A. (1955). Cross sections of the mean zonal flow and temperature along 80°W. *J. Meteorol.* **12**, 95–106.

Krishnamurti, T. N. (1959). A vertical cross section through the "polar night" jet stream. *J. Geophys. Res.* **64**, 1835–1844.

Kulcke, W., and H. K. Paetzold. (1957). Über eine Radiosonde zur Bestimmung der vertikalen Ozon-verteilung. *Ann. Meteorol.* **8**, 47–52.

Labitzke-Behr, K. (1960). Über markante Erwärmungen in der Stratosphäre. *Meteorol. Abhandl. Inst. Meteorol. Geophys. Freien Univ. Berlin* **9**, 57 pp.

Labitzke-Behr, K. (1962). Beiträge zur Synoptik der Hochstratosphäre. *Meteorol. Abhandl. Inst. Meteorol. Geophys. Freien Univ. Berlin* **28**, 32–72.

Labitzke-Behr, K., K. Petzoldt, R. Scherhag, and G. Warnecke. (1960). Tägliche Höhenkarten der 10 mbar Fläche für das Internationale Geophysikalische Jahr 1958. Teil. I–IV. *Meteorol. Abhandl. Inst. Meteorol. Geophys. Freien Univ. Berlin* **13**, 100 pp.

Labitzke-Behr, K., Z. Petkovsek, and R. Scherhag. (1962). Preliminary daily Northern Hemisphere 10 millibar synoptic weather maps of the year 1962, Part I, January–March. *Meteorol. Abhandl. Inst. Meteorol. Geophys. Freien Univ. Berlin* **26**, 110 pp.

Lee, R., and W. L. Godson. (1957). The Arctic stratospheric jet stream during the winter 1955–1956. *J. Meteorol.* **14,** 126–135.

Lettau, H. (1956). Theoretical notes on the dynamics of the equatorial atmosphere. *Beitr. Physik Atmosphare* **29,** 107–122.

Libby, W. F., and C. E. Palmer. (1960). Stratospheric mixing from radioactive fallout. *J. Geophys. Res.* **65,** 3307–3317.

London, J. (1952). The distribution of radiational temperature change in the Northern Hemisphere during March. U. S. Air Force, Cambridge Research Laboratories, Geophysical Research Papers No. 18.

London, J. (1963a). The distribution of total ozone in the Northern Hemisphere. *Beitr. Physik Atmosphare* **36,** 254–263.

London, J. (1963b). Ozone variations and their relation to stratospheric warming. *Proc. Intern. Symp. Stratospheric Mesospheric Circulation* pp. 299–311. Inst. Meteorol. Geophys. Freien Univ. Berlin.

London, J., and C. Prabhakara. (1963). The effect of stratospheric transport processes on the ozone distribution. *Proc. Intern. Symp. Stratospheric Mesospheric Circulation* pp. 291–299. Inst. Meteorol. Geophys. Freien Univ. Berlin.

Lowenthal, M. (1957). Abnormal mid-stratospheric temperatures. *J. Meteorol.* **14,** 476.

McCreary, F. E. (1961). Variation of the zonal winds in the equatorial stratosphere. Joint Task Force Seven, Meteorological Center, University of Hawaii, JTFMC TP-20.

Mac Dowall, J. (1960). Distribution of atmospheric ozone; a preliminary analysis of some International Geophysical Year observations. *Nature* **187,** 382–383.

Machta, L., and R. J. List. (1959). Analysis of stratospheric strontium measurements. *J. Geophys. Res.* **64,** 1267–1276.

Maeda, K. (1962). On the heating of the polar upper atmosphere. *NASA* (*Natl. Aeron. Space Admin.*), *Tech. Rept.* **TR-S-141.**

Maeda, K. (1963). Occurrence of abrupt warming in the stratosphere and solar activity. *Proc. Intern. Symp. Stratospheric Mesospheric Circulation* pp. 451–506. Inst. Meteorol. Geophys., Freien Univ. Berlin.

Manage, S., and F. Moller. (1961). On the radiative equilibrium and heat balance of the atmosphere. *Monthly Weather Rev.* **89,** 503–531.

Mantis, H. T. (1963). Note on the structure of the stratospheric easterlies of mid-latitude. *J. Appl. Meteorol.* **2,** 427–429.

Mason, R. B., and C. E. Anderson. (1963). Summer 100 mb anticyclone over Southern Asia. *Proc. Intern. Symp. Stratospheric Mesospheric Circulation* pp. 153–161. Inst. Meteorol. Geophys., Freien Univ. Berlin.

Mastenbrook, H. J., and J. E. Dinger. (1960). The measurement of water-vapour distribution in the stratosphere. U. S. Naval Research Laboratory, Washington, D. C. Report No. 551, 16.11.60, AD 247 760.

Mastenbrook, H. J., and J. E. Dinger. (1961). Distribution of water-vapour in the stratosphere. *J. Geophys. Res.* **66,** 1437–1444.

Masterson, J. E. (1956). Rocket photography of hurricanes. *Office Naval Res., Res. Rev.* pp. 4–7.

Mateer, C. L., and W. L. Godson. (1960). The vertical distribution of atmospheric ozone over Canadian station from *umkehr* observations. *Quart. J. Roy. Meteorol. Soc.* **86,** 512–518.

Michelson, I. (1957). Ultimate design of high altitude sounding rockets. *Jet Propulsion* **27,** 1107–1108.

Moreland, W. B. (1960). Antarctic stratospheric circulation and ozone observations. *In* "Antarctic Meteorology," pp. 394–409. Pergamon Press, Oxford.

Murcray, D. G., J. Brooks, F. Murcray, and C. Shaw. (1958). High altitude infrared studies of the atmosphere. *J. Geophys. Res.* **63**, 289–299.

Murcray, D. G., F. H. Murcray, W. J. Williams, and F. E. Leslie. (1960a). Water-vapour distribution above 90,000 ft. *J. Geophys. Res.* **65**, 3641–3649.

Murcray, D., J. Brooks, F. Murcray, and W. Williams. (1960b). Atmospheric absorptions in the near infrared at high altitudes. *J. Opt. Soc. Am.* **50**, 107–112.

Murgatroyd, R. J. (1955). Wind and temperature to 50 km. over England. Anomalous sound propagation experiments 1944/45. Meteorological Office, Great Britain, Geophysical Memoirs, No. 95.

Murgatroyd, R. J. (1957). Winds and temperatures between 20 and 100 km—a review. *Quart. J. Roy. Meteorol. Soc.* **83**, 417–458.

Murgatroyd, R. J., P. Goldsmith, and W. E. H. Hollings. (1955). Some recent measurements of humidity from aircraft up to heights of about 50,000 ft. over Southern England. *Quart. J. Roy. Meteorol. Soc.* **81**, 533–537.

Newton, C. W., and A. V. Persson. (1962). Structural characteristics of the subtropical jet stream and certain lower-stratospheric wind systems. *Tellus* **14**, 221–241.

Oort, A. H. (1962). Direct measurement of the meridional circulation in the stratosphere during the IGY. Planetary Circulations, Project, Dept. of Meteorology, MIT, Cambridge, Massachusetts, Sci. Report 6.

Paetzold, H. K. (1961). Messungen des atmosphärischen Ozone. *In* "Handbuch der Aerologie" (Hesse, ed.), pp. 458–528. Leipzig.

Palm, E., and A. Foldvik. (1960). Contribution to the theory of two-dimensional mountain waves. *Geofys. Publikasioner, Norske, Videnkaps-Akad. Oslo* **21**, 1–30.

Palmer, C. E. (1954). The general circulation between 200 mb. and 10 mb. over the equatorial Pacific. *Weather* **9**, 341–349.

Palmer, C. E. (1959). The stratospheric polar vortex in winter. *J. Geophys. Res.* **64**, 749–764.

Palmer, C. E., and R. C. Taylor. (1960). The vernal breakdown of the stratospheric cyclone over the South Pole. *J. Geophys. Res.* **65**, 3319–3329.

Panofsky, H. A. (1961). Temperature and wind in the lower stratosphere. *Advan. Geophys.* **7**, 215–247.

Panofsky, H. A., L. Krawitz, and P. R. Julian. (1958). A study and evaluation of relations between tropospheric and stratospheric flow. Pennsylvania State University, Final Report N189 (188) 34785A.

Pavlovskaya, A. A. (1961). Interseasonal variations of atmospheric circulation in the upper troposphere and lower stratosphere. Trudy Central Festg. Inst. Moscow, No. 104, pp. 54–88.

Plass, G. N. (1959). Carbon dioxide and climate. *Sci. Am.* **201**, 41–47.

Pressman, J. (1954). The latitudinal and seasonal variations of the absorption of solar radiation by ozone. U. S. Air Force, Cambridge Research Laboratories, Geophysical Research Papers No. 33.

Ramanathan, K. R., and K. P. Ramakrishamn. (1933). Distortion of the tropopause due to meridional movements in the substratosphere. *Nature* **132**, 932.

Reed, R. J. (1962a). Evidence of geostrophic motion in the equatorial stratosphere. *Quart. J. Roy. Meteorol. Soc.* **88**, 324–327.

Reed, R. J. (1962b). Some features of the annual temperature regime in the tropical stratosphere. *Monthly Weather Rev.* **90**, 211–215.

Reed, R. J. (1962c). Wind and temperature oscillations in the tropical stratosphere. *Trans. Am. Geophys. Union* **43**, 105–109.

Reed, R. J. (1962d). Energy changes and transformations in the sudden warmings of late January 1957 and 1958. Arctic Meteorol. Research Group, Published in Meteor. No. 47, McGill University, Montreal, Canada, pp. 19–21.

Reed, R. J. (1963a). On the cause of the stratospheric sudden warming phenomenon. *Proc. Intern. Symp. Stratospheric Mesospheric Circulation* pp. 315–335. Inst. Meteorol. Geophys., Freien Univ. Berlin.

Reed, R. J. (1963b). On the cause of the 26-months periodicity in the equatorial stratospheric winds. *Proc. Intern. Symp. Stratospheric Mesospheric Circulation* pp. 245–259. Inst. Meteorol. Geophys., Freien Univ. Berlin.

Reed, R. J., and D. G. Rogers. (1962). The circulation of the tropical stratosphere in the years 1954–1960. *J. Atmospheric Sci.* **19**, 127–135.

Reed, R. J., W. J. Campbell, L. A. Rasmussen, and D. G. Rogers. (1961). Evidence of a downward-propagating annual wind reversal in the equatorial stratosphere. *J. Geophys. Res.* **66**, 813–818.

Reed, R. J., J. M. Mercer, and R. V. Cormier. (1964). A climatology of wind and temperatures in the tropical stratosphere between 100 mb. and 10 mb. U. S. Navy Weather Research Facility, Norfolk, Virginia, NWRF 26-0564-092, May, 56 p.

Regener, V. H., H. K. Paetzold, and A. Ehmert. (1954). Further investigation on the ozone layer. *J. Atmospheric Terrest. Phys.* **1**, Spec. Suppl., 202–207.

Regener, V. H. (1960). On a sensitive method for the recording of atmospheric ozone. *J. Geophys. Res.* **65**, 3975–3977.

Reiter, E. R. (1962). The atmospheric micro-structure and its bearing on clear air turbulence. Colorado State University, Fort Collins, Colorado, Dept. of Atmospheric Science, Tech. Paper No. 39, 20 pp.

Riehl, H. (1950). Variations in the structure of high-level cyclones. *Bull. Am. Meteorol. Soc.* **31**, 291–294.

Riehl, H., and R. Higgs. (1960). Unrest in the upper stratosphere over the Caribbean Sea during January 1960. *J. Meteorol.* **17**, 555–561.

Rubin, M. J. (1953). Seasonal Variation of the Antarctic tropopause. *J. Meteorol.* **10**, 127–134.

Sawyer, J. S. (1961). Quasi-periodic wind variations with height in the lower stratosphere. *Quart. J. Roy. Meteorol. Soc.* **87**, 24–33.

Scherhag, R. (1952). Die explosionsartigen Stratosphärenerwärmungen des Spätwinters 1951/52. *Ber. Deut. Wetterdienstes U. S. Zone* **38**, 51–63.

Scherhag, R. (1959). Über die Luftdruck-, Temperatur- and Winschwankungen in der Stratosphäre. *Abhandl. Math.-Naturw. Kl. Akad. Wiss. Manz* pp. 86–91.

Scherhag, R. (1960). Stratospheric temperature changes and the associated changes in pressure distribution *J. Meteorol.* **17**, 575–582.

Scherhag, R. (1961). Die weltweite Stratosphärenerwärmung in der zweiten Dezemberhälfte 1960. *Meteorol. Abhandl. Inst. Meteorol. Geophys. Freien Univ. Berlin* **11**, B 1–6.

Schwerdtfeger, W. (1960). The seasonal variation of the strength of the southern circumpolar vortex. *Monthly Weather Rev.* **88**, 103–110.

Schwerdtfeger, W. (1963). The southern circumpolar vortex and the spring warming of the polar stratosphere. *Proc. Intern. Symp. Stratospheric Mesospheric Circulation* pp. 207–221. Inst. Meteorol. Geophys., Freien Univ. Berlin.

Scorer, R. S. (1949). Theory of waves in the lee of mountains. *Quart. J. Roy. Meteorol. Soc.* **75**, 41–56.

Scorer, R. S. (1950). The dispersion of a pressure pulse in the atmosphere. *Proc. Roy. Soc.* **A201**, 137–157.

Shellard, H. C. (1950). Simultaneous aircraft soundings of temperature and frost point. *Meteorol. Mag.* **79**, 355–356.

Sion, E. (1955). Time constants of radiosonde thermistors. *Bull. Am. Meteorol. Soc.* **36**, 16–21.

Stormer, C. (1948). Mother-of-pearl clouds. *Weather* **3**, 13.

Teweles, S. (1958). Anomalous warming in the stratosphere over North America in early 1957. *Monthly Weather Rev.* **86**, 377–396.

Teweles, S. (1961). Time section and hodographic analysis of Churchill rocket and radiosonde winds and temperatures. *Monthly Weather Rev.* **89**, 125–136.

Teweles, S. (1963). Reduction of diurnal variation in stratospheric radiosonde data reported by some countries of the Eastern Hemisphere. *Proc. Intern. Symp. Stratospheric Mesospheric Circulation* pp. 507–517. Inst. Meteorol. Geophys., Freien Univ. Berlin.

Teweles, S., and F. G. Finger. (1958). An abrupt change in stratospheric circulation beginning in mid-January, 1958. *Monthly Weather Rev.* **86**, 23–28.

Teweles, S., and F. G. Finger. (1960). Reduction of diurnal variation in the reported temperatures and heights of stratospheric constant pressure surfaces. *J. Meteorol.* **17**, 177–194.

Teweles, S., and M. Snidero. (1962). Some problems of numerical objective analysis of stratospheric constant pressure surfaces. *Monthly Weather Rev.* **90**, 147–156.

Teweles, S., L. Rothenberg, and F. G. Finger. (1960). The circulation at the 10 millibar constant pressure surface over North America and adjacent ocean areas, July 1957 through June 1958. *Monthly Weather Rev.* **88**, 137–149.

Van Mieghem, J. (1950). L'equation aux dérivées partielles de la pression de perturbation associée aux ondulations de grande longueur d'onde du courant geostrophique zonal. *Inst. Roy. Meteorol. Belg. Mem.* **39**, 1–45.

Van Mieghem, J. (1951). Hydrodynamic instability. *Compendium Meteorol.* pp. 434–453.

Van Mieghem, J. (1956). The energy available in the atmosphere for conversion into kinetic energy. *Beitr. Physik Atmosphare* **29**, 129–142.

Van Mieghem, J. (1963). New aspects of the general circulation of the stratosphere and mesosphere. *Proc. Intern. Symp. Stratospheric Mesospheric Circulation* pp. 5–63. Inst. Meteorol. Geophys., Freien Univ. Berlin.

Veryard, R. G. (1961). Fluctuations in stratospheric winds over Australia. *Meteorol. Mag.* **90**, 295.

Veryard, R. G., and R. A. Ebdon. (1961a). Fluctuations in equatorial stratospheric winds. *Nature* **187**, 791.

Veryard, R. G., and R. A. Ebdon. (1961b). Fluctuations in tropical stratospheric winds. *Meteorol. Mag.* **90**, 125–143.

Veryard, R. G., and R. A. Ebdon. (1963). The 26-month tropical stratospheric wind oscillation and possible causes. *Proc. Intern. Symp. Stratospheric Mesospheric Circulation* pp. 225–245. Inst. Meteorol. Geophys., Freien Univ. Berlin.

Warnecke, G. (1956). Ein Beitrag zur Aerologie der arktischen Stratosphäre. *Meteorol. Abhandl. Inst. Meteorol. Geophys. Freien Univ. Berlin* **3**, 60ff.

Warnecke, G. (1962a). Der Ablauf der Nordhemisphärischen Stratosphärenzirkulation im Jahre 1958 anhand monatlicher Mittelkarten der 100-, 50-, 25- und

10-mbar-Fläche. *Meteorol. Abhandl. Inst. Meteorol. Geophys. Freien Univ. Berlin* **28,** 74 pp.

Warnecke, G. (1962b). Über die Zustandsänderungen der nordhemisphärischen Stratosphäre. *Meteorol. Abhandl. Inst. Meteorol. Geophys. Freien Univ. Berlin* **28,** 33 pp.

Warnecke, G. (1963). Discussion remark—on the accuracy of various radiosonde data of the Northern Hemisphere during 1962 at the 100 mb. level. *Proc. Intern. Symp. Stratospheric Mesospheric Circulation* pp. 517–526. Inst. Meteorol. Geophys., Freien Univ. Berlin.

Wege, K. (1957). Temperatur- und Strömunsverhältnisse in der Stratosphäre über der Nordhalbkugel. *Meteorol. Abhandl. Inst. Meteorol. Geophys. Freien Univ. Berlin* **5,** 145 pp.

Wege, K. (1958). Druck- und Temperaturverhältnisse der Stratosphäre über der Nordhalbkugel in den Monaten März, Mai, September, November und Dezember. *Meteorol. Abhandl. Inst. Meteorol. Geophys. Freien Univ. Berlin* **7,** 20 pp.

Wege, K., H. Leese, H. V. Groening, and G. Hoffman. (1959). Mean seasonal conditions of the atmosphere at altitudes of 20–30 km and cross sections along selected meridians in the Northern Hemisphere. *Inst. Meteorol. Geophys. Freien Univ. Berlin* ASTIA No. 157-154.

Wexler, H. (1951). Spread of the Krakatoa volcanic dust cloud as related to the high-level circulation. *Bull Am. Meteorol. Soc.* **32,** 48–51.

Wexler, H. (1959). Seasonal and other changes in the Antarctic atmosphere. *Quart. J. Roy. Meteorol. Soc.* **85,** 196–208.

Wexler, H., and W. B. Moreland. (1958). Winds and temperatures in the Arctic stratosphere. *In* "Polar Atmosphere Symposium, July 1956," Part. 1, Meteorol. Section, pp. 71–84. Pergamon Press, Oxford.

White, R. M. (1951). The meridional eddy flux of energy. *Quart. J. Roy. Meteorol. Soc.* **77,** 188–199.

Wilkes, M. V. (1949). Oscillations of the earth's atmosphere. "Cambridge Monograph on Physics," 74 pp. Cambridge Univ. Press, London and New York.

Wilson, C. V., and W. L. Godson. (1962). The stratospheric temperature field at high latitudes. Arct. Met. Res. Gp. Publ. in Met., 46, McGill University, Montreal, Canada.

Witt, G. (1960). Polarization of light from noctilucent clouds. *J. Geophys. Res.* **65,** 925–934.

4

The Upper Stratosphere

Introduction

The upper stratosphere bears the distinction of being the most stable region of significant mass in the earth's atmosphere. This stability is thermal in origin, with a negative lapse rate of approximately 0.25°C per 100 meters in tropical regions, that is, roughly one-fourth the magnitude of the positive lapse rate of the troposphere. While mixing from the surface heat source establishes the tropospheric temperature structure through adiabatic processes, the upper stratospheric temperature structure is the result of internal heating through absorption of solar radiant flux in the ultraviolet region of the spectrum. The radiant flux acts as a catalyst in the dissociation of ambient molecules, the oxygen atom products of which in turn combine to form stable ozone molecules (O_3). This element is highly efficient in the absorption of electromagnetic energy in the 2000–3000 angstrom wavelength region and in transfer of energy into kinetic motions of air molecules through collision processes.

While it is by no means certain that ozone absorption is the only significant mechanism for internal heating of the upper stratosphere, it is probable that this mode of energy conversion is dominant. For instance, water vapor may be instrumental in establishing the thermal structure, although its contribution might well be in the negative sense, with the water molecules radiating and absorbing most effectively in the infrared spectral region. It would then be most closely related to the earth's upward radiant flux. A major problem for the atmospheric scientist concerns the abundance of water vapor in the upper stratosphere. Most current analyses of the stratospheric radiation equilibrium have considered the water vapor concentration to be essentially negligible, although such is not necessarily proved by the data.

As has been illustrated in Figs. 3.45–3.48, the peak concentration of ozone is located at near 20-km altitude. Relatively little is known about the distribution above 30 km, and the assumption of a constant mixing ratio with height is generally made for the upper stratosphere. This is equivalent to assuming an exponential decrease in ozone concentration with height where the exponent is the same as that for air, such as would be produced in an inert gas of the molecular weight of air in a gravitational field. Our picture of the generating and destruction mechanisms for ozone does not point toward such a distribution as the most likely, although there are no real disagreements with observations at this time. It is probably a fair criticism to point out that we are not familiar with the ozone concentration in the upper stratosphere. We do know, however, the gross temperature structure of the upper stratosphere with a reasonable accuracy. Assuming that ozone is the principal ultraviolet absorber and that other heat exchange mechanisms are reasonably negligible, it is probable that the ozone mixing ratio would increase with height in the upper stratosphere. The reduced relative specific heat capacity at higher levels thus produces the temperature maximum which we have designated the stratopause and the negative lapse rate which is a principal feature of the structure of the region. The horizontal distribution of ozone concentration in the stratopause region is not clearly defined either. Reaction rates are quite high at these altitudes, and they are not known precisely enough to estimate the equilibrium concentration under the strong diurnal change in solar flux. At first look it appears reasonable to assume that a maximum ozone concentration will be obtained under the subsolar point at the stratopause and that lesser concentrations would be found on the dark side of the earth and in the diminished solar flux regions about the poles. This is essentially the opposite distribution from what has been observed in the lower stratosphere, at least in the meridional case.

It has been implied in the above discussion that the heat input to the upper stratosphere will be sensitive to the aspect of the solar radiation as it is incident on the earth's atmosphere. As a result of the gross vertical density gradient, the heat disposition will be concentrated in a minimum height range where the incident radiation is normal to the spherically stratified atmosphere if the absorber mixing ratio is uniform; the level of maximum heating should be a minimum at that point. As the incident angle is increased (i.e., by moving away from the normal position) the absorbed radiation will be distributed through a greater volume and the level of maximum input will be higher than in the normal case. Thus, over the poles at equinox times or over the sunrise and sunset lines, the region of maximum heat input should be at higher altitudes and less intense than at the normal incidence point.

This is the general result noted from observational data. That is, a cold, low-pressure system is found at both poles in the upper stratosphere for periods about the times of the equinoxes. Winds are, therefore, westerly on a global scale during this interim period as the normal incidence point crosses the equator to shift hemispheres. The winds are light, however, so the effect is relatively small in comparison with the annual range of meridional temperature gradients to be observed during the summer and winter seasons. As the subsolar point moves away from the equator, a complication is introduced in that the length of the periods of illumination and shadow will be altered by the geometry of the earth-sun position relationship. These points are illustrated by Fig. 4.1, which indicates that certain physical changes are to be expected in the stratospheric heat input distribution as a result of changes in solar aspect angle.

Even during the equinoctial periods, the extreme upper portions of the ultraviolet absorbing regions over the poles are subjected to continuous irradiation when that level is reached where the radiation penetrates beyond the axis of rotation before it is absorbed (probably in the mesosphere). There probably exists, then, a small, warm, high-pressure region of inverted conical shape at high levels, roughly symmetrical about the earth's rotational axis. Because a greater amount of heat will be deposited on the subsolar side of the conic, a diurnal wobble can be predicted. The subsolar side of the conic will advance slightly toward lower latitudes and the antisolar side will retreat toward the pole. This wobble should become more pronounced in amplitude (possibly not in intensity) in the lower portions of the conic until near the vertex the axis of the conic will be deflected away from the earth's axis of rotation, always toward the subsolar point. A sort of diurnal whip action should result.

Dimensions of this anticyclonic vortex are currently unexplored, but it appears likely that the lower tip would be at mesospheric altitudes during equinoctial periods and would be constrained to positions of only a few degrees latitude from the rotational axis. At ionospheric altitudes, however, the horizontal extent of the easterly current would be enlarged, and the diurnal variation would become less pronounced as increased transparency resulted in a more even heat disposition across the polar region. The anticyclone should be largely symmetrical at the level of the earth's "limb" for the controlling wavelength. Thus, a portion of the vortex will be shaped by the ultraviolet-ozone interaction in the meso-

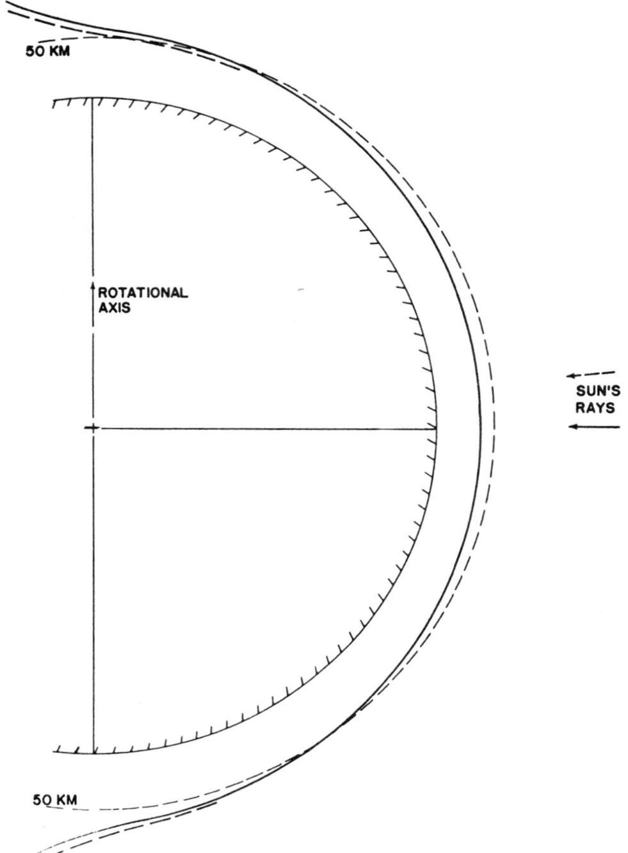

Fig. 4.1. Idealized cross section of solar ultraviolet heat input into the stratosphere (exaggerated vertical scale) in the plane which includes the earth's rotational axis and the subsolar point. The solid curve indicates depth of penetration of a particular radiation at the equinox, and the dashed curves illustrate the changes that are to be expected as the solar aspect changes.

sphere, a second portion by the far ultraviolet–atomic oxygen interaction in the lower ionosphere, with additional modes of formation at higher altitudes. Our conical surface is thus corrugated along the vertical axis, and possibly even divided into segments.

Now as the sun begins to rise over the hemisphere which will next experience summer, the geometrical relationship changes. Rays which had traversed the polar regions tangent to spherical surfaces are now tilted to penetrate into the more dense medium of lower altitudes on the side of the polar vortex opposite the sun. A diurnal variation should persist with a maximum in the subsolar region, but it should diminish in intensity. As the sun rises over a hemisphere, the region of continuous illumination should increase in size. The latitudinal extent of the conic will increase and the altitude will lower.

Again considering a homogeneous absorbing medium with an exponential vertical distribution, we would expect a maximum heat deposit at a minimum altitude on the subsolar point. More importantly, the basic assumption of uniformity leads one to expect a certain symmetry between hemispheres in a meridional plane, particularly at times of the equinox. At the times of the solstice this symmetry should still exist, but would be biased by a general thermal gradient, positive in the direction of the summer pole.

The latter item is a most important aspect of the solar heat input distribution. Continuous insolation in the summer polar regions results in rather warm temperatures, while the lack of heat input produces a cold center in the winter polar region. The distribution of the meridional temperature gradient would not be uniform even in the homogeneous case, and simple inspection of the problem indicates the probable existence of a peak in the negative meridional temperature gradient at upper middle latitudes in the winter hemisphere and a rather weak gradient elsewhere, except for a moderate positive meridional gradient in summer high latitudes. Dominance of the geometric relationship, resulting in longer periods of insolation in the summer hemisphere, produces a positive meridional temperature gradient from the equator to the pole.

All of the above remarks are based on the assumption of a homogeneous absorbing medium. Our experience with atmospheric phenomena does not support the likelihood of such a quiescent distribution. For instance, it is possible that some minor constituent such as water vapor or carbon dioxide has a higher concentration in tropical regions. If that were so, the equilibrium heat condition might be significantly altered. Most important would be the gradients which would be established by changes in concentration of the contaminant. It is possible, then, for meridional temperature gradients to be generated by these and other factors, and thus to introduce detail structure in the temperature and wind fields of the stratosphere.

Observation and analysis of the 20- to 30-km layer from balloon data had determined the monsoonal character of the stratospheric circulation prior to the establishment of the MRN. From several sources it was known that winds are easterly in summer and westerly in winter, and balloon sounding of the lower stratosphere had determined by the middle 1950's that gross variations occur in winter. These phenomena appeared to propagate downward, with the resulting indication that they originated in the upper stratosphere.

It was to explore those predictable and puzzling aspects of the stratospheric structure and to analyze deviations from these simple patterns that the Meteorological Rocket Network (MRN) was initiated. The MRN had its birth at the missile ranges of the United States because the prime users of the data are concerned with the space-age expansion of our sphere of operations, and they are located principally at these sites. There was no real program designed to provide for systematic analysis of the stratospheric structure but simply a group of individual projects which took data as needed. The general features of the MRN development have been outlined by Webb *et al.* (1962).

There existed a coordination forum for missile range meteorologists in the Inter-Range Instrumentation Group (IRIG) of the Range Commanders Council, the Secretariat of which is located at White Sands Missile Range, New Mexico. These scientists engage in developing meteorological techniques to the point of providing adequate environmental data for their respective ranges. Since their problems were similar, it was natural that a considerable interchange of ideas and equipments should result. The MRN fell into that category since in this new area of endeavor maximum progress and efficiency could be achieved by close coordination. The Meteorological Rocket Network Committee (MRNC) was formed in April 1959 as a subcommittee of the Meteorological Working Group (MWG) of IRIG to achieve the desired coordination. The MRNC has been instrumental in assembling and publishing the data in a series of volumes, at first on a quarterly basis and, after the summer of 1962, on a monthly basis. These volumes have become the focus of the MRN, welding data from diverse observational systems into a unified format which is compatible with synoptic analysis.

Since the inception of the development of the Meteorological Rocket Network, the Inter-Range Instrumentation Group's (IRIG) Meteorological Working Group (MWG) has acted as a coordination point for the activities of the several participating organizations. This coordination has concerned such items as the transfer of information relative to system techniques, range firing problems, and schedules, as well as the collection, reduction, and publication of the resultant data. IRIG Document 105-60 "Initiation of the Meteorological Rocket Network" (1961) was designed to make available a record of the many details concerning the implementation of the Meteorological Rocket Network for use in evaluating the data available and in indoctrinating new participants in the

MRN. The document was first released in July 1960 and was revised to depict the rapidly expanding Meteorological Rocket Network in August 1961. Subsequent progress has made it desirable to produce a second revision of this document, which was released on 1 March 1965 as IRIG document 111-64 titled "The Meteorological Rocket Network."

Until the advent of rocket systems, the stratosphere was one of the relatively unexplored regions of the earth's atmosphere. Balloon probes barely penetrated the stratosphere during the initial attempts to extend our observational systems to higher altitudes in the atmosphere. In addition, the sensing techniques employed by these early balloon systems were inadequate for accurate probing of stratospheric structure even in these lower regions. Early investigations into the lower margins of the stratosphere erroneously indicated it to be a relatively uninteresting volume of the atmosphere, and many early investigators were accordingly diverted into other areas of research. Subsequent improvements of the balloon techniques have resulted in information of great interest relative to the circulation and structure of the lower stratosphere.

The relatively slow progress made in understanding stratospheric processes is in part due to the fact that this is a transition region where observational and analysis techniques which are applicable to the tropospheric phenomena are beginning to give way to the decidedly different techniques required to observe the atmospheric media where electromagnetic phenomena predominate. The contrast between physical processes in the troposphere and in the ionosphere is rather tremendous by any standards, and a working knowledge of the interim region separating (or coupling) these distinct atmospheric layers is of prime importance to a comprehensive understanding of the earth's entire atmosphere. Rocket vehicles have provided a means for obtaining data in this very difficult observational region; specifically, small rockets have proved to be particularly efficient in providing the synoptic atmospheric data which are traditionally within the field of the meteorologist. A steady reduction in the size and cost of the vehicles and improvements in techniques has resulted in the development of a Meteorological Rocket Network which has the objective of adequate synoptic observation of the earth's stratosphere.

The modern development of flight, rocket, and space activities has resulted in a dramatic change in the sphere of influence for which the meteorologist is held accountable. The obvious progression of air transportation toward higher speeds which, in general, specifies higher altitudes, has directed the attention of the aviation industry and its associated supporting activities toward the practicability of supersonic flight in the stratosphere. The development of rocket systems with their regular traverse of the stratospheric region requires a working knowledge of the physical structure of this region as it affects the flight of high-speed probes. In addition, we are currently faced with the problem of reentry vehicles which, as they become more sophisticated, require more intimate knowledge of the detailed features of the stratosphere and its interaction with the flight problem.

The Meteorological Rocket Network has developed as a result of the exploratory activities of a number of research groups. It awaited the development of suitable rocket vehicles, a research item which was very intensively pursued during the late 1950's. The resulting vehicles, while far from optimum, were adequate to initiate an early synoptic exploration of the stratosphere. The firing of even small rockets requires certain safety precautions such as the availability of adequate real estate, and the effective safety of air space above the launch site and in the reentry zone. These parameters are characteristic of missile test ranges, so it is

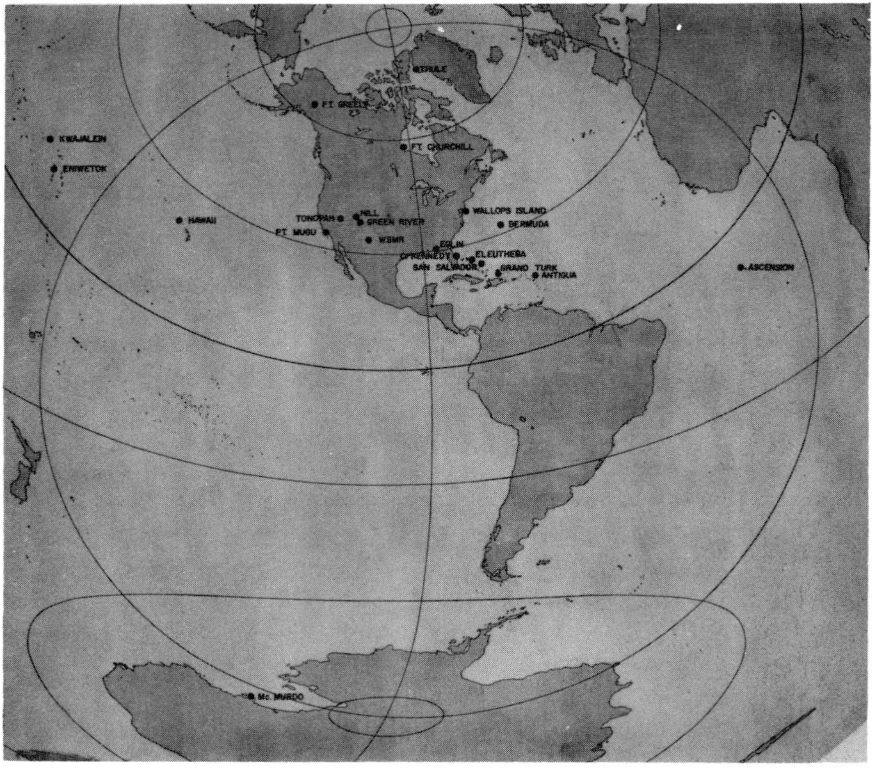

Fig. 4.2. Meteorological Rocket Network sites equipped to take stratospheric observations as of 1 January 1965.

not unexpected that the initial development of the Meteorological Rocket Network has occurred at the sites illustrated in Fig. 4.2 (see also Table 4.1).

TABLE 4.1
Meteorological Rocket Network Stations on 1 July 1964

Station	Call letters	International index number	Latitude	Longitude	Altitude (meters)	Total MRN firings
Antigua	MKPA	78861	17° 09′ N	61° 47′ W	3	
Ascension (AS)	FHAW	61902	07° 59′ S	14° 25′ W	30	76
Barking Sands (BKH)	BKH	91162	22° 03′ N	159° 47′ W	5	178
Cape Kennedy (CK)	COF 1	74794	28° 27′ N	80° 32′ W	3	678
Eglin AFB	VPS	72221	30° 23′ N	86° 42′ W	3	83
Eleuthera	MYEM	78076	25° 16′ N	76° 19′ W	6	
Eniwetok	PKMA	91250	11° 26′ N	162° 23′ E	6	14
Fort Churchill	YQ	72913	58° 44′ N	93° 49′ W	23	114
Fort Greely (FG)	FGJ	70266	64° 00′ N	145° 44′ W	450	244
Grand Turk	MKJT	78118	21° 26′ N	71° 09′ W	5	6
Green River, Utah	GRV	72477	38° 56′ N	110° 04′ W	1372	14
Keweenaw, Michigan			46° 26′ N	87° 42′ W	329	5
Kindley AFB, Bermuda	MYKF	78016	32° 22′ N	64° 40′ W	6	
Kwajalein, Marshall I.	PKWA	91366	08° 44′ N	167° 44′ E	5	26
McMurdo, Antarctica	NGD	89664	77° 53′ S	166° 44′ E	70	26
Point Mugu, Calif. (PM)	NTD	72391	34° 07′ N	119° 07′ W	3	560
San Nicolas, Calif.	NSI	72291	33° 14′ N	119° 25′ W	213	
San Salvador	MYSM	78089	24° 07′ N	74° 27′ W	12	10
Tonopah, Nevada	TTR	72485	38° 00′ N	116° 30′ W	1631	80
Wallops Island, Virginia (WI)	WAL	72402	37° 50′ N	75° 29′ W	3	366
White Sands Missile Range, N.M. (WSMR)	WSD	72269	32° 23′ N	106° 29′ W	1210	905

While rocket vehicles and the required personnel, equipment, and real estate provide proper background for stratospheric exploration, the principal problem rests in the availability of stratospheric observational systems which can adequately provide information relative to the more important atmospheric parameters of that region. The development of these sensing systems has been a major problem in the development of the MRN. The desire for synoptic analysis of time and space variability in the stratosphere precludes the expansion of each individual observation to include all possible parameters of interest. The decision to restrict the MRN observational systems to the minimum acceptable observed data was, in large part, dictated by the desire to increase the frequency of observations and the number of points from which data were obtained. A second major consideration was the fact that sensors for the observation of additional atmospheric parameters are frequently quite complex and raise significant economic questions relative to their application. Most successful measurements which have thus far been connected with synoptic stratospheric exploration have concerned observation of the structure of wind, temperature, and density. In general, observations of the wind system must necessarily be made in the accomplishment of other observations, so it is essentially true that all instances of reported data in the MRN will contain a wind profile. In addition, the profiles of measured temperature or measured density are occasionally obtained, and, in general, when one of these parameters is available the remaining thermodynamic parameters are evaluated through hydrostatic and perfect-gas assumptions. Other observational parameters which are sacred to meteorological analyses, such as water vapor and ozone concentration or the absorption of incident radiant energy, have not yet proved feasible for Meteorological Rocket Network application.

The total observational rate of the MRN is illustrated in Fig. 4.3.

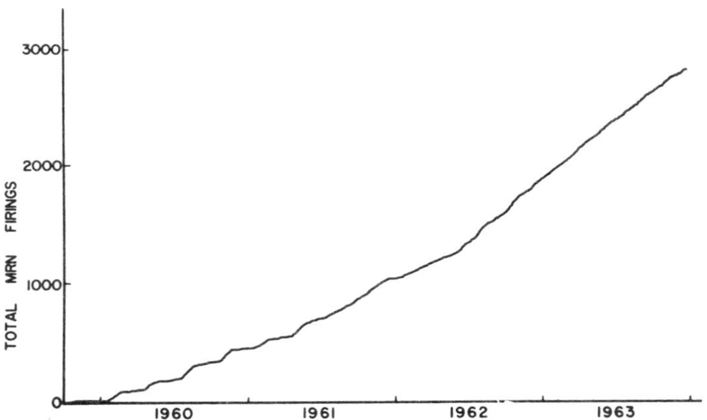

Fig. 4.3. Cumulative totals of Meteorological Rocket Network observations that have been published in the MRN data reports.

These data illustrate a steady growth in the observational system and point out that the early phases from October 1959 through April 1961 were conducted on a seasonal basis with frequent firings during a month's period. A steady firing schedule was initiated in the summer of 1961 with an objective of regular soundings from each participating station on every Monday, Wednesday, and Friday.

It is obvious from the map of the Meteorological Rocket Network stations (Fig. 4.2) that certain regions even of the North American continental area are relatively unobserved by the current distribution of firing sites. The particularly important omissions are the tropical regions, those near 50 degrees north latitude, and the arctic area. Regardless of the large number of important small-scale phenomena which have been observed by the sensitive observational systems of the MRN, it is clear that simple observation in a single quadrant of the globe will not satisfactorily define the large-scale circulations that are apparent in the stratosphere. Inspection of the data substantiates the need for expansion of the MRN with regard to the number of sites, the frequency of firings, and the peak altitude to which data are obtained.

One of the most important aspects of the IRIG coordination and the MRN has been the systematic collection and publication of the data obtained by MRN participants. All data obtained by meteorological rocket systems in cooperation with the MRN have been assembled, evaluated by common reduction techniques, and published in a standard format under the heading of IRIG Document 109-62. The first volumes of this document contain data for each of the seasonal periods except for Volume 4, which contains 20 soundings obtained in cooperation with the International Rocket Interval of 16–23 September 1960. The large amount of data acquired in the summer of 1962 required the issuance of two volumes. Subsequently, the MRN data have been published on a monthly basis. A total of 38 volumes of Document 109-62 are thus available, containing graphical and digital presentations of all data obtained in the Meteorological Rocket Network through December 1964.

This technique for assembling and publishing the data was selected principally to make sure the data were in the hands of users at early dates for more pressing research investigations. It necessarily required the use of less sophisticated analysis techniques than those that could be used on some of the data. The investigator who is attempting to look at the detailed structure of the atmosphere or to extract special information from the meteorological rocket soundings must contact the originator of each item of the data, since the MRN data reduction procedures are most suited to general synoptic studies. Several of the sensors, data acquisition techniques, and data processing techniques that were used to obtain the data from which the MRN publications were derived are amenable to much more sophisticated analyses.

The development of the Meteorological Rocket Network has been brought to the attention of scientists in general through the publication and presentation of numerous technical papers in a variety of journals and at numerous national and international meetings. The first meeting devoted specifically to objectives of the MRN was an American Meteorological Society meeting held on 17, 18, and 19 October 1958 at Texas Western College in El Paso, Texas. The meeting, titled "Conference on the High Atmosphere," was designed to bring together those scientists who had previously been engaged in rocket research of the upper atmosphere and the Meteorological Rocket Network scientists to assure maximum application of this early knowledge. It was followed in 1960 with the joint Instrument Society of America–American Meteorological Society meeting in Los Angeles, relative to meteorological rocket instrumentation, and again in 1961 by a meeting of the American Rocket Society on meteorological rocket systems at Baylor University in Waco, Texas. The American Meteorological Society again cosponsored a meeting at Texas Western College in El Paso in 1961 to review the state of development of stratospheric observation systems.

The first meeting devoted specifically to the presentation of analyses of stratospheric synoptic data obtained by balloon and rocket systems was held at the Free University of Berlin, Germany, during the period 20–31 August 1962. On 19, 20, and 21 November 1963, the American Meteorological Society again sponsored a meeting at Texas Western College in El Paso. This conference had the triple purpose of summarizing the information on the accuracy and sensitivity of stratospheric measurements, presenting the results of analyses of stratospheric synoptic studies, and considering possible future directions of rocket synoptic studies. The development of the Meteorological Rocket Network has been principally the result of a very healthy spirit of cooperation among the several scientific groups employing small rocket vehicles for upper atmosphere exploration. The impetus for this growth has come mainly from missile range meteorologists, stemming from their very pressing need for a better understanding of the environment in which their ranges conduct flight tests. It is clear, however, that synoptic study of the stratosphere has far wider applications to the advancing research program of our modern technology and should be intensified in response to the many requirements for data.

4.1 Stratospheric Circulation

The upper stratosphere is defined as that region of the atmosphere bounded beneath by the stratonull surface and above by the stratopause. The stratonull is located roughly at the 24-km altitude level and the stratopause is nominally located at the 50-km level. The upper stratosphere is characterized by an increasing temperature with height, and generally by increasing wind speed with height. Thus, the upper stratosphere is a very stable region in the atmosphere relative to vertical motions and isolates the upper atmosphere from the troposphere and lower stratosphere. This separation of the tropospheric and stratospheric circulations would be almost complete if the total atmospheric structure were only that exhibited by the equatorial or midlatitude regions; this is not the case, however, since the polar regions represent at least a partial breakdown in the rigidity of the stable layer. The long residence times that were first pictured for the stratosphere appear to be excessive even for this upper portion of the stratosphere, and only in equatorial regions does it appear likely that stratospheric storage times might be measured in months.

A well-developed zonal circulation is one of the most obvious properties of the upper stratosphere. As is illustrated in the winter case of Fig. 1.2, winds of the upper stratosphere increase with height from weak westerly or even easterly to strong westerly at the stratopause level. In general the maximum speed in the winter seasons occurs at about the stratopause level, and decreases somewhat above. While there is much detail observed in these profiles, the mean curves illustrate a steady strengthening of the flow with altitude throughout the upper stratosphere. The stratospheric circulation reverses in the spring, and a moderate easterly flow develops to a maximum in late July and then decays during the late summer season. It is not uncommon to have easterly tropospheric thermal winds, so the zonal wind at the stratonull level is frequently a few tens of meters per second from the east. The easterly thermal wind which is characteristic of the upper stratosphere is added to this value, generally with a stronger positive wind gradient in the upper stratosphere. Peak value of the easterly summer circulation in the upper stratosphere is of the order of 50 meters/sec in middle latitudes at the stratopause, and the wind profiles above the stratopause level generally show still increasing easterly winds with height.

The rather uniform vertical gradients in zonal wind speed have been adopted to obtain a simple measure of the stratospheric circulation which is informative for analysis purposes. This parameter is referred to as the Stratospheric Circulation Index (SCI) and is the mean component flow (meters per second) in a 10-km-thick layer centered at the 50-km level. Thus, the SCI is essentially the mean wind at the stratopause and is commonly reported in zonal and meridional components. The SCI is a comparatively conservative measure of the circulation and is useful in picturing the general state and trends of the stratospheric circulation. The results of zonal SCI observations over White Sands Missile Range since the beginning of MRN firings are illustrated in Fig. 4.4. It is obvious from these data that there is a very regular pattern to the winter westerlies and summer easterlies of the stratosphere.

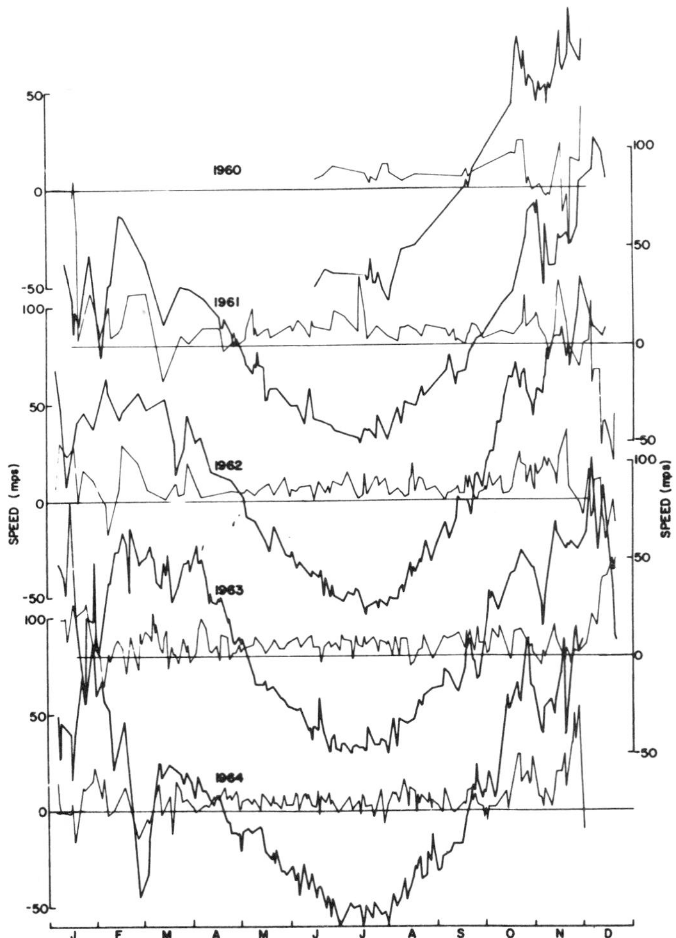

Fig. 4.4. Zonal and meridional Stratospheric Circulation Index (SCI) data for White Sands Missile Range, New Mexico, are presented by the thick and thin lines, respectively. The data points represent individual mean wind components in a 10-km-thick layer from 45- to 55-km altitude.

Winter westerly winds appear at about the date of the fall equinox and develop rapidly into a strong flow which circles the hemisphere in two to three days. After the initial headlong development of the first few weeks of winter, the westerly flow increases more slowly to reach its peak during the first half of December. The zonal winds show a marked change in the middle winter period, with rapid decreases in speed occurring over a few days' time. In special cases the westerlies characteristic of a weak summer situation put in an appearance. These interruptions of the winter westerly flow have a short life over White Sands Missile Range, and it is generally only a week or so before the westerly circulation is re-established. It is never quite the same again, however, with the mean speed of the variable flow showing a much lower value than that evidenced during the early winter period. This highly variable period of the stratospheric circulation is termed the "winter storm period" and it is during this time that gross circulation disturbances of the entire upper atmosphere on occasion result in changes in atmospheric structure which Scherhag (1952) called "explosive warmings."

The winter storm period may exhibit none or any number of these major changes in stratospheric circulation during any one year, and the dates on which these events occur vary over the years of data thus far available. This phase of the annual cycle is usually over by mid-February, and the stratospheric circulation stabilizes to a rather steady circumpolar flow of about one half the speed that was evidenced during the early winter season. About the first of April this weak westerly circulation starts to deteriorate, and the spring reversal in stratospheric circulation begins. The spring reversal is executed quite smoothly, with the date of SCI crossover of the zero circulation line near 1 May, which is some 40 days after the spring equinox. It is obvious from the data that the spring of 1964 deviates significantly from the data for the previous years of record in that the speed of the flow of the late winter season was only a third of the usual for that season and that the reversal occurred quite early over White Sands Missile Range. The reason for this different behavior during the 1964 spring reversal period is uncertain, but the data point out the fact that the data sample thus far collected is too small to assume that it is entirely representative.

The summer season stratospheric circulation exhibits a very conservative pattern, both in the symmetry of the development and decay of the zonal flow and in the lack of drastic day-to-day changes in the general flow structure. After a late start in spring, the summer easterlies tend to maintain their development until a late date so that the peak of the summer stratospheric circulation does not generally occur until well into the month of July. Decay of the summer easterly winds is rather rapid, although not as precipitous as is the buildup of the winter westerlies after the fall equinox. The summer season of the stratosphere is about four and one half months long and exhibits considerably less variability than does the winter season.

The above discussion illustrates the reasons for subdividing the stratospheric circulation into six definite seasons for analysis purposes. The four most stable periods are the "late winter season," the "summer season," the "early winter season," and the "winter storm period," which are designated to occur from 15 February to 1 April, 15 June to 15 August, 15 October to 15 December, and 15 December to 15 February, respectively. The usual mean values of wind and temperature and their standard deviations provide relatively meaningful information on the stratospheric structure during these intervals. The remaining two periods are characterized by gross systematic changes which require more detailed treatment to provide adequate representation. Our seasonal breakdown of the stratospheric circulation, except for the winter storm period, follows closely the regimes which would be predicted by a direct response to the annual variation in heat input resulting from a changing earth-sun geometry. In the case of the winter storm period there is no obvious solar cause, and one is led to inspect the data in a more global sense for dynamic factors which could precipitate the dramatic effects noted in the winter westerly circulation.

Inspection of the meridional component data in Fig. 4.4 brings out the rather interesting fact that there is a southerly flow at the stratopause over White Sands Missile Range year around. This south wind is strongest and most variable in the winter seasons but is of the order of a few meters per second throughout the year. It is clear that there must be a return flow, and a crucial question concerns whether that return flow occurs at other longitudinal stations at other times, or at other levels of the atmosphere. The latter case will imply ascending and (or) descending motions, and is thus of gross import to problems involving residence times or contaminant transport. Since observation of the probable vertical flows is beyond the resolution of current systems, adequate global observation of the continuity of this meridional flow characteristic will be required to establish the significance of these data.

While the data obtained over White Sands Missile Range may be representative of the stratospheric circulation in lower midlatitudes, it is not so likely that the same conditions exist in polar and equatorial regions. SCI data which have been obtained by the MRN station at Fort Greely during the past several years are illustrated in Fig. 4.5. These data point out the fact that the zonal circulation is generally weaker at high latitudes than at lower middle latitudes. Easterly winds are relatively common at Fort Greely during the winter season, and the meridional flow there is significantly greater than at White Sands Missile Range (Fig. 4.4). North winds are persistent and strong during the winter season but reverse and become light southerly during the summer season.

It appears, then, that at high latitudes the strong monsoonal characteristic of the zonal flow is weaker than that found at lower latitudes,

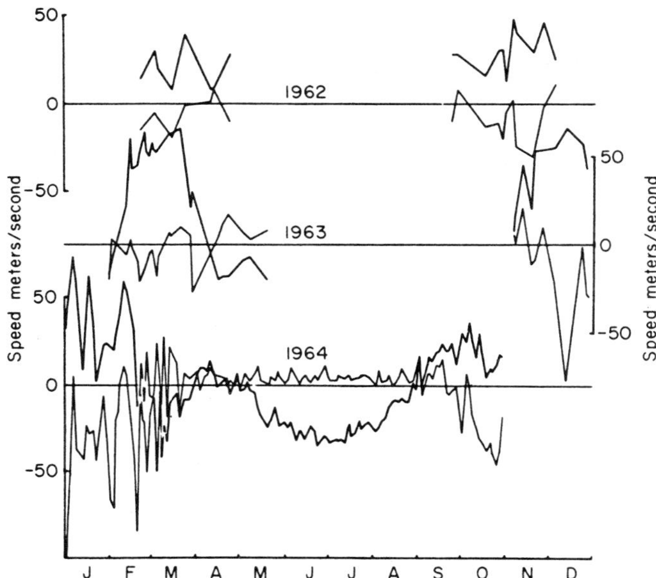

Fig. 4.5. Zonal (thick) and meridional (thin) SCI data for Fort Greely, Alaska.

while a new meridional monsoon is substituted at high latitudes for the monodirectional flow which has been found to be the annual distribution at lower latitudes. The reversal of the Fort Greely meridional flow occurs in the spring on about 17 May and in the fall on 25 August. The summer southerly winds are quite uniform for the entire season, with only minor variations. During the winter, on the other hand, large changes occur over a period of a few days in both the zonal and meridional components of the flow. To illustrate the nature of the vertical profiles from which the SCI data of Fig. 4.5 were obtained, two examples of the winter MRN data are presented in Fig. 4.6 and represent a rather typical change in the Fort Greely winter circulation.

In view of the likelihood that the stratospheric circulation is complex to the extent that the SCI data from White Sands Missile Range (Fig. 4.4) and Fort Greely (Fig. 4.5) are not representative of the hemispheric circulation, summaries of the SCI data for selected stations are presented in Figs. 4.7 and 4.8. The data for this presentation were designed to illustrate the meridional variations that are encountered in the MRN data. Ascension Island (7° 59′ S) data, although very limited in amount, is included to depict the stratopause circulation system in equatorial regions; Barking Sands, Hawaii (22° 3′ N) and White Sands Missile Range (32° 23′ N) to represent lower midlatitudes; Wallops Island

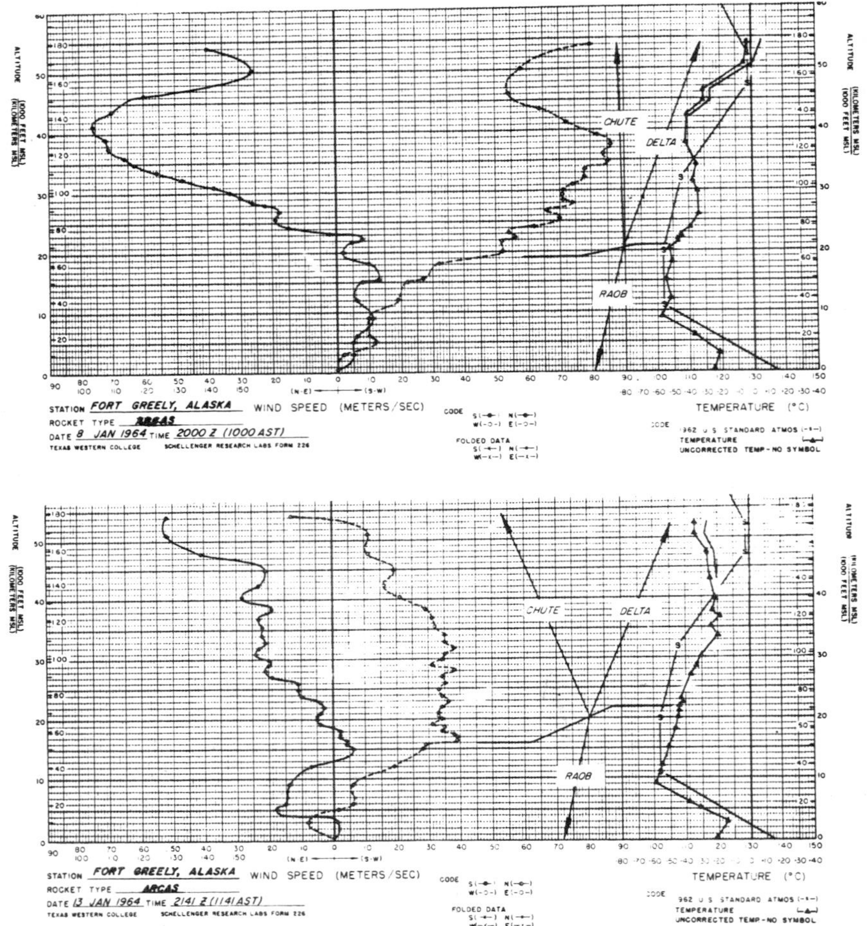

Fig. 4.6. Zonal (dashed) and meridional (solid) vertical wind and temperature (solid) profile data obtained at the MRN station at Fort Greely, Alaska on 8 January 1964 and 13 January 1964 as published in the MRN data reports.

(37° 50′ N) to represent middle latitudes; and Fort Greely, Alaska (64° N) to indicate the high-latitude flow. The curves of Figs. 4.7 and 4.8 were obtained by averaging all MRN data for each station through October 1964 at semimonthly intervals, and the number of observations included in each data point are presented in Table 4.2. As has been indicated by the White Sands Missile Range data presented in Fig. 4.4, a seasonal breakdown of the stratospheric annual cycle of

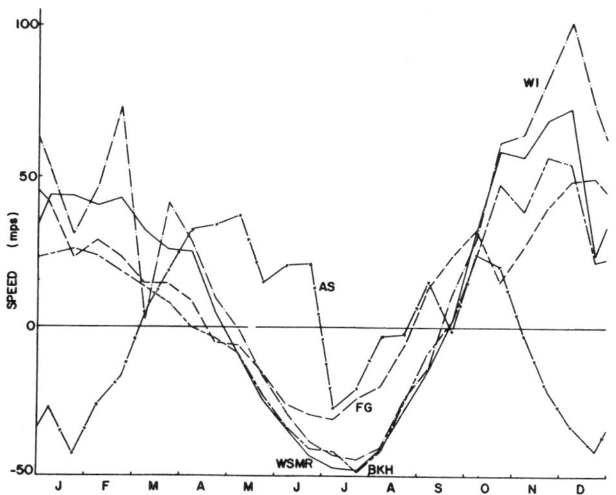

Fig. 4.7. Mean zonal SCI data for selected stations plotted to illustrate the meridional structure of the stratospheric circulation.

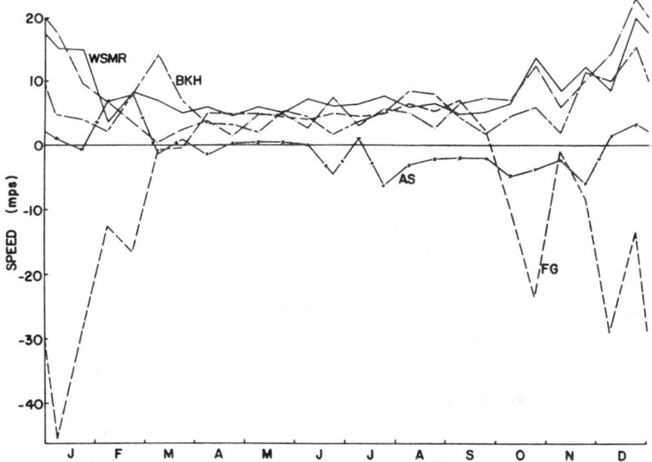

Fig. 4.8. Mean meridional SCI data for the cases reported in Fig. 4.7.

the type illustrated in Fig. 4.7 provides a meaningful climatological subdivision for all locations except polar and equatorial regions.

A most impressive bit of information contained in these curves is the considerable regularity of the annual progression in the stratospheric monsoonal circulation. Particularly in the development of the summer easterly circulation there is a striking similarity over a large portion of

TABLE 4.2

Number of Observations Included in Each Data Point of Figs. 4.7 and 4.8

	Ascension Island	Barking Sands	Cape Kennedy	Fort Greely	Point Mugu	Wallops Island	White Sands Missile Range
January	1	4	13	6	11	0	18
	5	7	28	12	21	11	26
February	3	6	25	13	15	13	18
	1	0	16	17	11	7	17
March	2	2	22	16	14	8	21
	3	1	19	8	14	11	24
April	6	3	22	8	15	6	22
	3	6	20	9	25	12	28
May	2	4	14	9	24	10	23
	2	7	32	7	27	6	24
June	2	4	24	7	27	7	27
	2	3	19	5	17	5	33
July	2	12	15	7	24	9	26
	3	13	34	7	29	15	38
August	2	7	23	8	26	11	31
	5	8	18	9	20	11	26
September	3	9	17	6	20	11	24
	8	9	13	11	23	4	38
October	8	13	15	7	30	13	24
	8	9	28	16	36	18	39
November	4	4	15	8	21	6	39
	1	7	13	5	11	0	33
December	4	4	11	3	10	6	20
	3	3	14	2	7	2	6

the stratopause region. Only in the polar and tropical regions does the sequence of events deviate significantly from that hypothesized for climatological purposes. In view of the need for a systematic pattern for analysis of the meteorological variability of the stratosphere, it would appear that these deviations are of minor importance. In polar regions the spring and fall reversals appear to be accomplished at an earlier date than is the case in middle latitudes. The spring reversal occurs only a short time after the spring equinox in polar regions, whereas the spring reversal is some 40 days after the equinox at middle latitudes. In tropical regions the stratopause is the site of a completely different type of circulation, with a semiannual cycle the principal feature. Again in the fall the polar region reversal occurs about one month earlier than at lower latitudes. It would appear that all of these data fall into a general circulation structure of the stratosphere which lends itself to simplified analysis along the lines discussed above.

The second item of importance in the zonal data presented in Fig. 4.7 is associated with the characteristics of the flow during the winter storm period. A gross reduction occurs in the zonal circulation in the latter part of December at all stations except in tropical regions. In this case the effect is noted to appear first at lower latitudes and work poleward. The circulation at the stratopause has its maximum value late in the early winter season, after which time the speeds drop to one half or less, with a partial recovery to moderate levels during the late winter season. There is little or nothing in the earth-sun heat input relationship to cause such an alteration in the stratospheric circulation. It would be expected that the winter circumpolar circulation would develop with increasing intensity as the polar night intensifies until a rising sun begins the reversal of the meridional temperature gradient. Clearly, that is not what is happening in the stratosphere during the central portion of the stratospheric winter.

A most likely second candidate for the cause of these fluctuations is that they are dynamic in origin, resulting from instabilities in the flow. This is a very interesting consideration and merits the full application of the synoptic capabilities of the MRN to acquire an understanding of the processes that can produce the results obtained thus far. Inspection of the data for Fort Greely in Fig. 4.7 points toward a second feature of the mean high-latitude circulation and shows a variation that is not obviously tied to the pattern of solar heat input. These features are the rather significant changes in slope of the zonal curves immediately after the spring and fall reversals. There appears to be no direct factor in solar control of the stratospheric circulation to explain these features, and again it is likely that the origin of this irregularity is dynamic and thus must be attacked on a synoptic basis. The White Sands Missile Range data of Fig. 4.4 show perturbations of this type in the individual data, particularly in the fall reversal period. These variations occur each year and are small enough that they are averaged out in the mean White Sands Missile Range curves of Fig. 4.7. All of these data point toward a disturbance which has its origin and greatest intensity in the polar regions, but which propagates toward lower latitudes with decreasing intensity.

A facet of the meridional SCI data of Fig. 4.8 requiring attention is the fact that these mean data show an equator-to-pole flow at all stations for the entire year except during the winter season at Fort Greely. If these data were actually representative for the various latitude bands about the entire hemisphere, it is clear that there would be a considerable convergence of stratopause air in high latitudes, particularly in the case of the summer season. The great stability of the summer circulation makes it probable that the data presented here are reasonably representative of the hemispheric circulation, and this meridional flow of mass at the stratopause level must necessarily be accounted for.

In the winter case, the data are less certain. It is entirely possible that a lack of symmetry in the winter circumpolar circulation could produce the observed oppositely directed meridional flows between high latitudes and middle and low latitudes. Synoptic charts of the lowest portion of the upper stratosphere indicate just such a circulation pattern, with a ridge over the eastern Pacific and a trough over the North American continental region a persistent feature of the winter season. If this circulation pattern should extend upward to the stratopause it could produce a compensating meridional flow at other longitude locations so that in the mean no vertical flow would be required by continuity considerations. This can only be established by extensive deployment of MRN-type observation systems on at least a hemispheric basis. On the other hand, investigators have found significant downflows of stratospheric air across the tropopause associated with tropospheric cyclonic circulation systems. Since these storms have their playground in middle latitudes, if the downward flow should come from the stratopause level there would be a significant converging flow at that level. The current MRN data coverage is probably inadequate for resolution of these points at this time.

A most interesting annual distribution of the stratopause zonal circulation for tropical regions is illustrated by the dash-cross curve of mean SCI data in Fig. 4.7. Limited data were available for construction of this curve, so the irregularities may or may not be indicative of the general circulation characteristics of the tropical region. These tropical data of Fig. 4.7 are based on MRN observations at Ascension Island, located at approximately 8 degrees south latitude, and thus the circulation patterns indicated by this curve are six months out of phase with respect to the other curves presented in the figure. A semiannual cycle is clearly present in the zonal winds of the tropical stratopause, although the two cycles are definitely not symmetrical at this 8-degree latitude station. There is good reason to expect the semiannual variation to be more nearly symmetrical at the equator and there only, since it is required by the characteristics of the solar heat input that the semiannual fade out at the maximum latitudinal displacement of the subsolar point, above which the annual cycle will dominate. According to these data, the reversal in fall from summer easterlies to winter westerlies occurs early in March, well ahead of the equinox. This reversal closes a well-developed summer circulation of approximately four months' duration in which maximum zonal speeds of 40 meters/sec were achieved. The winter westerlies at 8 degrees latitude of the Southern

Hemisphere appear to be definitely weaker than are observed at middle latitudes in the Northern Hemisphere, with a maximum of only 38 meters/sec observed in these mean data. These speeds are slightly lower than the maximum winter speeds observed at Fort Greely.

There are valid questions about the representativeness of the decrease in zonal circulation that is an obvious feature of the curve, with a minimum in the first half of May. The amount of data is small, and concerns only one year. On the other hand, our experience with stratospheric data, particularly in the case of the SCI measurement, is that most of the data points are highly representative, and should be ignored only after considerable deliberation. In this case, the perturbation is in exact agreement with the fall perturbation characteristic of all of the SCI data curves of Fig. 4.7 in late October and early November. The data sample in these curves is adequate (Table 4.1) to assure that this feature actually appears each year. The data indicate that the disturbance initiates at high latitudes, as is evidenced by its appearing first at Fort Greely during the last half of October, and works equatorward rapidly during early November. Apparently it crosses the equator and disturbs the winter hemisphere flow in tropical regions as is indicated by the Ascension data of late May.

The easterly winds of the Southern Hemisphere winter evidenced by data from Ascension are in effect for only a short period, of the order of two months. This variation corresponds in phase with the winter storm period of the high latitudes of the winter hemisphere except that it occurs slightly later, by the equivalent of two weeks. Consideration of the SCI zonal curve for Fort Greely (Fig. 4.7) indicates that the perturbation that is such a prominent feature of the middle latitudes arrives at high latitudes weak and late (23 January). It would appear that the disturbance which produces these results in the SCI data has its maximum impact at middle latitudes and extends this influence toward the winter pole and the tropics with time. Both of these events occur at the beginning and at maximum points in the development of the easterly summer circulation, and thus there is reason to suspect that these events are associated.

Weak westerlies of the order of 25 meters/sec are established again after the winter storm period in tropical regions for a period of approximately two months before the spring reversal of the Southern Hemisphere. The spring reversal at Ascension Island is quite similar in rate of change with that of higher latitudes in the Northern Hemisphere, but there is a significant difference in the tropical SCI of the summer in that a rather large depression occurs in the zonal stratopause speed during the middle of the summer season. It amounts to about 16 meters/sec and is centered on the first half of January, or a corresponding period of the first half of July in the Northern Hemisphere. The Barking Sands (22° N) SCI data of Fig. 4.7 show a similar type of variation, although of much smaller amplitude. These data would lead one to expect a general perturbation of the tropical stratopause during the winter storm period characterized by a reduction of the zonal easterlies which are resident there during this period, probably having a maximum effect at the equator.

Possibly the most interesting features of the zonal circulation of the stratopause region are illustrated by the curves presented in Figs. 4.9 through 4.20. These curves were obtained from the data presented in Fig. 4.7, but in this case the zonal wind speed was plotted as a function of latitude instead of time. To obtain the maximum latitudinal expanse of zonal wind profile from the relatively meager data available, the hemispheric zonal circulations were assumed to be similar. Data are available to substantiate the validity of this assumption in the form of scattered soundings at points such as the Woomera Range in Australia, although the only data comparable with the MRN data are those taken in October and November of 1963 at Cape San Lorenzo, Sardinia (Attmannspacher *et al.*, 1964). Mean SCI values determined from the soundings obtained nearest noon on the observation days are included in the mean time distribution data in Figs. 4.7 and 4.8. In any case, only additional MRN data on a global scale can satisfactorily resolve this problem. The data presented in this series of graphs represent the mean zonal flow in a 10-km-thick layer centered at 50-km altitude as a function of latitude, with a curve for each semimonthly period, and therefore illustrate the sequence of events for the entire year; that is, one period (summer) between equinoxes is presented in the upper portion of the graphs and the other (winter) period between equinoxes is illustrated by the lower portion of the graphs.

These data provide a fresh look at the sequence of events which surround development of the monsoonal circulations of the stratospheric system. The fact that the winter westerlies have begun on about the fall equinox and thus are getting under way while the westerlies are still dying out in the hemisphere that is approaching summer means that the entire globe has westerly winds at the 50-km level until the reversal starts in early April as is illustrated in Fig. 4.9, although the sunshine has already begun to heat the upper polar atmosphere and develop an easterly thermal wind in the Northern Hemisphere. With time the sunshine expands the warmed region around the north pole deeper into the polar atmosphere, so that rapidly this fledgling easterly circulation system embedded in the westerly counterflow extends past Fort Greely to herald the start of the summer season.

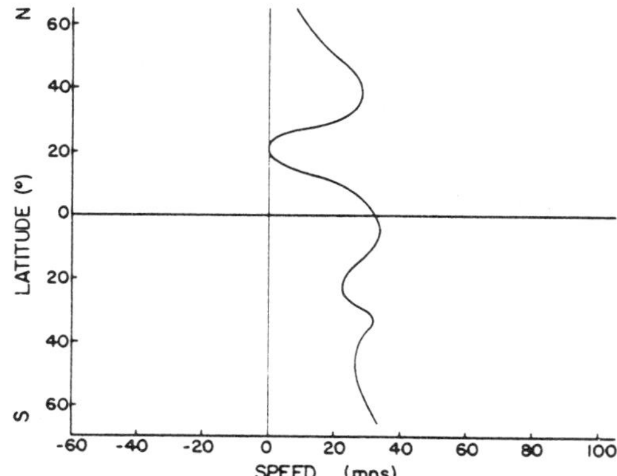

Fig. 4.9. Latitudinal distribution of mean zonal SCI data for the first semimonthly period after the spring equinox in the Northern Hemisphere. Data used for construction of the curve were obtained from the curves of Fig. 4.7 for the first semimonthly periods of April and October.

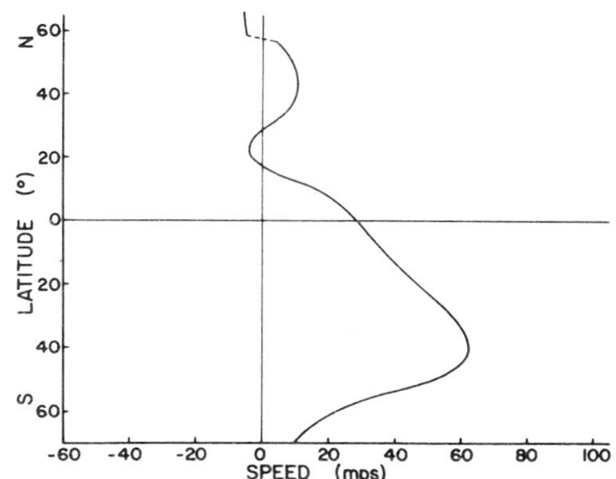

Fig. 4.10. Latitudinal distribution of mean zonal SCI data for the second semimonthly period after the spring equinox in the Northern Hemisphere. Data used for construction of the curve were obtained from the curves of Fig. 4.7 for the second semimonthly periods of April and October.

This polar start of the summer circulation is only a precursor, however, and the main easterly current has its origin at low latitudes. It expands rapidly on a broad front to cover the entire summer hemisphere by early May (Fig. 4.11). As the easterly circulation of lower latitudes expands poleward it encounters the polar cell of easterlies and a consolidation of the two circulation systems occurs as is indicated by the dashed portion of the curve of Fig. 4.10. Most of the energy of the easterlies is invested in the lower-latitude flow since a much larger mass

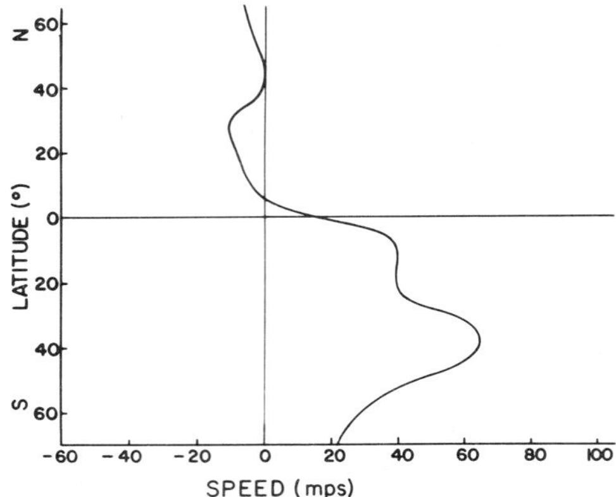

Fig. 4.11. Latitudinal distribution of mean zonal SCI data for the third semimonthly period after the spring equinox in the Northern Hemisphere. Data used for construction of the curve were obtained from the curves of Fig. 4.7 for the first semimonthly periods of May and November.

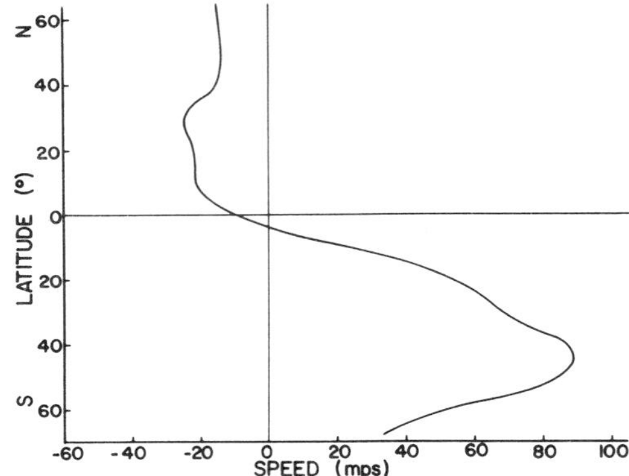

Fig. 4.12. Latitudinal distribution of mean zonal SCI data for the fourth semimonthly period after the spring equinox in the Northern Hemisphere. Data used for construction of the curve were obtained from the curves of Fig. 4.7 for the second semimonthly periods of May and November.

is involved, both because the dense lower portion of the upper stratosphere is involved and the volume is greater by a large factor. It is obvious that the polar circulation will be incorporated into the general summer easterlies when these systems merge. Thus, the speed of the polar easterly current, which has gained more speed than the lower-latitude circulation, will be diminished as the circulations merge (see Fig. 4.7). This locking of the two systems is clearly demonstrated by the change in speed in the Fort Greely SCI data during early May.

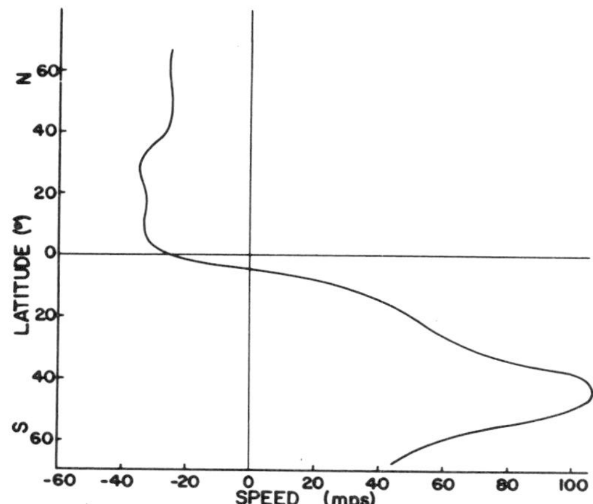

Fig. 4.13. Latitudinal distribution of mean zonal SCI data for the fifth semimonthly period after the spring equinox in the Northern Hemisphere. Data used for construction of the curve were obtained from the curves of Fig. 4.7 for the first semimonthly periods of June and December.

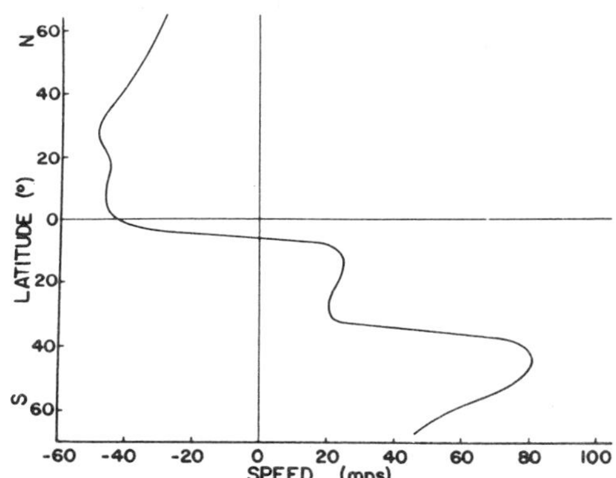

Fig. 4.14. Latitudinal distribution of mean zonal SCI data for the sixth semimonthly period after the spring equinox in the Northern Hemisphere. Data used for construction of the curve were obtained from the curves of Fig. 4.7 for the second semimonthly periods of June and December.

While the events discussed above are proceeding, another very special event is in progress in equatorial regions. Almost as rapidly as the easterly circulation spreads poleward it also expands equatorward, and by mid-May (Figs. 4.11 and 4.12) has crossed the equator. In a matter of a few weeks the easterlies strengthen and expand well into the winter

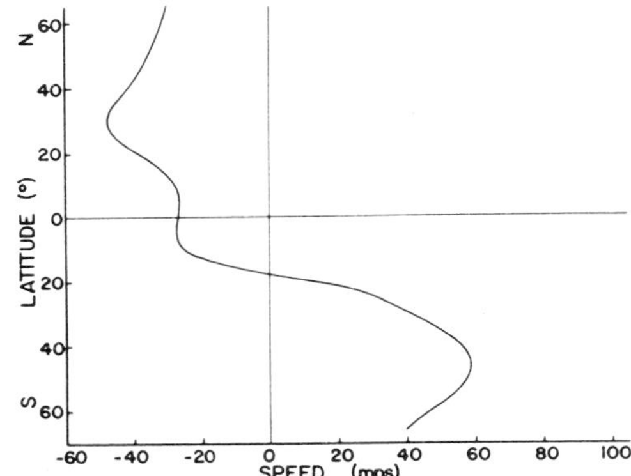

Fig. 4.15. Latitudinal distribution of mean zonal SCI data for the seventh semimonthly period after the spring equinox in the Northern Hemisphere. Data used for construction of the curve were obtained from the curves of Fig. 4.7 for the first semimonthly periods of July and January.

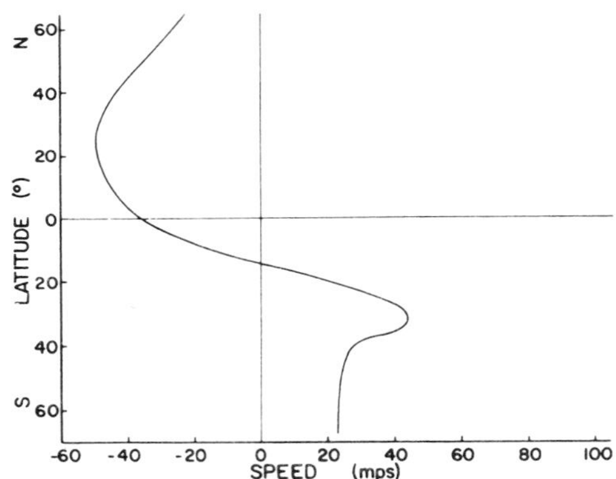

Fig. 4.16. Latitudinal distribution of mean zonal SCI data for the eighth semimonthly period after the spring equinox in the Northern Hemisphere. Data used for construction of the curve were obtained from the curves of Fig. 4.7 for the second semimonthly periods of July and January.

hemisphere. Such a result can occur only if the positive meridional temperature gradient of the summer hemisphere has the same slope in equatorial regions as the winter hemisphere. This is rather difficult to explain on straight geometric lines, and it appears certain that inhomogeneous structure in the composition or heat transfer mechanisms of the stratosphere must exist in low latitudes. The winter westerlies exhibit their maximum development in the curve presented in Fig. 4.13,

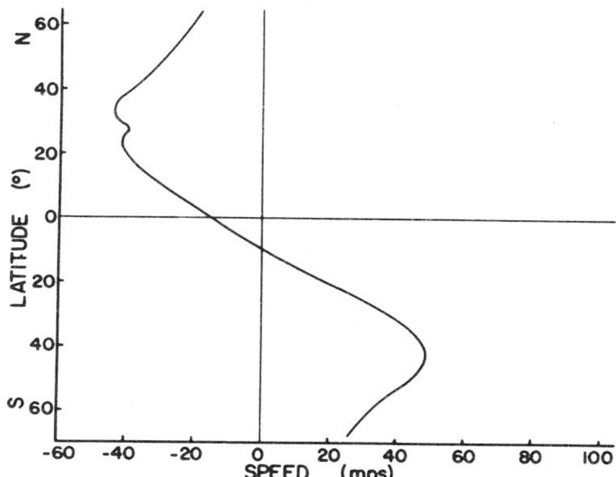

Fig. 4.17. Latitudinal distribution of mean zonal SCI data for the ninth semimonthly period after the spring equinox in the Northern Hemisphere. Data used for construction of the curve were obtained from the curves of Fig. 4.7 for the first semimonthly periods of August and February.

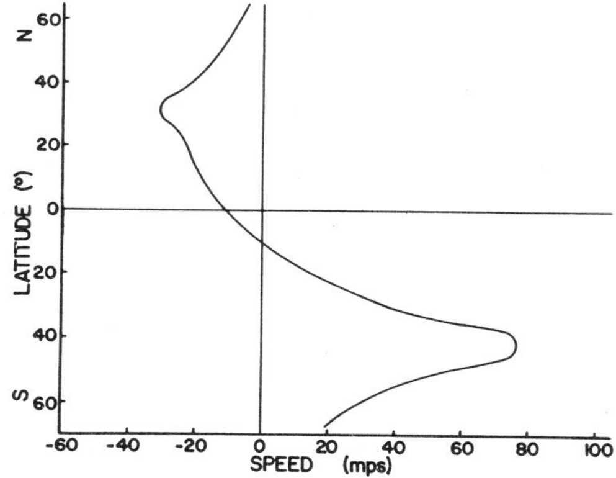

Fig. 4.18. Latitudinal distribution of mean zonal SCI data for the tenth semimonthly period after the spring equinox in the Northern Hemisphere. Data used for construction of the curve were obtained from the curves of Fig. 4.7 for the second semimonthly periods of August and February.

which is for the end of the early winter season before the winter storm period starts.

A principal result of this invasion of summer easterlies into low latitudes of the winter hemisphere is that at the 50-km level a semiannual oscillation is established in the zonal flow. This oscillation will not be sinusoidal at all equatorial stations owing to the asymmetrical character of the annual stratospheric circulation pattern. For instance, a station

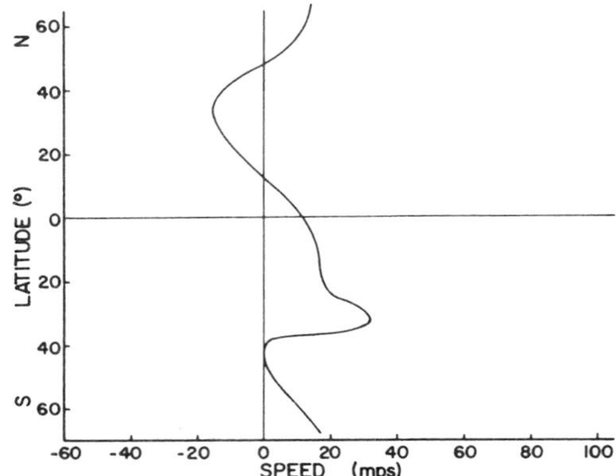

Fig. 4.19. Latitudinal distribution of mean zonal SCI data for the eleventh semimonthly period after the spring equinox in the Northern Hemisphere. Data used for construction of the curve were obtained from the curves of Fig. 4.7 for the first semimonthly periods of September and March.

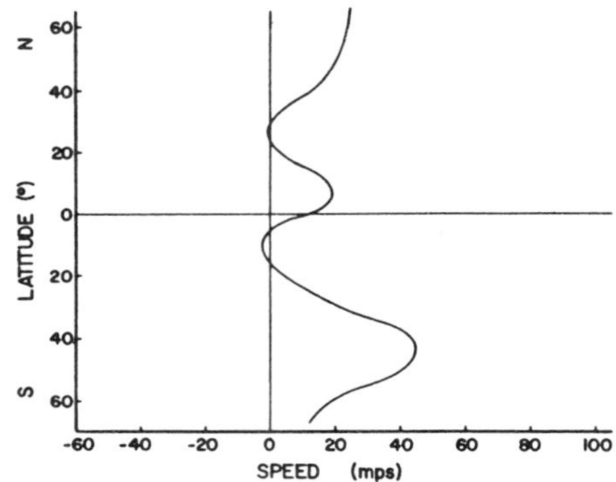

Fig. 4.20. Latitudinal distribution of mean zonal SCI data for the twelfth semimonthly period after the spring equinox in the Northern Hemisphere. Data used for construction of the curve were obtained from the curves of Fig. 4.7 for the second semimonthly periods of September and March.

located just outside the maximum penetration line (15–20 degrees latitude) of the easterly invasion of the winter hemisphere will experience the usual annual stratospheric monsoon characterized by an approximate four and one half months summer ending at the fall equinox and a seven and one half months winter season extending past the spring equinox. Inside tropical regions, however, the length of time during which this easterly flow will be evidenced over a particular station during the winter season will vary with latitude, becoming shorter as latitude increases. Inspection of the data in Figs. 4.9 through 4.20 indicates that the time interval during which this invasion is across the equator is almost exactly four months. Since the principal westerly circulation is always found at higher latitudes and occurs in tropical regions only when both hemispheres are experiencing westerlies during the seasonal reversals when the westerlies of the fading winter season are decreasing and the westerlies of the fresh winter season are increasing, it is probable that the westerly winds of the equatorial stratosphere are rather light.

The curves of Figs. 4.9 through 4.20 emphasize the basic differences in latitudinal distribution of the zonal winds of the summer and winter seasons. Winter westerlies have their maximum development in the region about 45 degrees latitude in the stratospheric circulation, while the summer easterlies show their peak development in the 20- to 30-degree latitude belt. The easterly surge starts at near 20 degrees latitude in a rather narrow belt as is indicated in Fig. 4.9, although by the time of its maximum development it is a consolidated stream ranging all the way from 14 degrees latitude in the winter hemisphere to the summer pole. Maximum speed in these mean data is 49 meters/sec at a latitude of 25 degrees. Westerly winds cover the globe for a short time about the equinoxes and dominate the winter hemisphere except for the low-latitude easterlies which develop during the other hemisphere's peak summer season. Their most distinctive feature is the very strong peak speeds in middle latitudes. This circulation reaches its maximum speed of over 100 meters/sec in these mean data in early December after a steady climb, rather precipitously immediately after the fall reversal.

It is with the encroachment of the summer hemisphere's easterly winds that the dynamic events of the midwinter season (the winter storm period) appear to be associated, and thus the principal deviation from solar control of the stratosphere is pin-pointed. If the stratosphere's heat-absorbing medium were uniformly distributed in the horizontal, a more likely dividing point between the hemispheric circulations, if it should be determined on other than rotational grounds, would be expected to shift with the subsolar point away from the equator into the summer hemisphere. The fact that the observations show this not to be the case and that the actual division, at least for a period, is found at roughly an equal angular distance on the opposite side of the equator is a matter of great interest. Obviously there must be a circulation-generating mechanism of the proper sign in the lowest latitudes of the winter hemisphere. The meridional temperature gradient must be positive. These considerations lead directly to the conclusion that there must be a horizontal gradient in stratospheric constituents or physical processes that are capable of producing this result. Ozone is a rather unlikely prospect for this job, since its concentration and ability to control the heat balance of the stratosphere are directly dependent on the intensity and aspect of the solar radiation, and with such tight control it is difficult to see how the required ripple in the meridional temperature gradient structure could develop. There are several other mechanisms that are quite capable of producing such a gradient. Vertical motions along a preferred longitude circle could easily produce the comparatively small temperature differences required to adiabatically yield the observed gradients. On the other hand, a general upwelling of stratospheric air in the region of the subsolar latitude would, through expansion, produce much the same result. The vertical velocities required in both cases are quite small. The known stability of the stratosphere makes both of these solutions unlikely, however, and leaves a more probable solution in the possible variation of other elements of composition of the stratospheric air.

Distribution of water vapor in the upper stratosphere is not well known. Direct measurements have led to conflicting results, with reports of increasing mixing ratios above the stratonull being questioned relative to their freedom from experimental contamination by water vapor which may have been carried upward from the troposphere by the balloon system. There have been measurements by indirect means (Murcray *et al.*, 1960), which indicate the presence of significant amounts of water vapor above the 30-km level, and it is in this area that further experimental work may have a profound influence. It is possible that residence time of the air in tropical regions is of sufficient magnitude to allow diffusive contamination of that region, whereas at higher latitudes the apparent presence of significant vertical motions could preclude the establishment of such an equilibrium. If such were the case, equatorial and summer hemisphere stratospheric regions would be the scene of enhanced moisture while the winter hemisphere, with the known shrinking of the atmosphere, and probable vertical motions, would offer a drier environment. Such a distribution of stratospheric circulation patterns could indeed produce a relatively sharp demarcation between the hemispheric circulation systems, and that break could fall in low latitudes of the winter hemisphere. An understanding of the spatial distribution of important atmospheric constituents such as ozone and water vapor in the upper stratosphere is imperative for our understanding of the stratopause circulation system. This knowledge can be obtained only by the acquisition of experimental data of sufficient reliability to permit resolution of the causing factor. It is already clear that this will be a very difficult problem.

Regardless of the causing element, the presence of these very strong opposing circulation systems in juxtaposition in the tropical regions of the winter hemisphere is the necessary ingredient for a spawning ground of smaller-scale circulation systems. The positive relative vorticity which will be a companion to these easterly circulation invasions of the winter hemisphere will assure the formation of cyclonic systems, and the general pressure gradient will assure that these systems will move toward the winter pole. A basic similarity exists between these subtropical whirls and their dynamic function and the Arctic cyclones of the troposphere which form along the wavy polar frontal zone and sweep into midlatitudes to perform an important heat exchange function. The equatorial front of the stratopause has a mean position in the 10- to 20-degree latitude region, but in view of our large deficit of detail data it is possible that its position is variable. The cyclonic whirls that form along this front will then perform the important function of transferring heat from the warm summer stratosphere into the cold winter hemisphere cyclone, but more importantly they will transfer mass across the proverbial equatorial boundary from one hemisphere to the other. This transfer will carry the composition of the residence environment into a new relative position, and the more stable elements such as radioactive materials should, with proper background information, provide a measure of the efficiency of this transfer process.

It should be realized that the cyclonic systems developed by this interaction of hemispheric circulation systems will evidence a wide range of sizes. If the usual rules of eddy exchange prevail, a majority of the whirls will be relatively small and will soon mix with the new environment as they move out of tropical regions. The result of such activity will be production of a zone of diminution of the winter westerly circulation such as that observed in Fig. 4.14 between 10 and 30 degrees latitude of the winter hemisphere. These smaller eddies will establish a mixing zone which, if it succeeds in becoming well established, will provide the observed brake on the winter westerly circulation during the winter storm period with no major disturbance in the general circulation of the winter hemisphere. On the other hand, a gross wave disturbance in the stratopause subtropical front may form and move toward the winter pole with sufficient stability to maintain itself well into the westerly circumpolar circulation. In this case the circumpolar circulation will be disrupted and very dramatic effects of this outbreak of summer easterlies will be evidenced. Again, the straightforward solution to our problem is the acquisition of sufficient data at the proper location to describe the dynamics of these subcirculation systems accurately.

The data presented in Figs. 4.9–4.20 were analyzed to obtain the cyclic variations in the zonal winds over the equator and at 10° N and 10° S latitude. These data are illustrated in Fig. 4.21. Over the equator

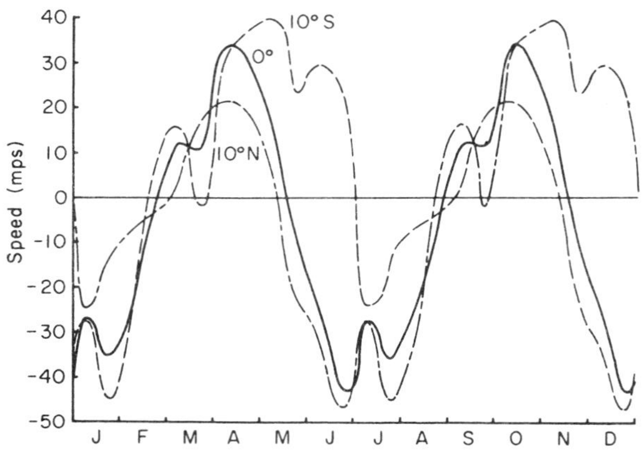

Fig. 4.21. Annual variation of the zonal SCI for 0°, 10° N, and 10° S latitudes. The SCI is the mean zonal wind speed between 45- and 55-km altitude.

there is a well-developed semiannual cycle which is skewed to an easterly mean value of approximately 5 meters/sec. According to these data, maximum westerly winds over the equator are approximately 33 meters/sec in April and October and maximum easterly winds are greater than 40 meters/sec in late June and July. Reversal dates are approximately 1 March, 15 May, 1 September, and 15 November. The equatorial stratosphere thus has westerly winds during the equinoctial periods, and has its principal circulation from the east during each of the hemispheric summers.

Since the heat input on a semiannual basis is only symmetrical at the equator, the zonal wind variation at 10 degrees north and south latitude can be expected to be asymmetrical. The greatest development of easterly winds will occur in a subtropical station during the period of the year when the subsolar point moves toward it; that is, during its summer season. Thus, a station at 10° N would experience a long period of easterlies starting in early May and lasting until late August, which is produced by the general easterly circulation of the Northern Hemisphere summer. The reversal to westerlies in late August heralds a period of strong winds as the westerly winds of the early winter season begin. In November the invasion of easterlies from the Southern Hemisphere summer circulation works northward across the equator and causes a short-period reversal to easterlies at 10° N, which starts about 1 January and is completed by the end of February.

As a result of our assumption of symmetry between the hemispheric circulation systems, the zonal wind oscillation at 10° S will be similar but shifted 180 degrees in phase. In addition to the component of tropical stratopause zonal flow which is tied to solar control, there appear to be certain special disturbances in the tropical circulation. As has been indicated in the discussion relative to Fig. 4.4, there are apparently perturbations generated in the stratospheric circulation as the general hemispheric circulation couples with the local circumpolar circulation, which is initiated early. In the fall case for the Northern Hemisphere this is illustrated by the September perturbation of the 10° N curve and by the 10° S curve in March. Also, the deep indention in the easterly circulations at the equator and the 10-degree latitude curve for the summer hemisphere would appear to be associated with the cross-equatorial invasion of easterlies into the winter westerly circulation, that is, the easterlies in the 10° S curve during July appear to result not only in a strong braking effect on the winter westerly circulation which they invade, but also in a reduction in speed of the tropical easterlies to which this event owes its genesis. The first of these disturbances extends to the equator with reduced amplitude as is illustrated by that data in March and September. It does not, however, show up in the 10-degree latitude curve of the alternate hemisphere. In the second case, there appears to be a very rapid damping of the hemispheric interaction as it is reflected in the summer hemisphere, as is evidenced by the summer curves of Fig. 4.4. This disturbance is strongly amplified in the winter hemisphere, sometimes to the complete disruption of the circumpolar westerly circulation.

The amount of penetration of the summer hemisphere into the winter hemisphere is a matter of keen interest. This information was obtained from Figs. 4.9 through 4.20 by noting the latitudes of the circulation separations with date and plotting them in Fig. 4.22. These curves show that the extreme penetration is about 18 degrees latitude, and is there for only a very short period. It would appear from these data that the major exchange of energy between these opposing circulations occurs in the earlier phases of the invasion at about 5 degrees latitude in the winter hemisphere. After this initial critical engagement of approximately one month the strength of the westerlies is broken and the easterlies push to their maximum extension. A fast withdrawal is effected as the sun's return to equatorial incidence removes the driving positive meridional temperature gradient.

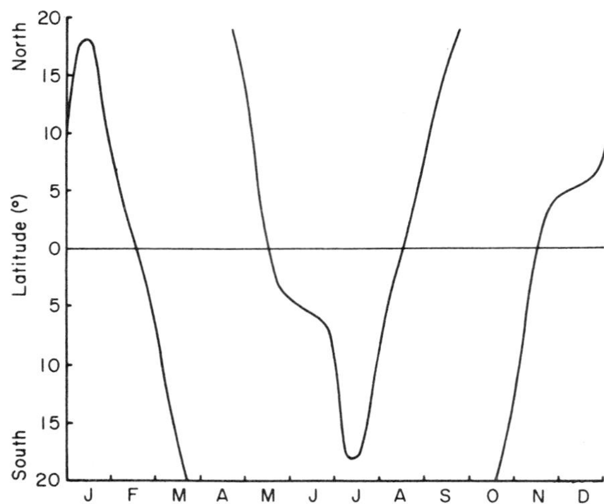

Fig. 4.22. Latitudinal invasion of summer easterlies into the winter hemisphere.

The data available thus far, therefore, indicate the presence of a meridional interhemispheric flow, from the Northern to Southern Hemisphere in July and the opposite in December, in the wind system over the equator at 50-km altitude. The MRN has not as yet obtained the required data, although the Ascension Island data at 8 degrees south latitude are very strong evidence that such winds must exist at the equator. It is very necessary that the required measurements be made, however, since the occurrence of such a flow is questionable and must be verified. From the data available today one must conclude that the appearance of westerly winds for a two-month period in spring and fall each year is a characteristic feature of the equatorial region.

The westerly circulation of the winter hemisphere also appears first in polar regions. Since the lower regions of the stratosphere are the first to be denied solar heating as the sun sets for the winter season, cooling and contraction begin there first. A cold, low-pressure center is established at low levels and the resulting westerly cyclonic flow expands up and away from the polar regions to reach Fort Greely in late August, approximately one month before the equinox. The principal winter westerly circulation has its origin in middle latitudes at about the time of the fall equinox and spreads rapidly to envelop the entire hemisphere. As is illustrated by the data of Figs. 4.12 and 4.13, strongest westerly winds of the winter season are usually observed at middle latitudes about two and one half months after the fall equinox.

Review of the solar heat input into the stratosphere according to assumptions of uniform distribution of the absorbing material and geometric considerations leads to the expectation of a peak in the westerly circulation at middle latitudes. The polar region is cooled by radiation roughly uniformly as a result of lack of incoming heat, so that a weak negative meridional temperature gradient is the rule at high latitudes. In the 60-degree latitude region incident radiation becomes appreciable, even in the winter season, and the gradient becomes rather strong across middle latitudes as temperatures along constant height meridians are raised to values more characteristic of the summer season. This abrupt fall in temperature as one enters the winter polar night zone is a standard feature of the atmosphere at all levels between the surface and the stratopause. It appears to be a pronounced feature of the stratopause level which, while the total temperature range from pole to pole is rather small, exhibits this very strong peak in the meridional profile of the zonal wind (Fig. 4.9). In connection with this point, it should be remembered that the thermal wind effect is cumulative, and thus the meridional temperature gradients of lower altitudes may have produced the principal characteristics of the wind pattern which we observe at the stratopause with only slight, or even possibly negative, contribution at the stratopause level.

Interaction between the hemispheric circulation systems as illustrated in Fig. 4.9 will result in the concentration of vorticity in low latitudes for an approximate two-month period centered at one month after the summer solstice. The strong meridional wind shears evidenced even in these mean data will produce vorticity values of at least double the largest values to be found in the strongest portions of the polar cyclone, and a full order of magnitude stronger than the general concentration of vorticity at the stratopause. The sense of this interaction will always produce positive vorticity, with the resulting probable development of cyclonic eddies in the shear zone. These mean data are probably valuable only in establishing the general location of the region of interaction and the gross features of the gradients. It is very likely that the interaction is highly dynamic, and that vortices of a wide range of dimensions will occur over a considerable latitude range centered as indicated in Fig. 4.9. It is probable that most of these eddies are small, and thus below the resolution of our current MRN observational system. In such a situation, it should not be too surprising if occasional very unusual rocket soundings are obtained as the sensors chance to fall into these small circulation disturbances. On the other hand, perturbations of this zone of interaction may well occur on a global scale in the low Rossby number range and produce cyclonic systems of many kilometers' horizontal dimensions.

In the case of these large circulation systems it should be possible, with a certain amount of luck, to chart the progress of the changes that are imposed on the stratospheric circulation as a result of these hemispheric interchanges. First, it would be essential that the disturbance originate in the North American region of the globe, since it is only there that synoptic data on the flow and temperature structure extend to sufficient heights to permit detection of the early phases of this circulation feature. Above about 50 degrees latitude progress could be observed by balloon systems, although a complete picture of the changes in the stratospheric circulation can be obtained only by a comprehensive MRN organized on a global scale.

Extension of the summer hemisphere's easterly circulation across the equator into the winter hemisphere presents some real problems in assessing the dynamic characteristics of the flow. In view of the reversal in direction of the coriolis parameter's horizontal component at the equator, a direct requirement for a stable flow in the winter hemisphere is a positive meridional temperature gradient in low latitudes. Now it appears unlikely that the geometric pattern of heat input can be called on to produce such a result. It would, instead, point toward a negative meridional temperature gradient from the subsolar point (low latitudes in the summer hemisphere) all of the way to the winter pole. We must look, then, to some other factor to produce these anomalous easterly winds in the winter hemisphere. The cause could be dynamic in nature, resulting from viscous entrainment of the tropical air into an easterly flow by close proximity of the strong easterly current of the summer circulation. Acceptance of such a mechanism for generation of these flows would require considerable enhancement of current concepts of the role of viscous forces, and would surely predict the existence of a turbulent tropical stratopause.

It seems far more probable, at this point, that the gross momentum of the tropical easterlies is stable, or that there actually exists a ridge of warm temperatures at 15–20 degrees latitude in the winter hemisphere. One of our basic assumptions in discussing geometric relationships of solar heat input into the stratosphere was that the absorber is reasonably uniformly distributed meridionally. Reflection would indicate that this is a rather broad assumption, and since it is based on an almost complete lack of data it should be carefully criticized for possible sources of error.

The discussion up to this point has concerned principally the meridional structure of available SCI data. As was indicated, an assumption of hemispheric similarity was required to achieve a global analysis of the stratospheric circulation. It is also true that the data currently available are inadequate for detailed evaluation of the longitudinal variability. MRN data have been obtained in significant quantities only in the Western Hemisphere. Even there adequate data in time have been obtained over only a limited longitudinal range. To illustrate the variabilities that one must expect along longitude circles, the annual variation of SCI determinations for Cape Kennedy, White Sands Missile Range, and Point Mugu are plotted for the zonal case in Fig. 4.23 and for the meridional case in Fig. 4.24. The number of cases included in each of the semimonthly data points of the curves of Figs. 4.23 and 4.24 is indicated in Table 4.2.

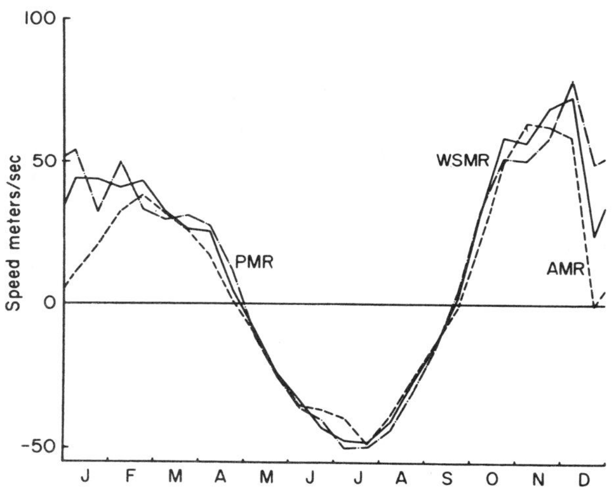

Fig. 4.23. Mean zonal SCI data for Cape Kennedy, White Sands Missile Range, and Point Mugu.

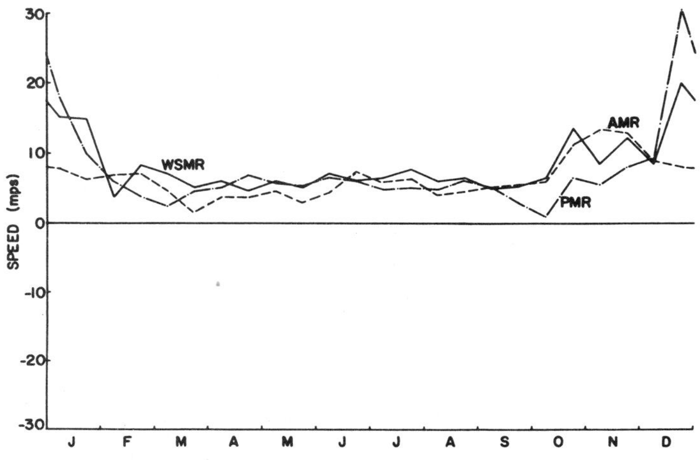

Fig. 4.24. Mean meridional SCI data for the stations of Fig. 4.23.

There is considerable similarity in the three curves presented here. This would lead one to expect that the data from an MRN station is at least roughly representative of the hemispheric circulation about a longitudinal circle. This is especially true of the zonal data for the summer season. The exception to this premise is found in late June and early July, when the Point Mugu and the Cape Kennedy data show strong evidence of reflection of the hemispheric interaction across the equator. It is interesting that these obvious decreases in mean zonal speed at Point Mugu and Cape Kennedy are not at all reflected in the mean data for White Sands Missile Range. These data indicate that the summers are slightly shorter at Point Mugu than at White Sands Missile Range and somewhat longer at Cape Kennedy. Reversal times vary by only a few days in these mean data, although more significant differences may exist in the individual data.

More pronounced differences are noted in the winter seasons. After a rapid climb into a strong westerly circulation the curves for Point Mugu and White Sands Missile Range show a sharp break and after mid-November a continued climb to peak values of the winter zonal circulation in early December. The curve for Cape Kennedy does not show such a variation in gradient, but climbs at a reduced rate to a maximum in early November. This maximum in the mean zonal flow at Cape Kennedy is significantly lower and occurs earlier than at the other stations. While it is to be expected that the circulation might be reduced at lower latitudes as a result of the fact that the peak of the winter zonal circulation occurs at upper middle latitudes, the difference appears to be excessive. This deduction results more from the shape of the curve than from the magnitude.

The pronounced reduction of zonal speed that characterizes the winter storm period is clearly illustrated by these data. Again the Cape Kennedy data show the greatest change, and the values of the mean zonal speed in late December for the three stations are distributed almost exactly in accord with their latitudinal position. In addition to its precipitous entry into the winter storm period, the Cape Kennedy curve exhibits a comparatively slow recovery to the moderate westerlies of the late winter season. It is clear from these data that there is a significant longitudinal variation in the characteristics of the stratospheric circulation.

Inspection of the meridional SCI data for these three stations as presented in Fig. 4.24 points out corresponding differences in that component also. At the spring equinox these meridional components range from 2 to 5 meters/sec and show a gradual increase to from 5 to 7 meters/sec at the height of the summer easterly circulation. There is a significant decrease in this southerly meridional flow as the summer circulation wanes, with a meridional component of 5 meters/sec in evidence by early September. The data indicate a correlation of the poleward flow with development of the summer circulation in these stratopause data.

There is an increase to more than 10 meters/sec in the fall at White Sands Missile Range and at Cape Kennedy. The data for Point Mugu, on the other hand, show a decrease in the early phase with a slow recovery, so that by early December, when the zonal winds have reached their peak, the three stations have nearly the same mean speed of approximately 10 meters/sec. With the winter storm period break in zonal circulation these mean data for Point Mugu and White Sands Missile Range show strong increases in the poleward flow. Cape Kennedy, on the other hand, shows a slight decrease in the meridional flow. Thus the south wind component speeds in these data range from 8 to 30 meters/sec from Cape Kennedy to Point Mugu. If this difference can be assumed to be due to longitudinal differences, it amounts to a gradient of about 0.5 meter/sec per degree longitude. This is quite small compared with a nominal value of 3 meters/sec per degree latitude zonal wind shear, but still requires attention for adequate resolution of many atmospheric problems.

These data show that the mean flow at the stratopause is directed away from the equator in both hemispheres throughout the year. Such a result could require a return flow for continuity considerations, and the geometry of such a flow is of prime importance. Because there is a strong heat source in tropical tropospheric regions it is reasonable to expect upward motions in that region. Analysis of the vertical meridional wind profile structure has shown that in the mean the speed is near zero in upper stratospheric altitudes except for the poleward flow indicated in the SCI data for the stratopause region. Considering only the summer side of the equatorial region, the required vertical flow at the tropopause to sustain the lateral outflow is of the order of 0.03 cm/sec. This vertical flow is small compared with most physical processes, and thus is evidence of a considerable stability of the tropical upper stratosphere, with residence times of the order of at least several months. Transport processes such as molecular diffusion cannot be judged insignificant in this quiescent situation.

4.2 Sudden Warmings of the Stratosphere

Research meteorologists have taken a lively interest in the so-called "explosive warmings" discovered by Scherhag (1952) through high-altitude balloon observations over Berlin during the winter of 1951–

1952. These events, more commonly called sudden warmings, are not of annual occurrence, but appear irregularly. Thus far, satisfactorily documented sudden warming events have been observed during the Januarys and Februarys of 1952, 1957, 1958, and 1963. Some definition is required to distinguish between the normal warming associated with return of solar radiant heating of the upper atmosphere with the spring equinox and these special events, which appear to have a very different origin. In general, changes in stratospheric temperature structure are classed as sudden warming events only if there is a temperature increase of several tens of degrees in a period of a few days and the timing of the event is significantly removed from the spring warmup. Actually, the sudden warming events are not at all difficult to spot. Much of the confusion centers on the fact that stations at different locations are affected differently, and only with a considerable amount of data is one sure to obtain a satisfactory picture.

SUDDEN WARMING: *A dynamic event in the stratospheric circulation which is principally characterized by a temperature increase in polar regions immediately above the stratonull level greater than 50°C over a period of ten days or less, accompanied by a disruption of the usual westerly zonal circumpolar flow of the stratospheric winter circulation.*

Stations centered under the warming region as it sweeps poleward provide a very dramatic record of the magnitude of events which must be in progress in the upper atmosphere. Temperature increases of from 30° to 90°C over a period of a week are the rule, and stratospheric wind changes of the order of 100 meters/sec out of the west to 50 meters/sec from the east are common. Such variations have all of the earmarks of the hurricane in the tropospheric system of weather patterns. It is a principal objective of the MRN to obtain a complete picture of the formation and propagation of these disturbances of very special form as well as to analyze the more symmetric general response of the upper atmosphere to the annual cyclic input of heat from the sun.

Progress of the January 1963 sudden warming of the stratosphere is charted in Fig. 4.25 by plots of mean temperature values at the 10-mb level for longitude circles from 40 degrees latitude poleward. These means were obtained by averaging the temperature reading at intersections of the specified longitude circle with the four cardinal meridians (0°, 90° W, 180°, 90° E). The data used were daily 10-mb charts prepared by the Institute for Meteorology and Geophysics of the Free University of Berlin (Kriester *et al.*, 1963). In addition, the light curve of Fig. 4.25 illustrates the daily temperatures observed at the 60° N latitude, 90° W longitude point to present the type of observational data

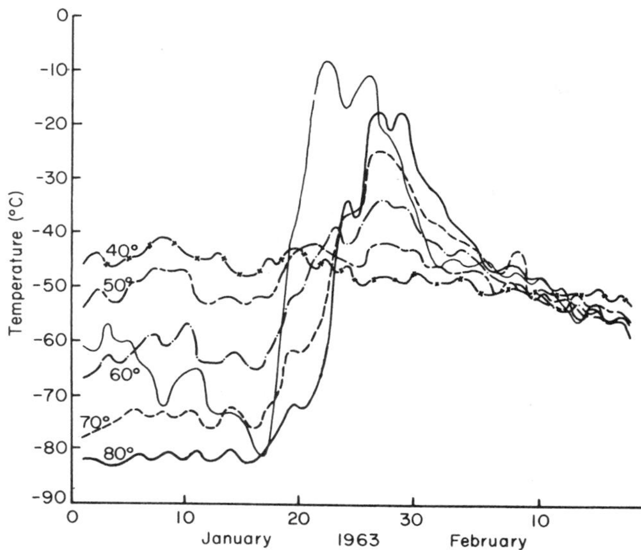

Fig. 4.25. Average temperature of various longitude circles on the 10-mb pressure surface of the upper stratosphere during the sudden warming event of January 1963.

obtained at an individual station as one of these phenomena passes overhead on the 10-mb pressure surface.

As is true in the general case, the cold circumpolar vortex characteristic of a strong winter stratospheric circulation is in evidence as the warming begins. Temperatures at the pole are colder than —80°C, and a positive meridional difference of roughly 40°C exists between midlatitudes and the pole. The meridional temperature gradient reverses at low latitudes owing to the cold tropical tropopause, which also has a temperature of the order of —80°C. These opposing gradients in meridional temperature structure confine the stratospheric circulation to the high-latitude regions of the winter hemisphere.

It is obvious that the "sudden warmings" constitute a dramatic change in the structure of the upper stratosphere. Not only is the high-latitude temperature gradient destroyed, but it is also replaced by a positive meridional temperature gradient of almost equal intensity. In this case the change was effectively accomplished in a 12-day period with an over-all mean rate of change of 5.3°C per day. During the period of maximum rate of change the gradient in these mean data shows strongest values of as much as 20° per day. The change in character of these mean temperature data first becomes evident at all latitudes on 17 January, and, with some detail structure in the early and late phases, the change is executed with a considerable uniformity.

It is known that the warming center formed in the Atlantic and moved northwestward across the Hudson Bay to a position northeast of Alaska. It then swung to the right to cross the pole and dissipate in the eastern Asiatic region. These obvious parts of the trajectory started on 14 January, crossed the pole on 1 February, and faded away on about 4 February. Since the gross dimensions of the warm sector were of the order of one quadrant in the 40- to 60-degree latitude regions, the effect on the mean temperatures of these latitudinal circles is proportionally smaller than at the 80-degree circle, where the dimensions of the central low-temperature cell were equivalent to the diameter of the region.

Central temperatures of the low center were warmer than —35°C on 18 January as the warm center started its northwestward motion, and had intensified to 0°C on 27 January as it turned to cross the pole. The warm center weakened rapidly as it moved poleward so that the warmest isotherm when it was centered over the pole was —30°C. During this movement away from the pole into the Asiatic mainland, the warm center continued to dissipate until at a temperature of approximately —40°C it became a part of the rather chaotic temperature structure of middle latitudes.

The light solid curve of Fig. 4.25 illustrates the temperature variations with time that a particular station observes at the 10-mb level as the phenomenon crosses it. Cold air of the polar cyclone was advected into the region of this station for a week prior to the event, so local temperatures were around —80°C as the warming sector advanced. In a five-day period the temperature climbed to above —10°C for an average change of some 14° per day and maximum rates approaching 20° per day. The sequence of events that will be observed at a particular station has a strong relationship to the position and paths of motion of the thermal centers. It is clear that we have picked a special case here, and that for many stations in less opportune locations the manifestations of a sudden warming might be far less obvious.

It is quite possible that the warming event discussed here resulted from the downward motion of stratospheric air over northern North America as the circumpolar low-pressure system split and the western low moved rapidly into the Pacific south of Alaska. Evaluations of the subsidence required to produce the observed temperature rises are of the order of 10 cm/sec (Kriester *et al.*, 1963). A composite picture of the events that precipitated this disruption of the winter circulation in the stratosphere is as yet not available, although evaluation of other possible energy sources such as auroral particles and meteoroid influx indicates that subsidence is the only adequate mechanism. It is of interest to note that a marked exchange of stratospheric air is implied here, with residence times of the upper stratosphere very greatly reduced during these periods of great temperature rise. This flushing of upper stratospheric air into the lower atmosphere would result in other structural changes, such as the increased ozone concentration produced by convective compression.

Recovery from the effects of the sudden warming is slow in comparison with the onset of the phenomenon. The early rapid recovery to a nearly homogeneous structure is probably the result of mixing, but the continued trend toward establishment of the typical winter meridional distribution proceeds rather slowly, and probably represents the insolation effect at that time. The proper latitudinal distribution of the data is again observed by 18 February, but the magnitude of the gradient is only one fifth of the prewarming gradient. It is clear from other data that the winter circumpolar circulation failed to regain its intensity after that particular warming.

Recent improvements in balloonsonding systems have made possible the exploration of the upper atmosphere to altitudes of the order of 30 km during the past several years. Thus penetrations into the extreme lower portion of the stratospheric circulation systems became possible and synoptic analysis of the circulation in this region of the atmosphere was achieved during the late 1950's. Ten-millibar charts have been prepared on a daily basis from radiosonde observations since 1958 by the Institute for Meteorology and Geophysics of the Free University of Berlin under the direction of Richard Scherhag. An example of the mean January 10-mb chart for the five-year period from 1958 through 1963 is illustrated in Fig. 4.26. The height contours for this 10-mb pressure chart are presented in kilometers at half-kilometer intervals, except in cases where more frequent intervals were desirable for illustration of the pressure surface slope. These data indicate that the pressure contours are not symmetrical about the polar region, but in general are elongated into the continental areas with a more pronounced extension over the Eurasian land mass in a distinct trough of low pressure extending across North America. A rather deep invasion into this low-pressure system is to be noted on the Pacific side of the pole by a comparatively weak high-pressure system. When we consider the fact that these are five-year mean data for the 10-mb level, it is comparatively obvious that the continental land masses have a pronounced effect on the circulation even at these high altitudes.

In the winter of 1951–1952, with these new upper-atmospheric sound-

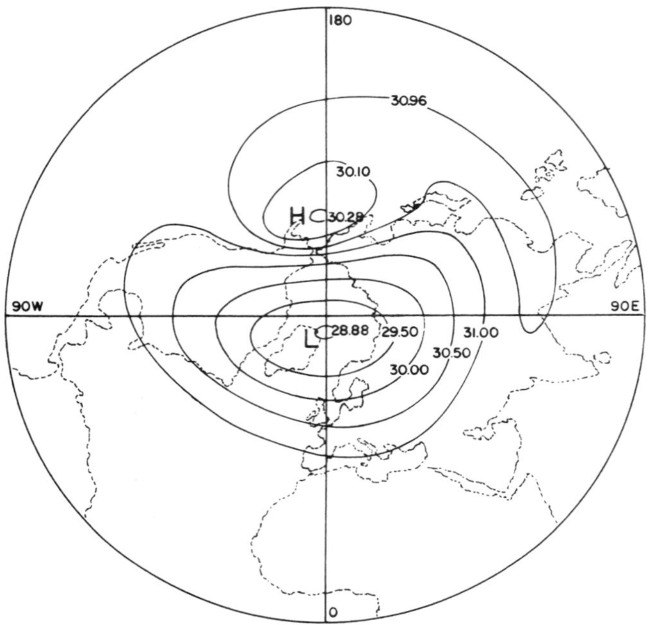

Fig. 4.26. Mean 10-mb chart based on data for the period 1958 through 1963 analyzed from daily radiosonde observations over the Northern Hemisphere. Contours are heights in kilometers.

ings, Scherhag (1952) noted the development of a disturbance in the upper reaches of radiosonde observation that constituted an almost complete disruption of the circumpolar circulation system which is characteristic of the early winter period in the stratosphere. He titled these disruptions in the circulation system "sudden warmings" because of the abrupt temperature rise that occurred at certain specific locations as a result of these shifts in circulation systems. There were indications in the meager data even then that the disturbance worked downward and poleward into the lower stratosphere. There have been numerous attempts to attribute these rather violent changes in the stratospheric circulation to various input sources of energy such as the precipitation of auroral particles or similar special devices which could result in the placement of these large heating values in the stratospheric night. In general, inspection of these various methods by which heat could be introduced into the winter stratosphere has indicated that there is insufficient energy available, and the suspicion has existed from the beginning that the variations were dynamic in nature and resulted from instabilities in the zonal circulation. With the development of a world-

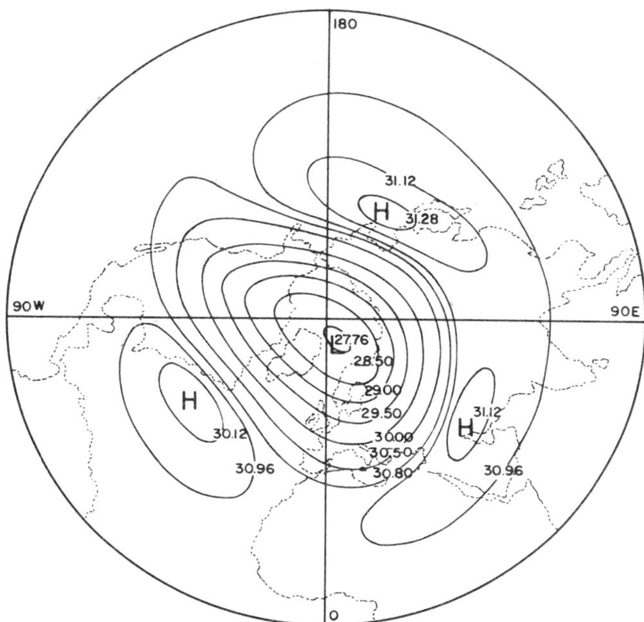

Fig. 4.27. Ten-millibar chart for 14 January 1963 over the Northern Hemisphere. Contours are heights in kilometers.

wide system of more adequate balloonsonding techniques, and with the development of the MRN, it became possible to document more adequately the events which attended the "sudden warming" events in the stratosphere. One of the best documented and strongest events to be observed occurred in January and February of 1963. In view of the rather complete documentation, this particular event will be considered here in detail from 10-mb charts illustrated in Figs. 4.27–4.36.

On 14 January 1963 (Fig. 4.27), the 10-mb map shows a well-developed zonal circulation centered almost on the polar region, exhibiting only slight deviations from the five-year mean data presented in Fig. 4.26. The egg shape of the low-pressure cell is rotated clockwise toward the west coast of North America and significant indentions are noted in both the Pacific and Atlantic Ocean areas. The circulation system is quite strong with a central 10-mb pressure altitude of 27.76 km as opposed to a central pressure altitude of 28.88 in the five-year mean data presented in Fig. 4.26. Ten-millibar winds are very strong in the Alaskan and European areas, and comparatively strong along the west coast of North America, but are lighter than normal for the season at other midlatitude stations due to the high-pressure invasions in the ocean

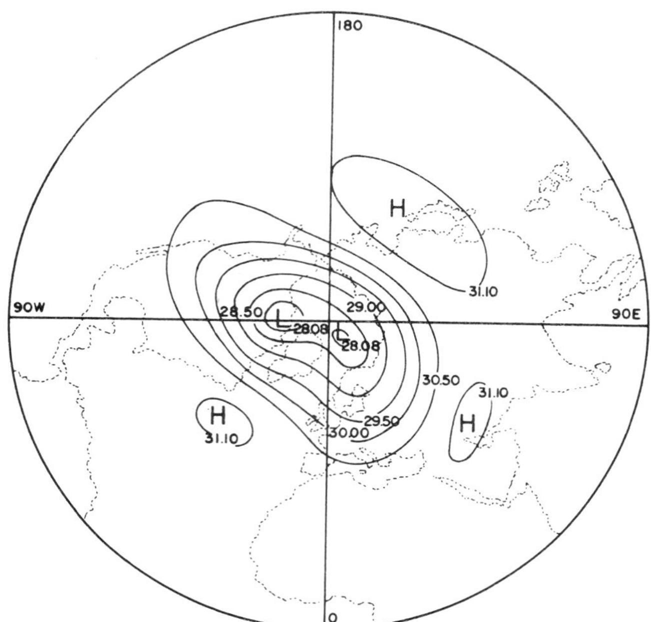

Fig. 4.28. Ten-millibar chart for 16 January 1963 over the Northern Hemisphere. Contours are heights in kilometers.

areas. As is illustrated in Fig. 4.29, by 18 January 1963 the 10-mb chart has assumed a different configuration from what is generally considered normal. Deep penetrations of the oceanic high-pressure region particularly in the case of the Atlantic Ocean have been initiated. Strong meridional circulations are to be noted over the European areas; and while the measurements are few, they probably exist also in the northwestern Pacific Ocean region. Strong zonal winds now dominate the North American continent area due to the elongation of the low-pressure center as it is compressed inward in the polar regions. The isotherms which had been roughly congruent with the height contours on 14 January are strongly compressed, particularly in the North American oceanic areas, where temperatures as high as —35°C have pushed into regions occupied four days earlier by temperatures of the order of —70°C. The —80°C isotherm which was initially displaced from the polar region toward the European area is found on the 18th to be displaced well to the Alaskan side of the North Pole. At the time of the 10-mb map of 18 January the warmest temperatures are centered over Nova Scotia.

Considerable filling of the deep low-pressure system had occurred

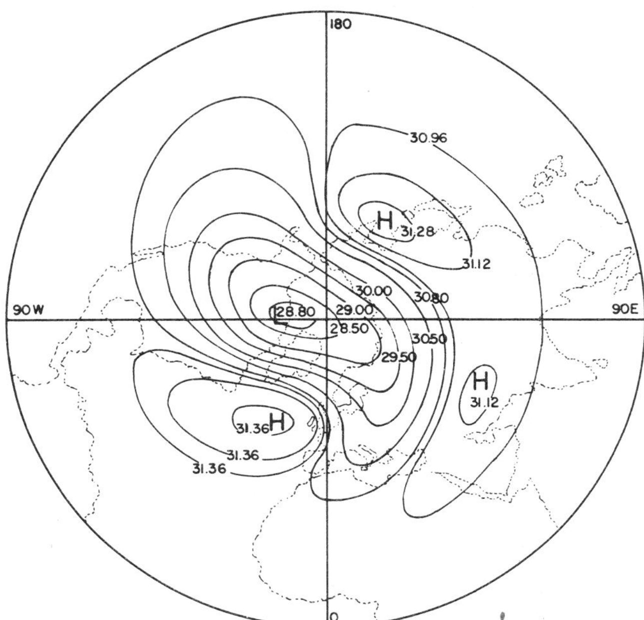

Fig. 4.29. Ten-millibar chart for 18 January 1963 over the Northern Hemisphere. Contours are heights in kilometers.

by the time these high-pressure cells began to move into the polar regions from the Atlantic and the Pacific so that by 20 January the lowest height observed is 28.21 km. The Atlantic high-pressure system, on the other hand, had grown from 31.12 km on the 14th to 31.36 on the 18th, and to 31.60 km on the 20th. The high-pressure system over the North Atlantic remained relatively stationary over these last two days, but increased in extent and intensity at the expense of the circumpolar low-pressure cell. The warming region intensified and moved toward the northwest from its location on the 18th near Nova Scotia and warmed to —20°C with the center of the warm air over the Hudson Bay area by the 20th. The —80°C cold core of the low-pressure system remained essentially stationary so that a very strong temperature gradient was established between the Hudson Bay and the extreme northwest Canadian region.

Except for the slipping of the —80°C isotherm to the Alaskan side of the pole, the isotherms had remained essentially circumpolar. By 23 January, the Atlantic high had continued to strengthen and the circumpolar low-pressure system had begun to divide into two cells, the one on the Alaskan side being the weaker with a central height of 28.89 km

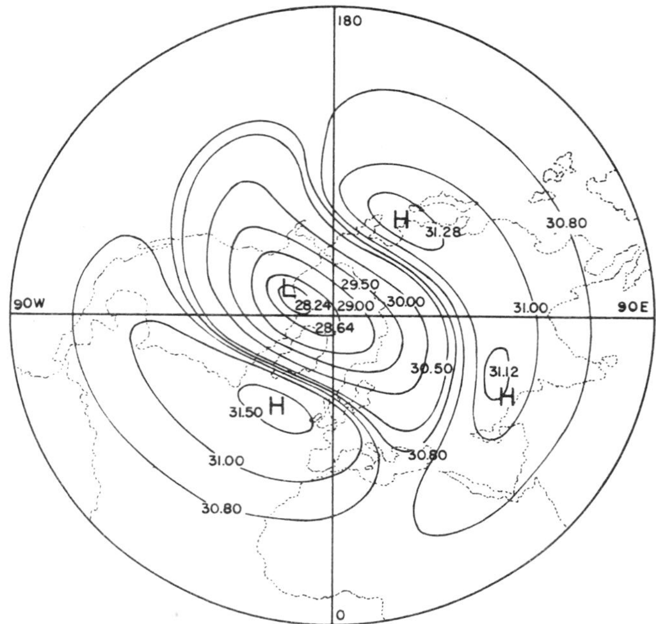

Fig. 4.30. Ten-millibar chart for 20 January 1963 over the Northern Hemisphere. Contours are heights in kilometers.

and the one on the European side stronger with 28.72 km as peak. The —80° isotherm had been eliminated with —70° isotherms. The coldest temperatures available separated into two cells with the European case centered on the constant-pressure height lines, and the cold air on the Pacific side displaced well into the Pacific from the center of the low-pressure system. The warm air center had continued to drift northwestward into the vicinity of the low-pressure system that was on that side of the polar region.

On the 24th, the low-pressure system continued to separate with rather well-defined circulations around each cell at this time. The low-pressure center to the Alaskan side of the pole had migrated more into the Alaskan region following the center of low pressure which was then located over the northern Pacific Ocean. The warm center had pressed poleward, forcing its way between the two low-pressure systems and at the same time the western Pacific high-pressure system had strengthened and pushed poleward.

It is clear from these data that the principal region of temperature change is found at high latitudes, centered at the pole. Maximum changes in the meridional temperature gradient are to be found in the vicinity

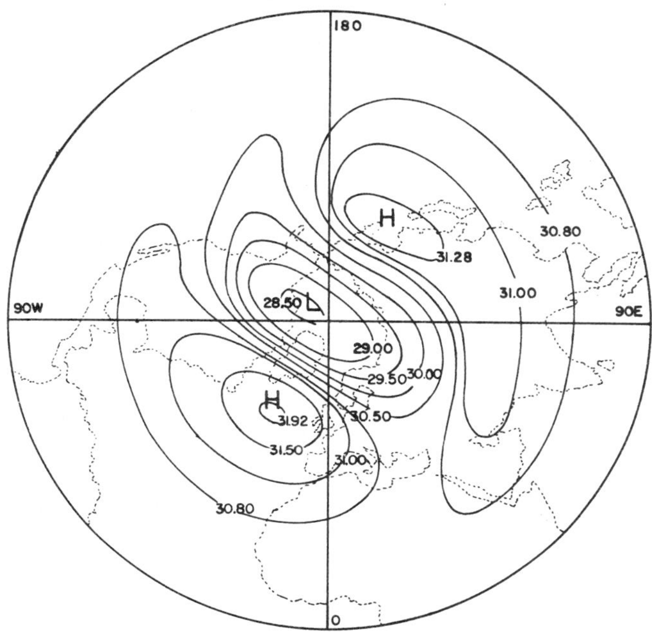

Fig. 4.31 Ten-millibar chart for 22 January 1963 over the Northern Hemisphere. Contours are heights in kilometers.

of the stratonull level of approximately 24-km altitude, and the meridional temperature gradient of the lower and central portions of the upper stratosphere at high latitudes is reversed to positive values from its usual negative sign. It should be remembered that these gradients are local to the region in which the warming is initiated during the early phases, and even during the peak of the disturbance, this area of the 10-mb surface continues to exhibit the cyclonic curvature of height contours characteristic of the usual winter circulation pattern. In this January 1963 case the polar low is elongated into the Alaskan and European regions (Fig. 4.32) and by 27 January (Fig. 4.35) is split by the ridge of high pressure between the two high centers located over the North Atlantic and Eastern Siberia. Consolidation of this high-pressure center from an original height of the 10-mb surface of less than 28 km, a full kilometer below the mean January height (Fig. 4.26), results in final heights of over 31 km by the end of January. Even in its principal development, the sudden warming anticyclone fails to achieve symmetry about the pole and thus the aspect presented at a particular station is a strong function of longitudinal location.

An illustration of a time cross section observed at White Sands Missile

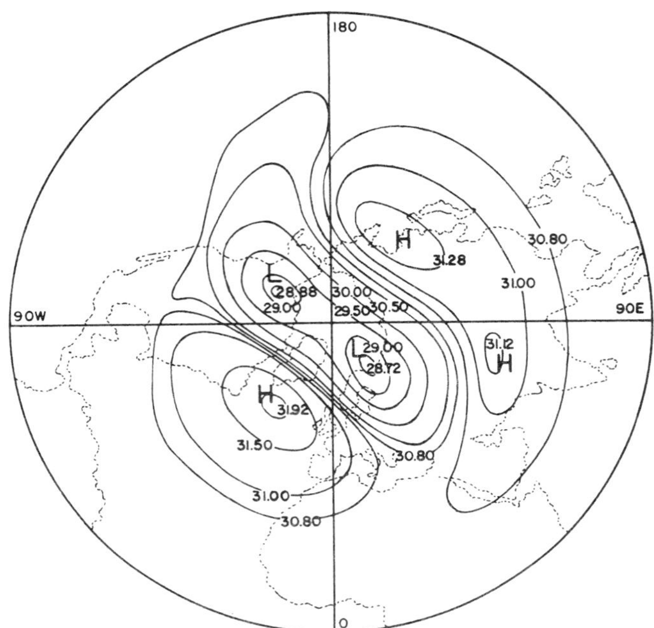

Fig. 4.32. Ten-millibar chart for 23 January 1963 over the Northern Hemisphere. Contours are heights in kilometers.

Range during the 1963 circulation disturbance is presented in Fig. 4.37. On 14 January the westerly zonal circulation increases steadily with height above the stratonull, indicating the presence of a negative meridional temperature gradient at all levels of the upper stratosphere to above the stratopause. Winds of 90 meters/sec in the 55- to 60-km height range are strong for the lower midlatitude location of White Sands Missile Range. The first changes appear near 50-km altitude, with a sharp drop in the zonal wind speed starting on 16 January and culminating in a reversal of the zonal flow on the 20th at an altitude of 46 km. A first principal surge of easterlies is observed at the 45-km level on the 22nd and 23rd as the White Sands Missile Range sector comes under domination of the Atlantic high-pressure center (Fig. 4.31). Withdrawal of this anticyclonic circulation to the east apparently was much more pronounced at the stratopause levels since there is a strong resurgence of westerlies on the 24th, building downward from the westerly flow of the lower mesosphere almost to the 40-km level. This change, which looks rather insignificant at the 10-mb level on the 24th (Fig. 4.33), is reversed so that a new strong jet of easterly winds is established just below the 40-km level by 26 January. The disturbance in the upper

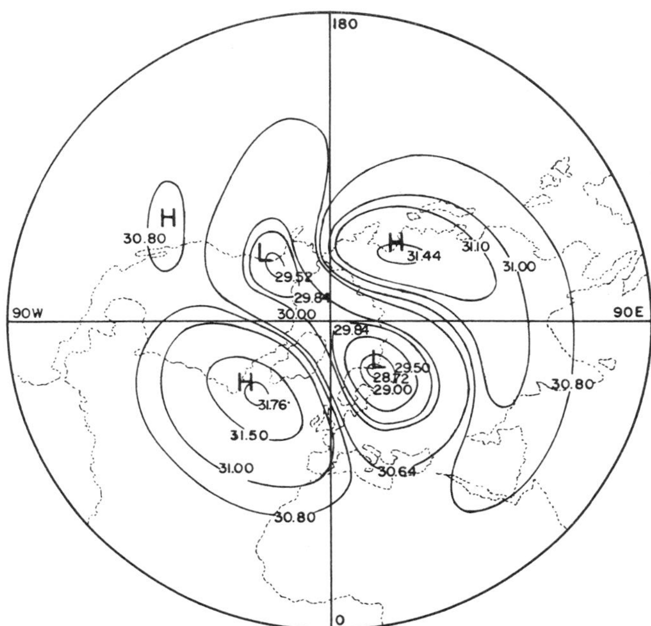

Fig. 4.33. Ten-millibar chart for 24 January 1963 over the Northern Hemisphere. Contours are heights in kilometers.

stratosphere begins to weaken after 27 January, over White Sands Missile Range at least, with a sharp decline about the end of January followed by a new strengthening after another downward surge of westerly circulation from mesospheric altitudes and a final decay of the circulation disturbance by mid-February.

A most obvious feature of the data presented in Fig. 4.37 is the apparent downward progression of events during this major circulation disturbance. The associated circulation changes progress from the stratopause to near the stratonull in a matter of some 25 days, with a roughly linear slope of the axis of the disturbance. Initial perturbations in the meridional temperature gradient must be the result of a sharp rise in polar stratopause temperatures which in a period of approximately one week produces an equal and opposite gradient. There are indications from the shift of circulation patterns at the 10-mb level that the surges of westerly winds downward from higher levels on the 24th and 30th are advective changes induced by rotation of the height contour patterns in the stratopause altitude zone back and forth across White Sands Missile Range. The first case was apparently associated with the movement of the warming center northwestward from the initial North

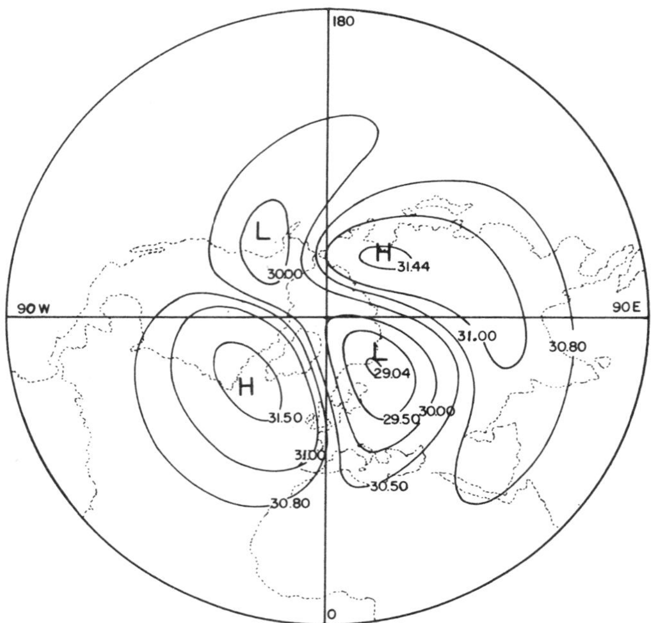

Fig. 4.34. Ten-millibar chart for 25 January 1963 over the Northern Hemisphere. Contours are heights in kilometers.

Atlantic position into the Hudson Bay area, and later by the poleward motion as the warming center crossed the pole into Asiatic high latitudes. Deterioration of the easterly circulation progresses steadily with time as the westerly zonal circulation strengthens in the lower mesosphere and works gradually downward.

Numerous investigators of the sudden warming phenomena have concluded that the disturbances also propagate toward higher latitudes in each of the cases of record as has been indicated by the data presented here. At the 10-mb level these motions are in no case advective transfer of mass along streamlines, but are rather a wave phenomenon that may propagate upstream or downstream, always with the poleward meridional motion. These facts are then in general agreement with the concept that these sudden warming events find their origin in subtropical latitudes of the winter hemisphere in the interaction zone between the hemispheric circulations and then extend their influence downward as they progress toward the winter pole. It is probable that at the stratopause level there is more or less strict advection of air mass of equatorial characteristics into the polar region in the form of strong cyclonic systems with dimensions in the thousands-of-kilometers range. The effects of these strato-

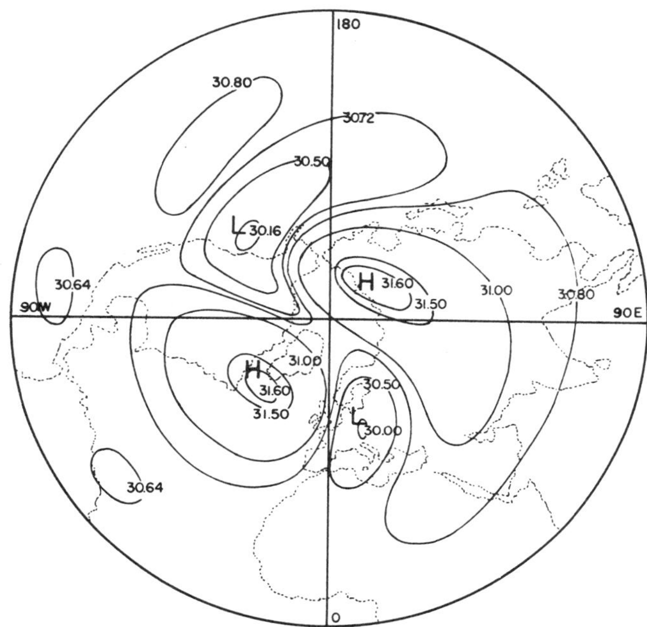

Fig. 4.35. Ten-millibar chart for 27 January 1963 over the Northern Hemisphere. Contours are heights in kilometers.

pause circulation systems would then be to alter the mass distribution above the central stratosphere which will in turn result in vertical motions. Temperature changes which have been observed strongly indicate subsiding motions which react to very efficient adiabatic heating to raise the temperature and expand the pressure surfaces upward. This accumulation of upper atmospheric air at the winter pole may well be a mesospheric circulation characteristic.

For instance, the strong wind system which is only partly observed over White Sands Missile Range in mid-January may well be a part of the sudden warming event. This jet of westerly winds in the mesosphere and its rapid decay during the period of the sudden warming and subsequent reformation in early February would indicate that possibly our SCI analysis of the interhemispheric conflict of easterly and westerly flows in tropical regions is only a sample of the base of a new circulation phenomenon which is characterized principally by a strong meridional component of flow. It has been understood for some time that the ionospheric regions were dominated by such circulations, and possibly the events which we are witnessing in the MRN data are induced at these lower latitudes by the confluence of the hemispheric circulations.

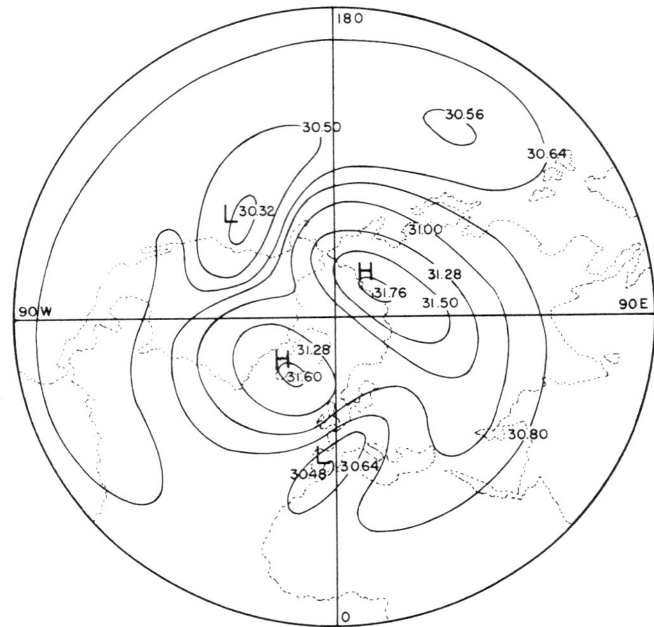

Fig. 4.36. Ten-millibar chart for 29 January 1963 over the Northern Hemisphere. Contours are heights in kilometers.

The few data already available at mesospheric altitudes show increasing easterly winds in the summer hemisphere well into the upper mesosphere, while the data have shown that, in general, peak westerly winds of the stratospheric circulation occur near the stratopause level. We may hypothesize, then, that the circulation changes which produce sudden warming events may originate near the mesopause at the very top of the stratospheric circulation over the tropical winter hemisphere and work down to the base of the stratospheric circulation, the level of the stratonull. From this point of view the slope of the axis of the event is seen to vary with time, exhibiting higher rates of descent at higher altitudes.

One is led to believe then that the initiation of the sudden warming events is associated with filling of the circumpolar low-pressure cell by strong meridional flows at mesospheric altitudes. Such a circulation system has already been predicted for ionospheric altitudes, with strong meridional winds moving from the summer pole to the winter pole. It is probable that the relatively small amount of mass transported in the very low-density ionosphere can do no more than dampen development of the winter stratospheric low-pressure center. The most obvious possi-

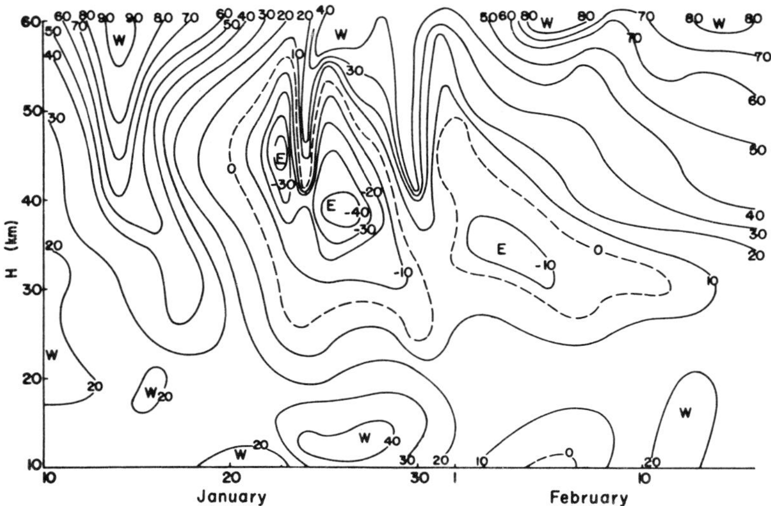

Fig. 4.37. Time cross section of the zonal wind structure over White Sands Missile Range during the sudden warming of 1963. Speed contours are in meters per second with west positive.

bility of generating the necessary circulation concentration is, therefore, the hemispheric circulation interaction zone of stratopause tropical regions where the density is sufficient to provide the mass transport. A most imperative objective of the MRN type of observation system should be a proper delineation of the physical characteristics of the circulation subsystems which accomplish this meridional transport of stratospheric mass from the summer hemisphere to the winter hemisphere.

The semiannual oscillation of the tropical stratopause which is associated with this interhemispheric exchange of mass fades out rapidly with decreasing altitude. The tropical stratosphere would appear to be isolated from the events discussed above, and there is some reason to believe that the residence times of the tropical stratosphere up to altitudes of approximately 40 km are very long, of the order of at least one half year. Observation of the meridional winds at low latitudes by the MRN during the summer season shows an almost negligible flow except at the stratopause. Consideration of Fig. 3.1 provides some reason for suspecting that there may be a significant interaction between the stratosphere and the troposphere in middle latitudes. Downward motions in the vicinity of tropospheric low-pressure systems have been identified by Danielsen (1964).

Synoptic analysis of the MRN data has developed comparatively slowly, in large part as a result of the very limited sample thus far obtainable of the gross features of stratospheric circulation. In special cases constant-pressure charts have been constructed at the 2-mb and the 0.4-mb levels through upward extension of radiosonde data by use of hydrostatic and geostrophic relationships with MRN wind and temperature profiles for a period of several days on either side of chart time. This work, carried out by the Stratospheric Research Laboratory of the U. S. Weather Bureau, is the forerunner of synoptic mapping of stratospheric circulation patterns on a routine basis. Maps were prepared by Teweles and Finger for the central period of the sudden warming event of January 1963 as illustrated in Fig. 4.38.

On 27 January there was complete separation of the stratospheric circumpolar low-pressure center of winter into two cells, one centered over Europe and the other in the Pacific just off the coast of North America (Fig. 4.35). A high-pressure system is located in the North Atlantic and imposes an anticyclonic circulation at the 10-mb level over the eastern half of North America. Retreat of the Western Hemispheric low into the Pacific was rapid during the latter part and the marked filling and weakening of its circulation as it moved into middle latitudes appears to be a standard part of the sudden warming event.

Inspection of the 2-mb and 0.4-mb charts of Fig. 4.38 points toward an earlier accomplishment of the breakdown of the polar winter vortex over North America. At 2 mb the continental regions are dominated by an easterly circulation with cyclonic curvature evidenced only along the west coast and in a weak trough which arcs southwestward across Canada into the lower midlatitude Pacific. On the 0.4-mb chart the low center is closer to the Pacific Coast area, and the height contours of this depression control the flow at the stratopause level over the United States. This trough line is oriented roughly east-west and the trough across central Canada, which is well defined at 10 mb (Figure 4.35) and is clearly present at 2 mb, is only faintly discernible at 0.4 mb.

Low-latitude MRN stations such as Point Mugu, White Sands Missile Range, and Cape Kennedy have southwesterly winds at the stratopause, while poleward from this midlatitude trough line the flow is easterly as is required by the gross expansion of the polar atmosphere attending temperature rises of the sudden warming in that region. It is unfortunate that data for hemispheric maps at 2 and 0.4 mb are not available, particularly in time sequence, since it appears that the circulation effects of the sudden warming event may be in a more advanced state at the stratopause. Such a conclusion is questionable, however, from the data at hand.

Height differences between the west coast midlatitude low and the high-pressure center north of Alaska range from approximately 1340

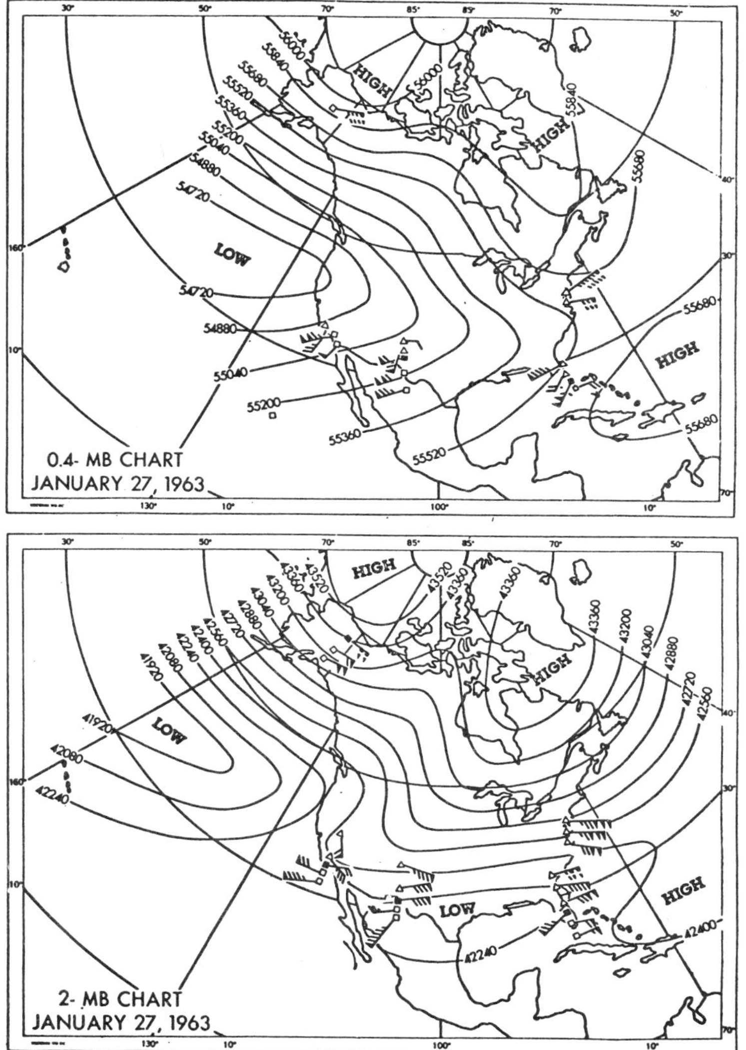

Fig. 4.38. *Two-millibar and 0.4-mb constant-pressure charts for North America on 27 January 1963, prepared by the Stratospheric Research Laboratory of the United States Weather Bureau. Contours are altitudes in meters and wind speed in knots where a barb represents 10 knots and a triangle represents 50 knots.*

meters at the 10-mb level to 1600 meters at the 0.4-mb level. These data would indicate that the level of maximum heating as a result of the sudden warming is to be found at the 40- to 45-km level in the upper stratosphere. This is not to say that the phenomenon was initiated there, but that it showed its maximum amplitude at that level.

4.3 Stable Stratospheric Winters

Sudden warmings of the stratosphere are an occasional feature of the winter stratospheric circulation which have been clearly identified only four times during the past 12 years in the Northern Hemisphere, and are yet to be observed in the Southern Hemisphere. They are, therefore, a rather special phenomenon in the stratospheric circulation, possibly on a level equivalent to that which hurricanes occupy in the tropospheric circulation system. The winter storm period, however, is a characteristic feature of the winter circulation pattern in the Northern Hemisphere as is evidenced by the curves of Fig. 4.7. A major problem of the MRN observation program is to establish the degree to which the two hemispheres are similar, and the lack of synoptic data from any Southern Hemisphere station precludes a firm stand at this time. If, however, hemispheric interaction between the opposing circulations in tropical latitudes is the generating cause of the slowdown in the winter westerlies during the winter storm period, there is little reason to doubt that a similar circulation phenomenon takes place in the Southern Hemisphere. A number of points relative to the global stratospheric circulation will necessarily remain unresolved until this question of similarity is settled, but to facilitate the intelligent exploration of the global circulation it is desirable to postulate the probable situation to establish meaningful checkpoints for early analysis.

As was pointed out in Chapter 3, the winter of 1963–1964 was without a major sudden warming event. The circulation patterns that existed on the 100-mb surface during the winter storm period were presented at five-day intervals in Figs. 3.31 through 3.38. It was noted that a basic difference exists in the sets of 100-mb charts which depict conditions under which a sudden warming occurs (Figs. 3.20 through 3.30) and the set mentioned above. This difference principally concerns the wave which develops over the Atlantic Ocean, and particularly the southern portion of that wave which extended into tropical regions in the case in which the sudden warming occurred. This is evidenced by the opening of the 16.50-km height contour over the South Atlantic on 20 January 1963 (Fig. 3.24), whereas the 16.50-km contour did not open in the latter case, although it did develop a considerable wave amplitude over Africa. Figures 4.4 and 4.5 very clearly indicate that while no discernible sudden warming occurred, the circulation of the stratosphere was far from quiescent, with gross variations over periods of a few days. Inspection of Figs. 4.39 through 4.46 indicates that these variations probably

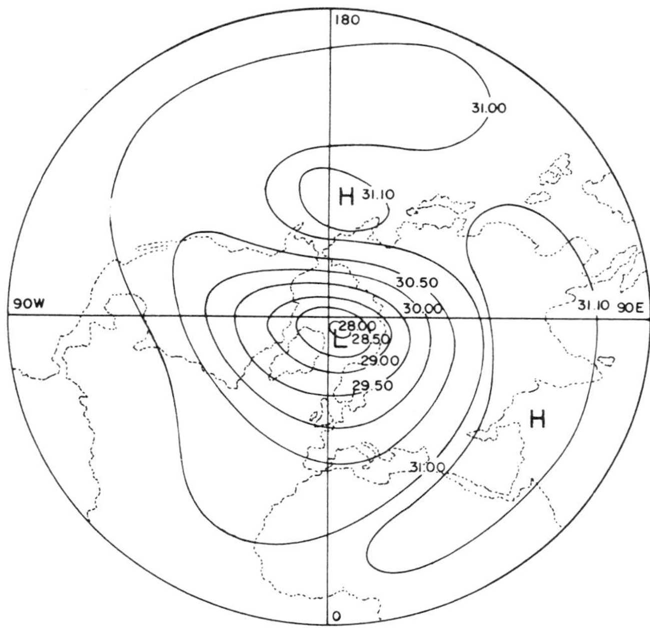

Fig. 4.39. Ten-millibar chart for 15 December 1963 over the Northern Hemisphere. Contours are heights in kilometers.

result from variations in dimensions of the central polar low-pressure system, and also from rotation and (or) transformation of the principal axis to different meridional locations.

On 15 December 1963 the 10-mb chart for the Northern Hemisphere is roughly typical of the contour patterns as the winter storm period gets under way. A well-developed low-pressure center is located almost at the pole with a minimum altitude of 28 km, and a rather steep gradient (negative meridionally) covers the high latitudes. The principal axis of this central low center arcs from Central Asia across the pole to western North America and into the eastern Pacific. This is approximately the mean position of the low system as is evidenced by the five-year mean data of Fig. 4.26. The westward hook in the principal axis in middle latitudes is somewhat unusual, however, and results in a reduced horizontal gradient of the height contours over White Sands Missile Range. Isotherms of the 10-mb surface generally follow the height contours in this 15 December data (Fig. 4.39), with a strong negative meridional temperature gradient in high latitudes and a thermal ridge in middle latitudes. The temperature gradient is rather flat in low latitudes, re-

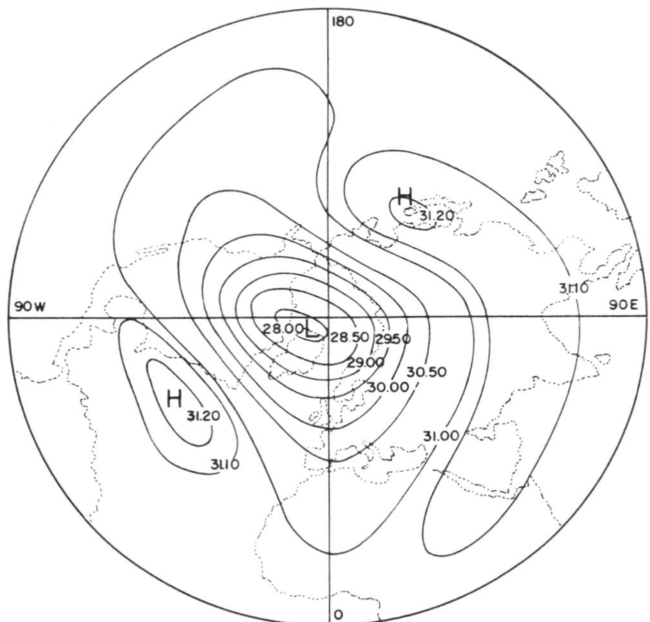

Fig. 4.40. Ten-millibar chart for 20 December 1963 over the Northern Hemisphere. Contours are heights in kilometers.

versing locally to provide a positive meridional temperature gradient in tropical regions.

Consideration of the data for 20, 25, and 30 December (Figs. 4.40 through 4.42) indicates a steady westward rotation of the Western Hemispheric end of the principal axis until by the end of the year it is oriented from Europe across the pole into the central Pacific. The low-pressure center has remained remarkably stable during this period in other respects, maintaining a central height of 28 km and roughly the same lateral dimensions. Temperature contours rotated with the height contours, so there was little change in the system other than the rotation of the axis.

On 4 January 1964 the contour pattern illustrated in Fig. 4.43 presented a rather curious change, probably representing a dynamic change in shape of the high-latitude low-pressure system which rotated the previous principal axis very rapidly eastward across North America into the North Atlantic. The result of this action, as far as the White Sands Missile Range data are concerned, was to diminish the zonal flow during early January as the low center deepened slightly to 27.5 km and withdrew slightly into the Eastern Hemisphere. The low center reorganized

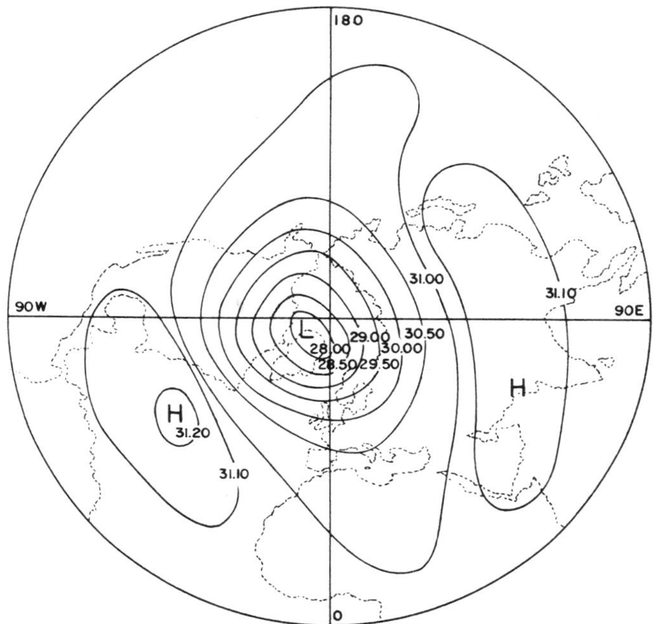

Fig. 4.41. Ten-millibar chart for 25 December 1963 over the Northern Hemisphere. Contours are heights in kilometers.

in its new position with a major axis extending from eastern Asia across the pole into eastern North America as is illustrated in Fig. 4.45 by the data for 14 January 1964. During these changes, as well as later variations of the stratosphere in the late winter of 1964, the meridional temperature gradient of the stratospheric circulation was relaxed through processes other than sudden warming.

The Stratospheric Research Laboratory of the U. S. Weather Bureau has analyzed the MRN data to determine North American constant pressure averaged from data obtained over several days during the winter of 1963–1964 as illustrated in Figs. 4.47 through 4.50. The chart for 4 December 1963 (Fig. 4.47) illustrates the more or less normal strong westerly circulation which characterizes the early winter period (see Fig. 4.7) of maximum annual westerly winds. Zonal winds of between 50–100 meters/sec are the rule over the United States, and it is clear that there is a lack of symmetry in the distribution about the hemisphere, with marked meridional flow in the Pacific and probably in the Atlantic. This is illustrated by the data for Fort Greely, where a strong north component is in evidence, as is indicated for that station by the meridional SCI data of Fig. 4.8. In part, the height contour configuration

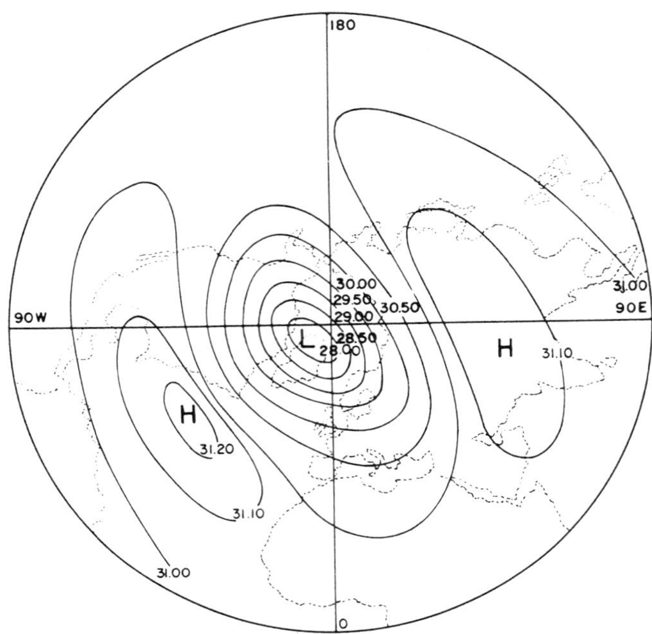

Fig. 4.42. Ten-millibar chart for 30 December 1963 over the Northern Hemisphere. Contours are heights in kilometers.

illustrated for this North American sector of the 2-mb and 0.4-mb constant pressure levels is derived from the 10-mb chart pattern. It could well be in error, therefore, and a clear picture of the hemispheric distribution can be established only by the operation of a global MRN. At this time it is reasonable to assume that higher levels in the upper stratosphere are characterized by a contour shape similar to that of the 10-mb level, possibly along the lines of the mean January map of Fig. 4.26.

Consideration of our annual progress charts of zonal SCI (Figs. 4.7 and 4.23) points out that the winter storm period occurs shortly after the example of Fig. 4.45, so the 2-mb and 0.4-mb charts in Fig. 4.48 for 29 January 1964 should represent the height contour pattern of the upper stratosphere after interaction between the hemispheric circulations has put an effective brake on the early winter westerlies. It is clear again in these data (Fig. 4.48) that the 2-mb level tends to dominate the upper stratospheric circulation system. A meridional height difference over the east coast of North America of 4.64 km from lower middle latitudes to the polar low-pressure center is noted on the 2-mb surface, while the

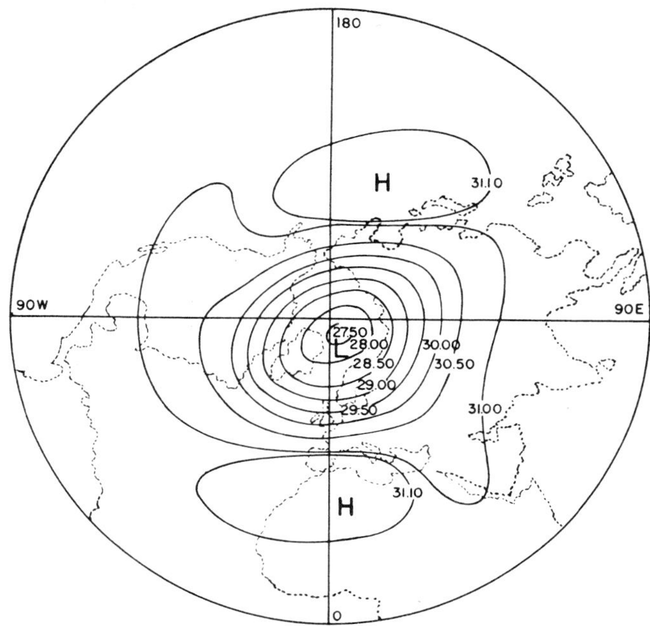

Fig. 4.43. Ten-millibar chart for 4 January 1964 over the Northern Hemisphere. Contours are heights in kilometers.

0.4-mb surface has a slightly smaller value of 4.32 km. This is essentially a 50% increase in over-all gradient over the 4 December case.

A principal difference concerns the latitudinal distribution of the gradient exhibited by the two cases. In December a strong meridional gradient exists in the 30-degree latitude zone (Fig. 4.47) at both the 2-mb and the 0.4-mb levels. On 29 January 1964 the meridional height gradient in lower midlatitudes has diminished by about one third. This decrease in low-latitude gradient is more than compensated by a gross strengthening of the high-latitude polar vortex. Maximum wind intensities, which were noted near 30 degrees latitude in December on the 0.4-mb surface, are found in middle latitudes in late January. Strongest winds were already at midlatitudes in early December on the 2-mb surface, but it is clear from the height contour pattern changes between the two cases that the strong winds have shifted poleward here also. Northwest winds of 100 meters/sec have appeared over Fort Greely, twice those reported in the December case. These data lead to the conclusion that the winter storm period, when there is no sudden warming, simply results in a shrinking of the winter circumpolar cyclonic vortex to high

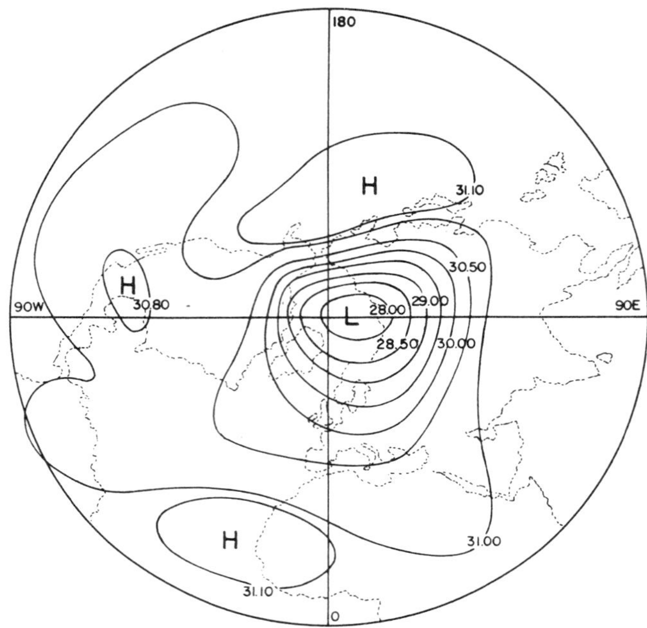

Fig. 4.44. Ten-millibar chart for 9 January 1964 over the Northern Hemisphere. Contours are heights in kilometers.

latitudes. It has actually strengthened in the process in that the height of the low center has dropped by over 2 km at the 2-mb level and by 1.3 km on the 0.4-mb surface.

These data would then indicate that the sharp reduction in zonal wind speed which is such a strong characteristic of lower-latitude stratospheric data is produced by encroachment of the summer hemispheric easterlies into the winter hemisphere. It is probable that, in the cases where no sudden warming occurs, a high-latitude station would observe a continuous buildup in the circumpolar circulation throughout the winter season until the spring reversal begins. Such a regime has been observed over McMurdo Sound during two winter seasons when MRN observations (Rotolante and Parra, 1965) have been conducted by personnel of the Schellenger Research Laboratories of Texas Western College, El Paso, sponsored by the National Science Foundation. These data raised the question of a possible lack of symmetry between the hemispheres since they did not exhibit the traditional winter storm period which is invariably a part of the Northern Hemisphere MRN data. It would appear that essentially the same result might well be obtained in the Northern Hemisphere above 80 degrees

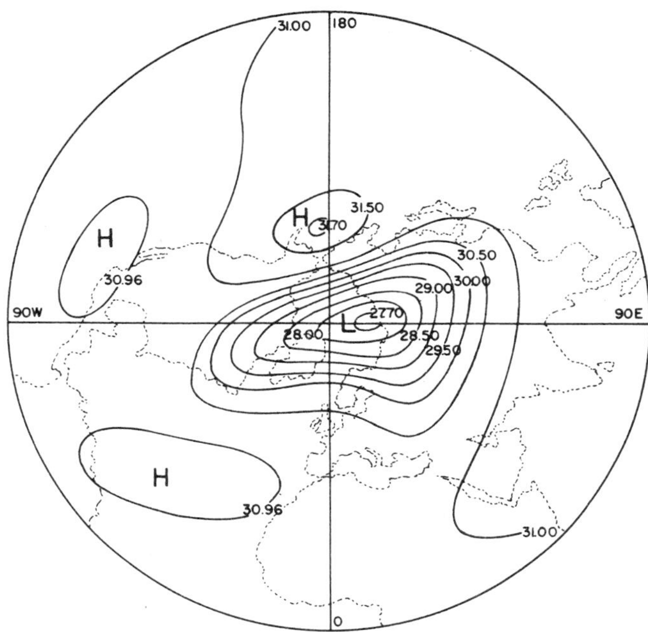

Fig. 4.45. Ten-millibar chart for 14 January 1964 over the Northern Hemisphere. Contours are heights in kilometers.

latitude during those years in which sudden warmings do not occur. The need for high-latitude data in the Northern Hemisphere is obvious.

Particularly in evidence in the 29 January data are strong longitudinal gradients in the circumpolar flow. On the 2-mb surface, a trough arcs southwestward in a characteristic fashion deep into the central Pacific. At the 0.4-mb level just above the stratopause the same type of arc exists, but it is foreshortened to influence only high latitudes. It is apparent that considerable portions of the lower and middle latitudes are dominated by easterly circulations. This is more clearly indicated by the 2-mb and 0.4-mb charts of Fig. 4.49 for 26 February 1964 where the Pacific high-pressure area has pushed into the MRN observation region to alter the circulation over North America. As was implied by the data in Fig. 4.48 the system is tilted poleward in the vertical, resulting in a more northward position of the high center on the 0.4-mb chart, and a resulting displacement of the polar low away from the pole into the Eastern Hemisphere. Considerable filling of the central low-pressure system has occurred, with height increases of almost 2 km in evidence on the 2-mb surface.

The high-pressure region of the North Pacific has height values which

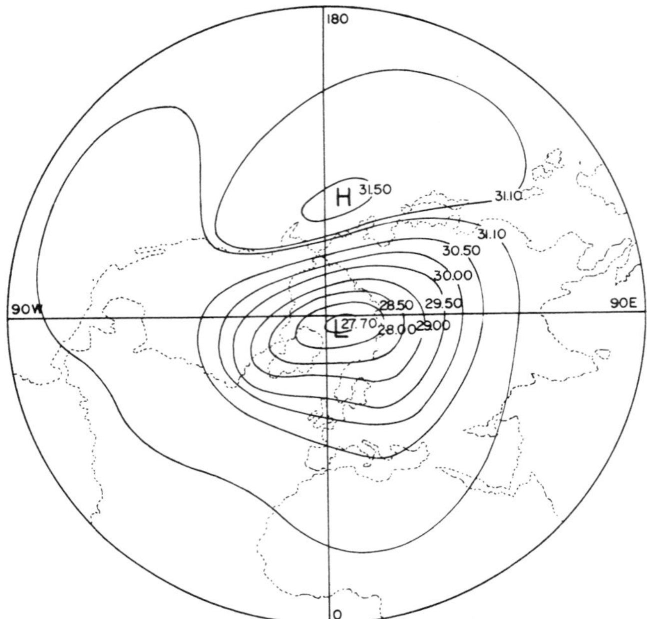

Fig. 4.46. Ten-millibar chart for 19 January 1964 over the Northern Hemisphere. Contours are heights in kilometers.

are in excess of those observed anywhere else in the hemisphere on both of the pressure surfaces. This can result only from a warming process, and thus differs from the sudden warming of the winter storm period only in the manner in which the disturbance is advanced into the winter polar regions. It is probable that, if interaction between the hemispheric circulations at low latitudes is the basic cause of this circulation change, the entry point of the easterlies into the westerly flow was in the western Pacific in this case.

Westerly winds of the stratospheric winter were largely dissipated during March as is evidenced by the charts for 25 March 1964 in Fig. 4.50. Meridional height gradients were weak on both charts, and are inconsistent with the usual winter pattern, at least on the 2-mb chart. Here we find a weak elongated low-pressure system along the 50th parallel across North America and the North Pacific, with a weak ridge of high pressure extending from the Hudson Bay area across the pole. Height of the 2-mb pressure surface at the pole fell during the early phases of the winter storm period as the circumpolar low tightened about the pole, but has climbed as the westerly circulation was eliminated from a low of less than 38-km altitude to a height of over 42.56 km

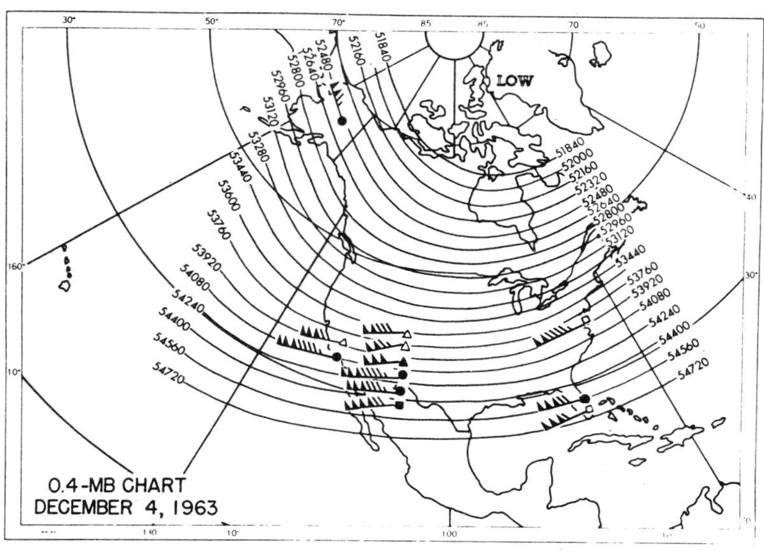

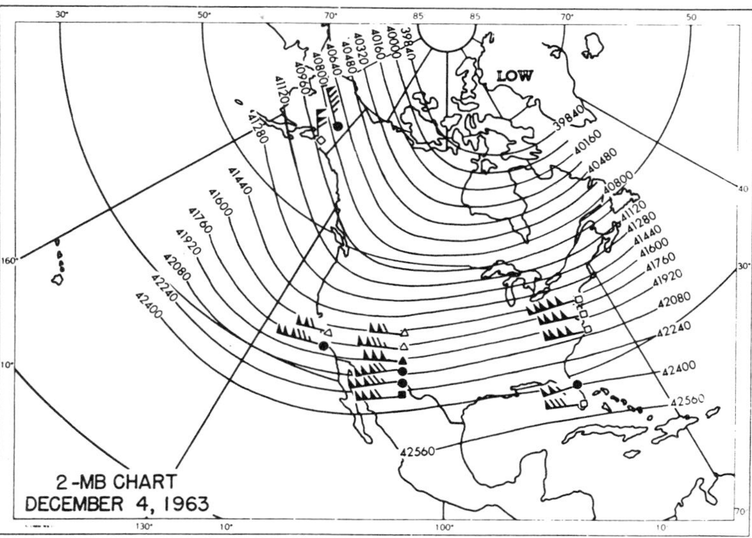

Fig. 4.47. Stratospheric constant-pressure charts for 4 December 1963 over North America prepared by the Stratospheric Research Laboratory of the United States Weather Bureau using MRN data. Wind speeds are in knots and heights are in meters, with wind barbs and triangles representing 10 and 50 knots, respectively.

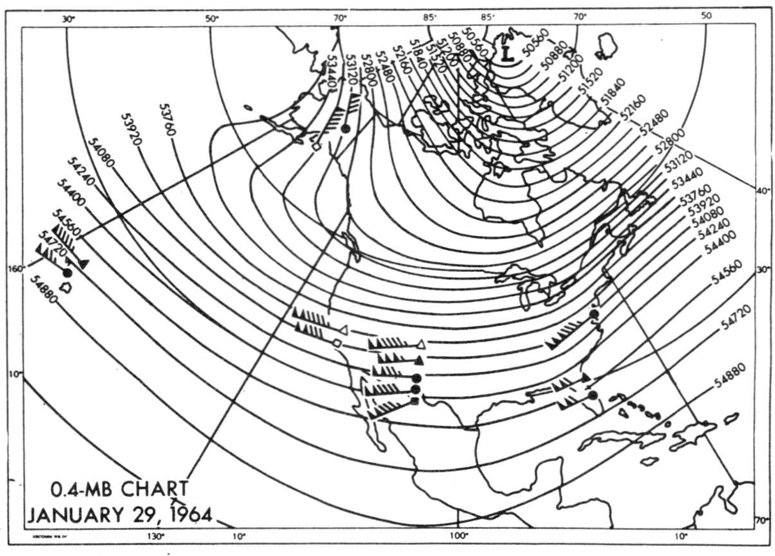

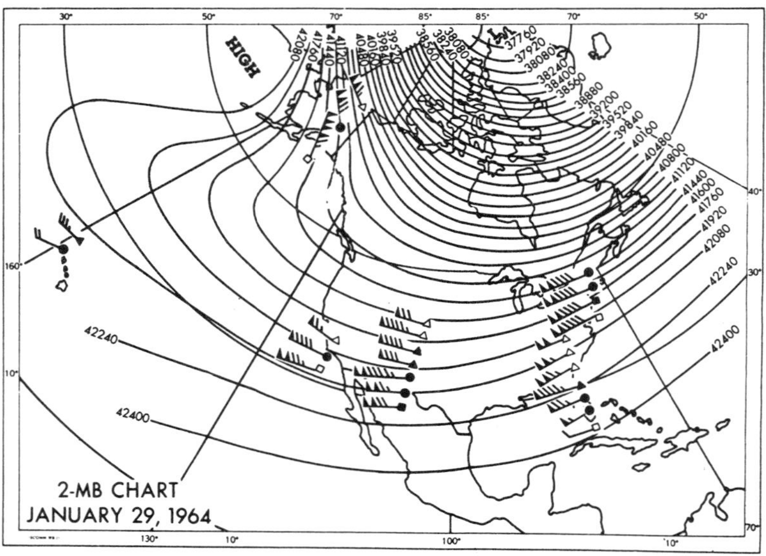

Fig. 4.48. Stratospheric constant-pressure charts for 29 January 1964 over North America prepared by the Stratospheric Research Laboratory of the United States Weather Bureau using MRN data. Wind speeds are in knots and heights are in meters, with wind barbs and triangles representing 10 and 50 knots, respectively.

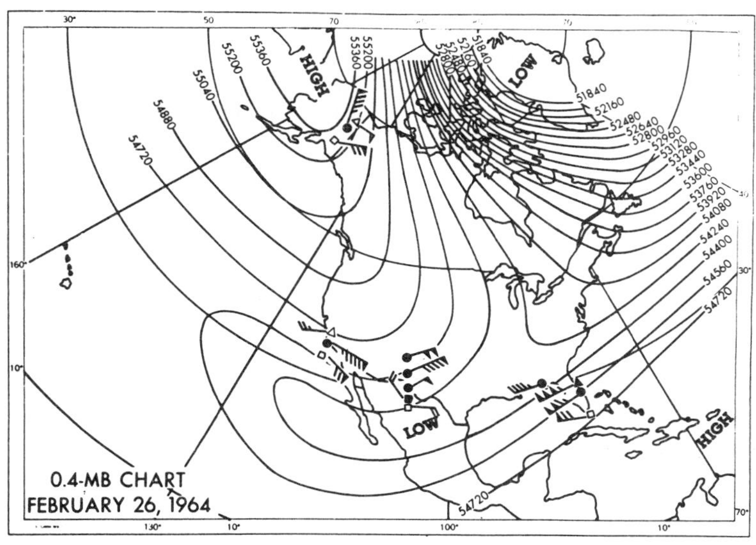

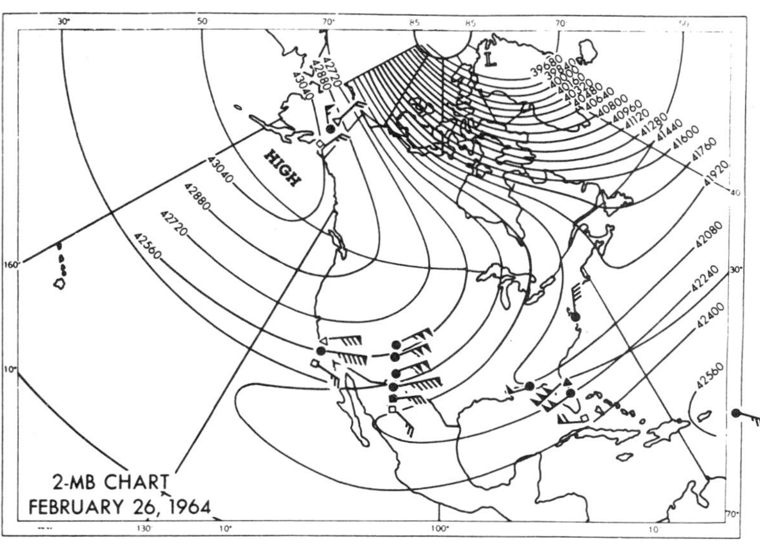

Fig. 4.49. Stratospheric constant-pressure charts for 26 February 1964 over North America prepared by the Stratospheric Research Laboratory of the United States Weather Bureau using MRN data. Wind speeds are in knots and heights are in meters, with wind barbs and triangles representing 10 and 50 knots, respectively.

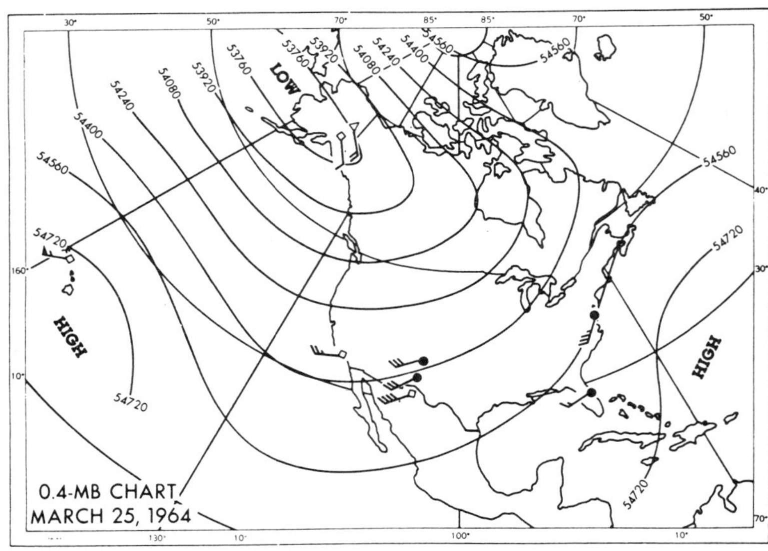

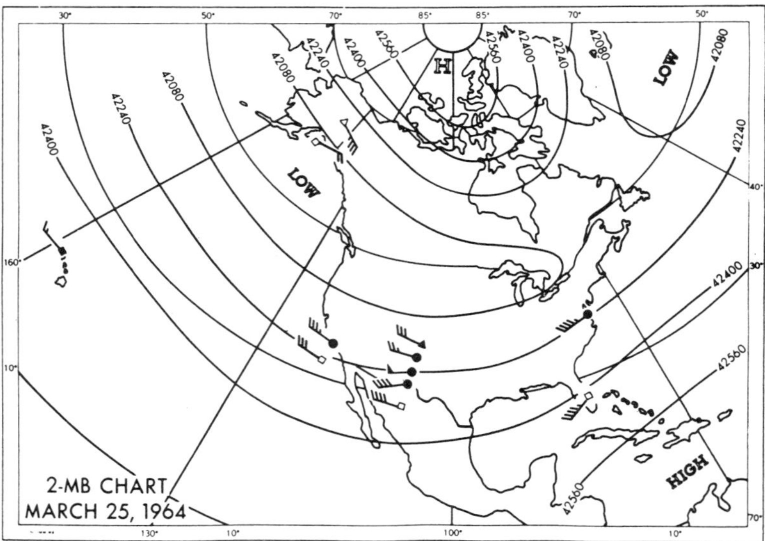

Fig. 4.50. Stratospheric constant-pressure charts for 25 March 1964 over North America prepared by the Stratospheric Research Laboratory of the United States Weather Bureau using MRN data. Wind speeds are in knots and heights are in meters, with wind barbs and triangles representing 10 and 50 knots, respectively.

on 25 March. An increase in height of similar magnitude also occurred at the 0.4-mb surface.

It is interesting to note that identifiable features of the upper stratospheric circulation structure tend to tilt toward the pole with increasing height. Such an effect could be the obvious reason for the known meridional motion of the upper stratosphere and mesosphere from the equator toward the poles that appears in the mean climatological data of the MRN.

4.4 Spring Reversal

Reversal of the stratospheric monsoonal flow in Northern Hemispheric spring from winter westerlies to summer easterlies is observed during April and May. The reversal has been described as initiating at the highest observed levels and working downward with time into the lower stratosphere. Such a program of events is in at least general agreement with theory, since as the sun rises for the summer season it will probably first become evident by establishing a positive meridional temperature gradient at the highest levels and impressing this structure downward as the incidence angle decreases. It is perhaps too much to expect that these events would occur uniformly, either in the detail which would be observed from day to day or in gross characteristics of each annual reversal. The MRN data provide an opportunity for a look at certain features of this rather prominent stratospheric event.

Inspection of Figs. 4.7 and 4.23 yields the information that stratopause reversals in the zonal flow occur around 1 May at low and middle latitude locations. High-latitude stations have gained easterly winds before the middle of April and thus are well ahead of lower-latitude locations in the reversal phenomenon. In addition, there is a special shape of the zonal SCI data curve for Fort Greely during the period after easterlies appear there and as the reversal gets under way at lower latitudes. It may be inferred from these data that there is a separate genesis of the summer easterly circulation in immediate polar regions and in lower latitudes.

Immediately after the spring equinox the sun's rays begin to illuminate the polar upper atmosphere in altitude regions that are shadowed except in the summer season by intervening medium. The influx of radiant energy is continuous and very rapidly heats the thin air of the mesosphere, causing it to expand and carry upward the constant-pressure surfaces to form a dome of warm air about which an anticyclonic circulation begins. The heated volume expands downward and equatorward as the solar elevation angle increases. Our only significant point of measurement thus far is that the polar anticyclone has spread downward and outward to 50-km altitude at Fort Greely (64° N) by early April. Mapping of this extremely interesting aspect of stratospheric circulation will undoubtedly be a major problem in future MRN activities.

The large mass of the upper stratospheric circulation is accelerated into easterly motion by meridional thermal gradients which begin and exhibit a maximum in lower midlatitudes. These easterly winds are first observed in the SCI data at 30 degrees latitude and spread rapidly poleward and equatorward, so that in a matter of a few days the entire summer hemisphere stratopause has an easterly flow. It is clear that the poleward spread of this wave of easterly winds will encounter the polar anticyclone as it expands toward lower latitudes. The rather curious behavior of the Fort Greely SCI data around 1 May is then possibly a measure of the coupling of these two circulation systems as they encounter each other just south of Fort Greely. Clearly, the polar anticyclone will have far less momentum than its midlatitude counterpart, and thus will be decelerated to the rotational speed of the general hemispheric circulation. This interaction zone at about 60 degrees latitude should be the scene of considerable turbulence as these flows are welded together.

A view of the mode of reversal is presented in Figs. 4.51 through 4.56. These figures represent time cross sections of data for White Sands Missile Range, New Mexico (32° 23′ N, 106° 29′ W) and Fort Greely, Alaska (64° N, 145° 44′ W). These data are characteristic of the information available on the two anticyclonic circulations of the summer season. In the lower latitude case the persistence of a stratum of easterly winds at the stratonull level after the winter storm period is a significant feature. Ignoring this complication, the easterly winds of the reversal are noted to appear first at the top of the cross section, generally during the month of April. Data for 1964 (Fig. 4.54) contain the earliest indications of easterly winds, with light variable easterlies occurring near the stratopause the last few days of March and in early April. The shift to steady easterlies is not effected until mid-April, however, which is in good agreement with an average appearance of the easterlies at 60 km on 20 April.

The reversal phenomenon works downward with time. There is a great deal of variability in the data, but a period of approximately one month is required before the upper stratospheric westerlies are completely eliminated at the 30-km level. Thus the spring reversal phenomenon in midlatitudes is a downward-propagating system with a velocity of a kilometer per day.

The matter of consistency from one year to the next is of some import. It may be noted that the development of easterly winds in the

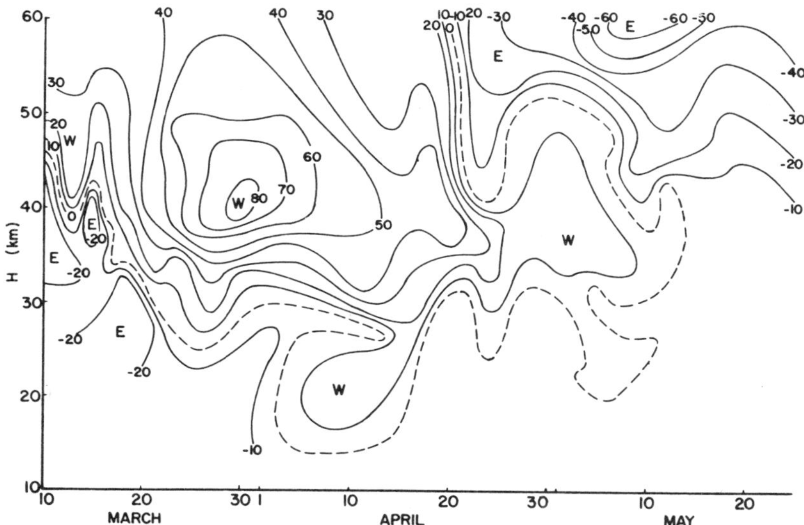

Fig. 4.51. Zonal wind structure of the stratosphere over White Sands Missile Range during the spring reversal season of 1961. Component speeds are in meters per second with winds from the west positive.

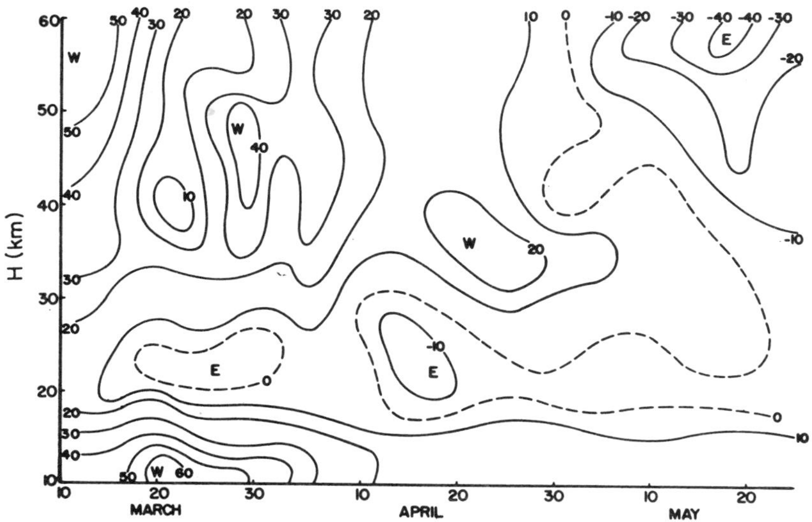

Fig. 4.52. Zonal wind structure of the stratosphere over White Sands Missile Range during the spring reversal season of 1962. Component speeds are in meters per second with winds from the west positive.

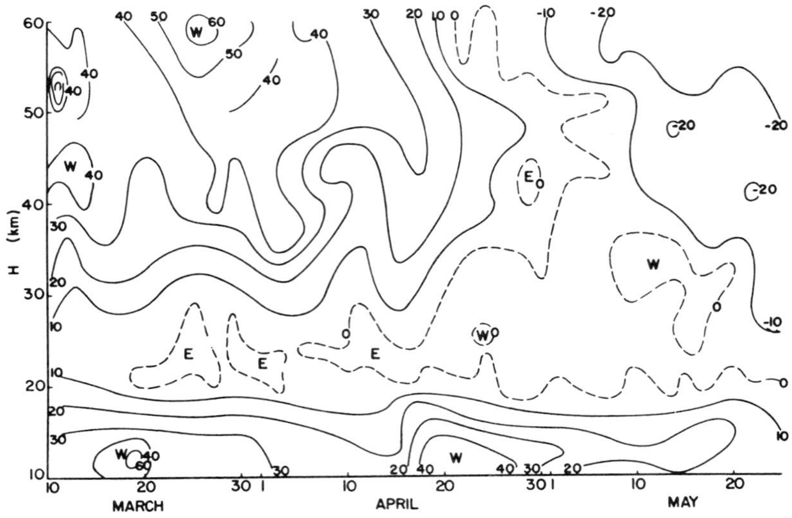

Fig. 4.53. Zonal wind structure of the stratosphere over White Sands Missile Range during the spring reversal season of 1963. Component speeds are in meters per second with winds from the west positive.

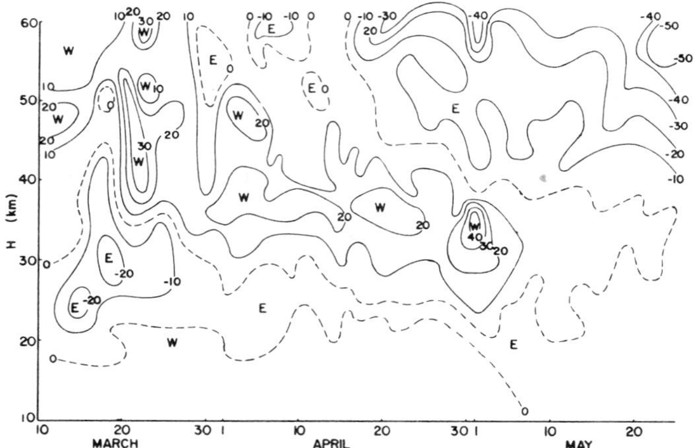

Fig. 4.54. Zonal wind structure of the stratosphere over White Sands Missile Range during the spring reversal season of 1964. Component speeds are in meters per second with winds from the west positive.

stratonull region is more pronounced in 1963 (Fig. 4.53) and 1964 (Fig. 4.54) than in previous years. This feature is characteristic of a weakened westerly circulation in the tropospheric and stratospheric westerly circulations which permits the easterly thermal wind of the lower stratosphere to play an enhanced role in establishing the vertical wind structure. This trend in the small amount of data available is uncertain, but it is in step with the sunspot cycle and thus could be a result of the diminished solar output. By far the most important part of the solar variability is found in the very high-frequency end of the radiation spectrum. It is generally believed that the 2000- to 3000-angstrom region of the solar ultraviolet is relatively stable over the 11-year cycle, but the data are limited in this area and the existence of an unvarying solar ultraviolet is not well established.

If there should be a variation in the intensity of the energy input to which the stratospheric circulation is sensitive it is probable that the first evidence that we would have available is a higher level for the base of that circulation; that is, the radiations that heat the stratosphere and drive its circulation system could not, with a reduced intensity, penetrate as deep into the atmosphere as usual and thus would not be capable of establishing the negative meridional temperature gradient of the winter westerlies down to the usual altitudes. There are, however, several other ways in which the depth of penetration of solar ultraviolet energy can be altered by strictly terrestrial means. The question of solar control, if the trends of stratospheric circulation should follow the sunspot cycle, will undoubtedly have to be resolved by precise measure of the solar ultraviolet stability.

The stratospheric circulation of high latitudes is quite variable, with large deviations from the generally zonal circulation of lower latitudes. The time cross sections available for Fort Greely (Figs. 4.55 and 4.56) illustrate this point for the years 1963 and 1964. If the 1963 data are presumed to be representative of the usual reversal picture in the high-latitude stratosphere, the reversal phenomenon appears at 60 km about the first of April and works downward to 40 km very rapidly, at a rate approximating 2 km per day. Completion of the reversal in the lower stratosphere is less decisive, and it takes the rest of April to establish an easterly flow at all stratospheric levels. It should be noted that the reversal process is completed at Fort Greely while just getting under way at White Sands Missile Range.

The time cross section for Fort Greely in 1964 (Fig. 4.56) is markedly different from that for 1963. It is surmised that the pronounced circulation event which dominates the upper stratospheric circulation pattern in early March is dynamic in origin and as such is not directly related to the nonseasonal monsoonal reversal. There is a likelihood that this disturbed zonal flow over Fort Greely is directly related to the "sudden warming" type of disturbance that was observed over White Sands Missile Range in late January (Fig. 4.4). Such events are believed to

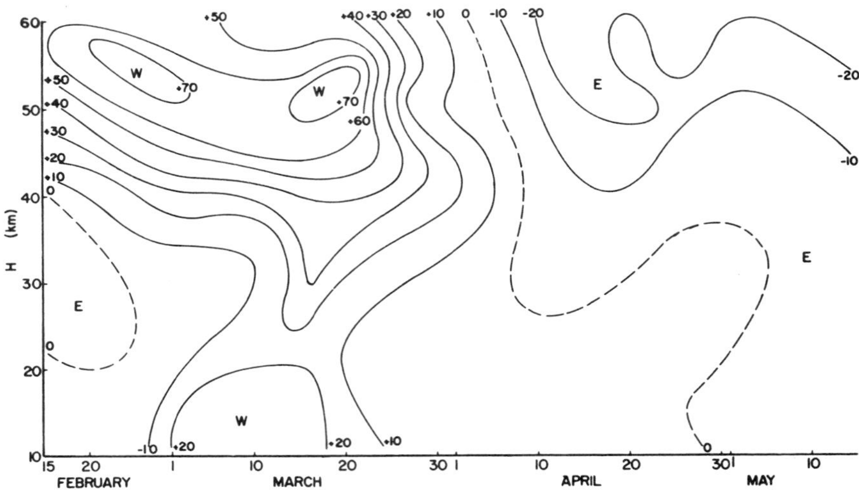

Fig. 4.55. Zonal wind structure of the stratosphere over Fort Greely during the spring reversal season of 1963. Component speeds are in meters per second with winds from the west positive.

originate in subtropical stratopause regions in the shear zone between the hemispheric flows and progress toward the winter pole, extending their influence downward into the lower stratosphere at high latitudes.

Events of this type more generally occur during the winter storm period from 15 December to 15 February. This late occurrence in 1964 could then be the source of the confusion which attended the spring reversal at all latitudes. The time for stabilization of the polar cyclonic vortex was simply inadequate for the reversal to be executed in a systematic fashion.

It is obvious from these data that a comprehensive picture of the spring reversal phenomenon is not available in the MRN data which have been compiled thus far. Additional spatial and temporal observations will be required, particularly in the form of complete hemispheric coverage of the event. It is possible, however, to trace one other aspect of the spring reversal with some measure of success. The SCI data indicate the mode of latitudinal development of the easterly circulation during the reversal period. These points are analyzed and presented in Fig. 4.7. These curves show that, while the high-latitude summer easterlies appear first, they are quickly incorporated into the general anticyclonic circulation. The greatest strength of this circulation system is to be found at low latitudes.

This is interesting since the meridional temperature gradient of the

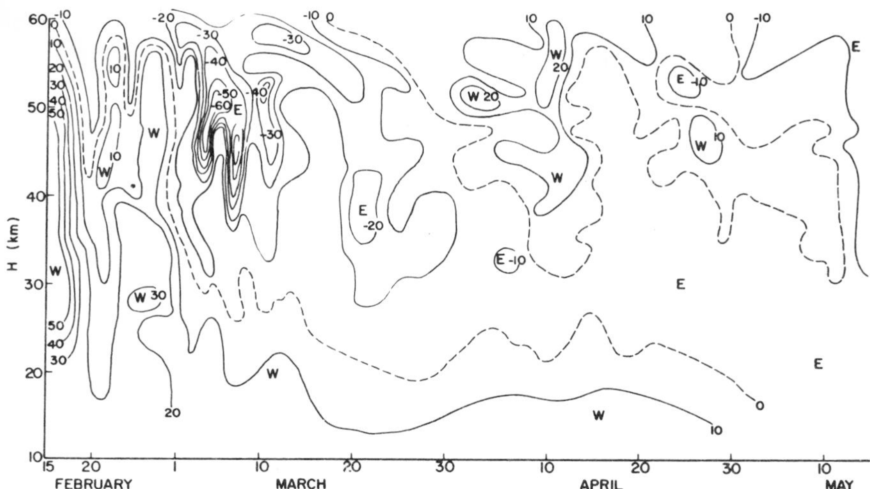

Fig. 4.56. Zonal wind structure of the stratosphere over Fort Greely during the spring reversal season of 1964. Component speeds are in meters per second wtih winds from the west positive.

stratosphere would not necessarily have been predicted to occur in that region. It appears essential that there be a considerable latitudinal gradient in stratospheric absorption of solar ultraviolet in the low latitudes of the summer hemisphere. It is known that the winter and spring storage of ozone at high latitudes (Figs. 3.45 and 3.46) couples with strong advective effects to produce strong gradients of ozone concentration in the lower stratosphere. It is not easy to see the relationship between this factor and the situation at 50-km altitude. A second possibility exists, in that there may be high concentrations of water vapor in the stratosphere where residence times are known to be quite long, and molecular diffusion may exert a role in establishing the vertical composition structure. Since residence times for stratospheric air above 20 degrees latitude are quite short and vertical motions will assure a dry stratosphere, there will probably be a strong meridional gradient in water vapor in the upper stratosphere. The considerations suggest that water vapor may play an important role in the spatial organization of the spring reversal.

4.5 Fall Reversal

Easterly winds of the summer stratosphere reach their peak speeds in late July and early August and begin to diminish as the favorable solar aspect angle which produces the warm summer fades back into equatorial regions. Figure 4.7 indicates a lag in the development of the summer easterlies, in that peak intensity follows the summer solstice (22 June) by approximately one month, with the latest occurrence in subtropical regions. These data also show that the fall reversal occurs earlier in high latitudes, following very closely the solstice in the stratosphere over Fort Greely. As in the spring, then, the fall reversal is complex, with an early shift to westerly winds in polar regions and a later and perhaps more enthusiastic turnaround in the mainstream of the stratospheric circulation.

From the SCI data of Fig. 4.7 it is determined that the actual reversal date of the stratospheric monsoon in the fall occurs almost on the date of the fall equinox at middle and low latitudes. Rate of decay of the summer easterlies is significantly more rapid in the late summer season than was the buildup in early summer, and development of westerlies in early winter immediately after the reversal is even faster. The mechanics of this fall reversal indicate a rather loose coupling to solar control, introducing a considerable asymmetry into the annual wind oscillation of the stratospheric circulation. These effects, coupled with the gross deviations that occur in the winter storm period, disrupt the smooth pattern of solar heat input and give the stratospheric circulation an individual character of considerable distinction. There are undoubtedly pressing reasons for these rather large differences in characteristics of the spring and fall reversals of the stratospheric circulation, and the observational program of the MRN is in part directed at evaluation of the sequence of events which accomplish these major changes in stratospheric structure.

Consideration of the geometry of solar aspect relative to the stratospheric absorbing medium suggests that denial of heating as the sun recedes over the equatorial horizon would first come to the lowest levels of polar regions. The positive meridional temperature gradient characteristic of stratospheric summer would thus first begin to decay and revert to the usual winter negative meridional temperature gradient at the base of the stratospheric solar heat sink and then gradually build upward during the latter part of the summer season. This shadow zone for solar ultraviolet should exhibit a strong diurnal oscillation since most of the heat would be deposited on the subsolar side of high-latitude longitude circles. The region affected would, in any case, expand upward and equatorward as the sun's withdrawal progresses to the equinoctial period. This denial of heating will reduce the temperature of this base layer of the stratosphere well ahead of the fall reversal so that a negative meridional temperature gradient will be established in the immediate polar region ahead of the scheduled circulation reversal date. No MRN stations are located in favorable positions for observation of such an event at this time, but there is reason to expect that the first evidence of fall reversal of the stratospheric monsoonal circulation is to be found in the 30- to 40-km altitude region of the immediate polar region.

Now heating and subsequent expansion of the upper portions of an atmospheric layer such as occurs in spring and a similar operation on the lower portions of a layer such as happens in fall are two decidedly different things. What goes on up above in such an environment may well be a matter of small concern to the lower layers, but the same can hardly be said in the reverse case. Cooling and contraction of lower layers in the stratosphere will result in subsidence of the entire stratosphere in polar regions with associated physical effects. One effect will be compressional heating of the descending air, which, in the upper layers, should result in an intensification of the summer easterly circulation. At high latitudes, then, the fall reversal should originate near the pole in the lower stratosphere. It is probable that the entire polar stratosphere becomes involved in this new circulation before it spreads significantly into the middle and lower latitudes, so the exact nature of the entire reversal phenomenon will necessarily be acquired from observational data.

An example of the sequence of events associated with a fall reversal at lower middle latitudes is illustrated in the data obtained at White Sands Missile Range for 1962, 1963, and 1964 as presented in Figs. 4.57 through 4.59. Development of the tropospheric westerly jets that are characteristic of the winter season is evident in early September of 1962, well in advance of the equinox. However, the fall seasonal reversal begins at above 60-km altitude and works gradually downward in the course of approximately five weeks' time, with the last of the easterly winds being eliminated from above and below at the stratonull level by mid-October. The driving mechanisms which accomplish the fall reversal thus appear to originate in the troposphere and in the mesosphere as far as lower-latitude regions are concerned, with the upper development gradually reversing the stratospheric circulation. A rather rapid intensification of the westerly zonal circulation at the stratopause level is indicated, with the center of the new wind system lowering to that altitude as the winter season gets under way.

A similar sequence of events is observed in the 1963 data for White Sands Missile Range that are presented in Fig. 4.58. A principal difference lies in the speed with which the reversal is accomplished, it taking approximately one month to complete the change in the stratosphere in 1963 compared with more than six weeks in the 1962 case. It is interesting to note that the westerlies of the tropospheric jet stream region

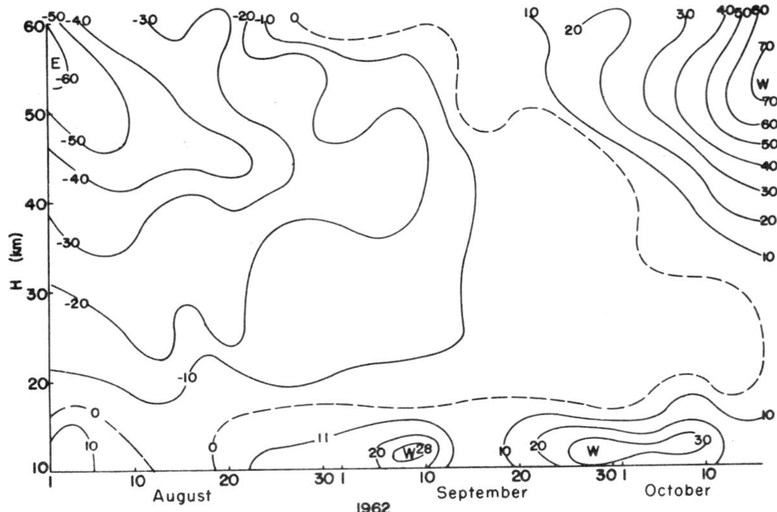

Fig. 4.57. Zonal wind structure of the stratosphere over White Sands Missile Range during the fall reversal of 1962. Speeds are in meters per second with winds from the west positive.

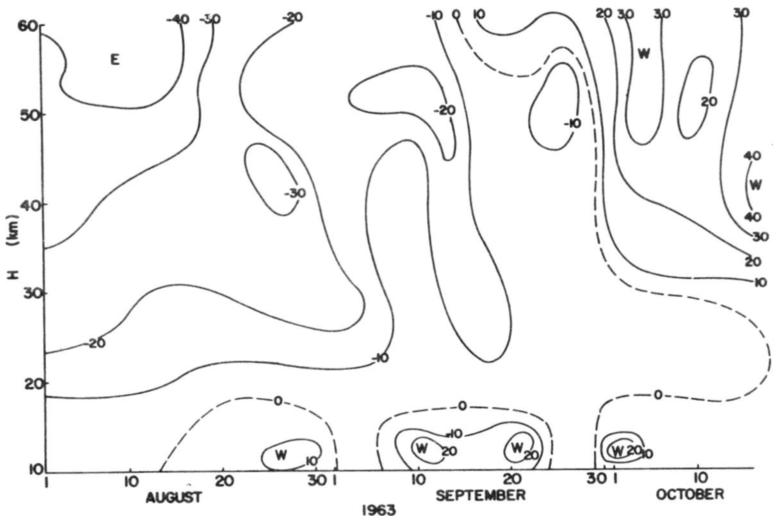

Fig. 4.58. Zonal wind structure of the stratosphere over White Sands Missile Range during the fall reversal of 1963. Speeds are in meters per second with winds from the west positive.

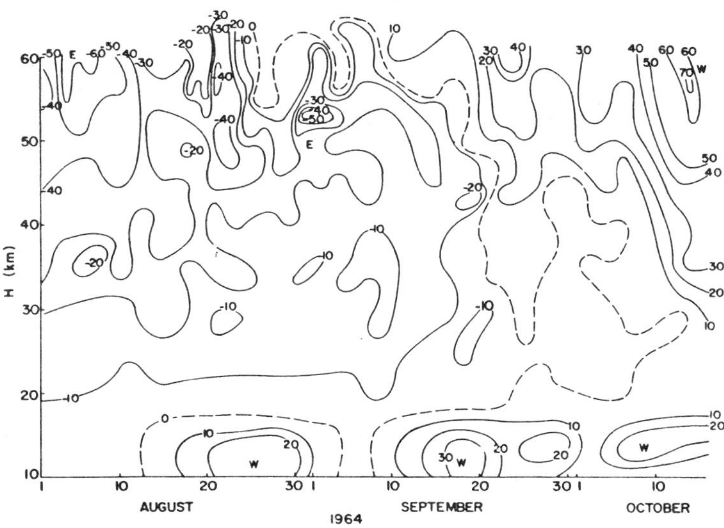

Fig. 4.59. Zonal wind structure of the stratosphere over White Sands Missile Range during the fall reversal of 1964. Speeds are in meters per second with winds from the west positive.

are also weaker in 1963, as is the general easterly circulation of the summer stratosphere just before the fall equinox. In addition, development of the winter westerly circulation in stratospheric altitudes is less well defined and precipitous in 1963 than it was in the 1962 case. All of these items point toward the fall reversal of 1963 being an example of an unusually mild fall reversal. If that is correct, apparently such reversals are characterized by a rather rapid switch from easterlies to westerlies as opposed to the prolonged reversal period exhibited by what must at this point be considered a strong reversal in 1962.

Stratospheric circulation data in the zonal plane over White Sands Missile Range in 1964 are shown in Fig. 4.59. This is a case that would undoubtedly fall in the strong category, with stratopause zonal winds changing from 50 meters/sec from the east on the first of September to 60 meters/sec from the west by mid-October. Again we see that the reversal sequence in the stratosphere is prolonged, starting in late August at the highest levels, but not completely eliminating the easterly summer flow at the stratonull level until well into October. It is clear from the strong resurgence of easterlies in the lower stratosphere in early October of the 1964 data that progress of the reversal is not always smooth, although it may well be that this is a local phenomenon in the White Sands Missile Range area.

These data point very clearly to a stratospheric circulation process in effect in lower latitudes, originating in the mesosphere and working downward at an average rate of almost 1 km per day, the stratopause reversal being accomplished on about the equinox, and termination of the process accomplished at the stratonull level about three weeks later. If additional data should bear this out, there appears to be a biannual cycle in the strength of the circulations associated with the fall reversals, with maximum intensities in the data of 1962 and 1964. The reversal rate is lower in the weak year reversals, amounting to only one half in the single available case of 1963.

In this connection, the strong fall reversals of record occurred when easterly winds dominated the biannual cycle of tropical regions in the lower stratosphere, as was the case in 1962 and 1964. The weak fall reversal of 1963 occurred when westerly winds were dominant at the stratonull level in tropical regions. Within the very tight limits of these few data, there is an implication that the two phenomena are related, and in view of the complete absence of a satisfactory explanation of either of these two items, it would appear that pursuit of research on MRN data of the global stratospheric circulation would be reasonably

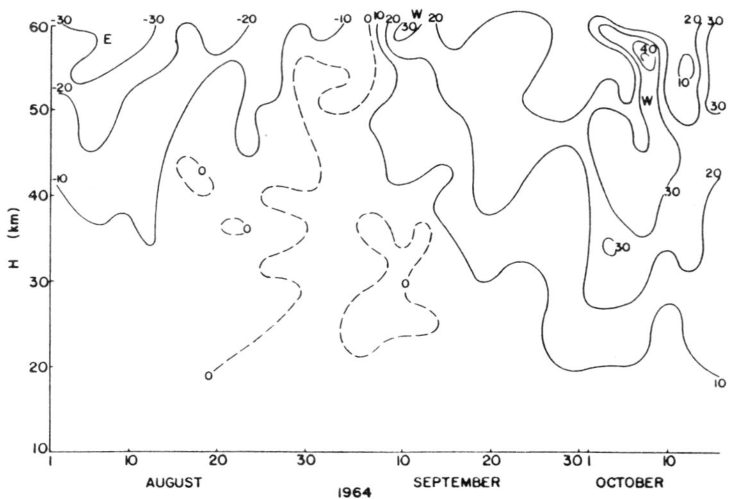

Fig. 4.60. Zonal wind structure of the stratosphere over Fort Greely during the fall reversal of 1964. Speeds are in meters per second and winds from the west are positive.

likely to provide at least an understanding of the structure of these phenomena, if not the cause.

Data on the fall reversal of the high latitude zonal circulation are relatively meager owing to a variety of circumstances. The data are inadequate for a firm determination at this time, but the data for Fort Greely obtained in the fall of 1964 (Fig. 4.60) bear out the concepts discussed in the opening paragraphs of this section. While the picture is far from clear, there is an indication that the reversal is decidedly different at high latitudes in contrast to that exhibited at low latitudes, with a slight trend for the reversal to start at low stratospheric levels and progress upward. The effect is not very marked in these data, as it should not be over a station at 64 degrees latitude. This trend for an upward progression of the reversal should be quite pronounced at higher latitudes, reaching a maximum in the vicinity of the pole. As was discussed above, the rate at which the phenomenon progresses upward through the stratosphere probably does not increase greatly over that observed at Fort Greely, even at the pole, since the upper layers cannot act independently of the lower layers in the fall reversal case.

4.6 Temperature and Density Structure

The temperature structure of an atmospheric environment exerts a controlling influence in determining the circulation patterns on a gross scale and is of significance in establishing the eddy diffusion structure, particularly in the vertical. The horizontal temperature structure at a particular level establishes the sources and sinks of atmospheric heat which the circulation system is obliged to attempt to eliminate. The earth's rotation complicates the circulation smoothing technique which, in essence, results in time delays during which further complications in the atmospheric thermal field are introduced by several sources. As a result of all of these factors, comparatively small temperature gradients are required to produce very gross global wind systems. As a result of the gravitationally structured atmosphere the vertical case is considerably different, with rather large temperature lapse rates the rule. A forced circulation results only if the lapse rates exceed the dry or moist adiabatic lapse rates for unsaturated and saturated cases, respectively. In the actual case, there is generally a considerable amount of turbulent energy available in the flow so that vertical motions may well become significant if the environmental lapse rate even approaches the dry or moist adiabatic lapse rate.

As was illustrated in Fig. 1.1, the troposphere and the stratosphere have decidedly different lapse rate structures in the vertical. The temperature lapse rate of the upper stratosphere is in almost all cases negative with a magnitude of about half the lapse rate observed in the troposphere. Exceptions to this rule are to be found only in the polar regions of the upper stratosphere. Here one may well find a very deep, nearly isothermal layer above the tropopause extending to well above the normal stratonull level before temperatures begin to increase with height. A typical case of this type is illustrated in Fig. 4.61 showing the

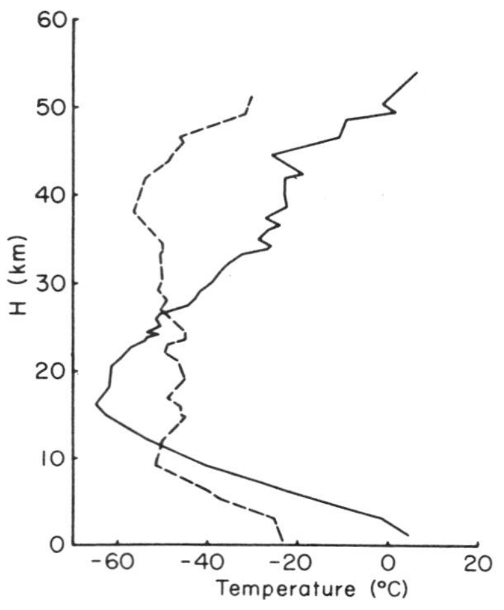

Fig. 4.61. Vertical temperature structure in the troposphere and stratosphere over Fort Greely (dashed) and White Sands Missile Range (solid) at 1017 AST, 18 November 1963 and 0730 MST, 19 November 1963, respectively.

very strong westerly circulation of the stratosphere that is typical of the early winter season and is directly dependent on a negative meridional temperature gradient. In this particular case of 18 November 1963, the negative meridional temperature gradient of the upper stratosphere begins at approximately 27 km, which is the level of the stratonull surface from these data for that date. The meridional temperature change between White Sands Missile Range (32 degrees latitude) and Fort Greely (64 degrees latitude) has its maximum at around 40-km altitude with some 32° difference at Fort Greely. Thus the meridional temperature gradient in this particular case has an average value from these data of —1°C per degree latitude. This meridional temperature gradient is essentially maintained in the 40— to 50-km altitude range in these particular data.

Now the thermal wind relationship, which is based on the geostrophic wind relation, states that a negative meridional temperature gradient will result in a westerly circulation, while a positive meridional temperature gradient will result in an easterly circulation. Thus, consideration of Fig. 4.61 indicates that westerly winds should dominate the troposphere, increasing in strength up to about 11 km with a positive wind lapse rate above that level, resulting from the positive meridional temperature gradient which is exhibited in the extreme upper troposphere and lower stratosphere. The easterly thermal wind which occurs between 11- and 27-km altitude is, in general, adequate to eliminate the westerlies of the lower troposphere so that the level of minimum wind between the tropospheric and stratospheric circulations is generally in the vicinity of 24-km altitude. These data would lead one to expect an approximate 40°C temperature difference between the equator and pole, which should, in the winter season, produce the strong westerly zonal winds that are characteristically a maximum during that early winter period. The largely isothermal region of the high-latitude data (Figs. 4.61 and 4.62) results from denial of solar input heating of ozone as the sun drops

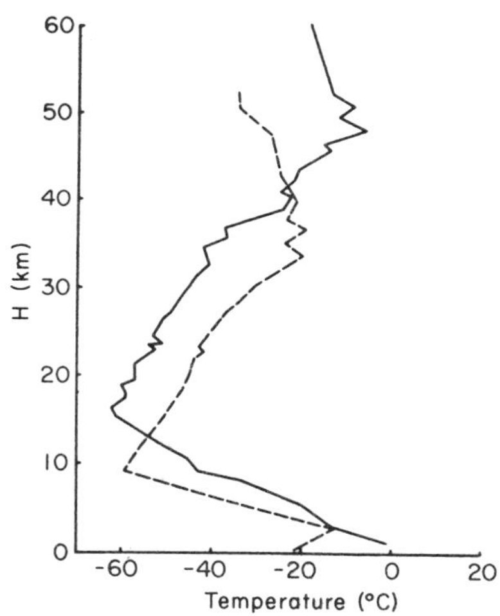

Fig. 4.62. Vertical temperature structure in the troposphere and stratosphere over Fort Greely (dashed) and White Sands Missile Range (solid) at 1141 AST, 13 January 1964 and 1100 MST, 13 January 1964, respectively.

below the horizon for the winter period. The temperature structure evidenced here for Fort Greely, then, builds upward from below as the westerly winds of the stratospheric winter circulation begin. By early December, the thermal structure illustrated here has reached its maximum amplitude and the westerly zonal circulation, which is largely circumpolar during this early winter season, has also reached rather high speeds, generally more than 100 meters/sec in the midlatitude region.

The data presented in Fig. 4.61 are not generally representative of the entire winter season since, beginning about mid-December, gross alterations occur in the simple circumpolar circulation, resulting in a comparatively complicated circulation pattern over the high latitudes of the winter hemisphere. Since there was no well-distinguished sudden warming event in the winter of 1963-1964, it is clear that the more normal winter storm period activities induced some rather profound changes in the upper stratospheric thermal structure. It must be remembered that, since the flow is not smoothly circumpolar in the winter circulation, during this period of the year one cannot use the temperature difference between White Sands Missile Range and Fort Greely at any given level as being representative of the general meridional temperature structure of the upper stratosphere. Even with these possible sources of error, however, it is obvious that considerable changes have occurred in that during the winter storm period over at least certain restricted areas a positive meridional temperature gradient must exist.

The reversal in meridional temperature gradient occurs at about 13 km in this case compared with 11 km in the November data. We see that the stratonull surface has been wiped out in these data, or at least, if we follow the definition we have advanced, it would appear at 42-km altitude. Near the stratopause the usual winter negative meridional temperature gradient is reestablished.

The rather large changes that are observed in the 35- to 45-km altitude range are noteworthy, in part produced by cooling at White Sands Missile Range principally at around 35-km altitude, with changes of the same order in the Fort Greely data. The tropopause is cooler by almost 10°C in January, while at 38-km altitude the temperature has increased between these two soundings by more than 30°C. Temperatures over Fort Greely at the stratopause remained essentially the same for these two soundings, although the altitude of the stratopause changed dramatically.

The 50-km level over Fort Greely in the November sounding was a full 30° cooler than midlatitude stratopause temperatures and the maximum temperature was located at about 50-km altitude. In January, however, the stratopause level, according to our general definition, has dropped to between 35- and 40-km altitude and has warmed to the point where it is only about 10° cooler than the midlatitude stratopause. It is from temperature data of this type that the inference has been drawn that many of the circulation events of importance in the stratosphere concern vertical motion. In particular, adiabatic heating due to compression resulting from downward motion of the general stratospheric environment can produce the results illustrated here; that is, the cold air of November at the 40-km level over Fort Greely will be heated along the lines indicated by the January data if subsidence occurs. This is not the only technique through which such a result can be obtained, however, but it has proved very difficult to find any other mechanism that might supply a significant portion of the required input heat. Even less correlation is to be found between the occurrence of such events and the variations in stratospheric circulation.

Since a large portion of the changes which result in the diminution of the westerly circulation of winter appeared to occur at the 35-km altitude level, the available data for Fort Greely, from December 1963 through December 1964, have been plotted in Fig. 4.63. From this it is

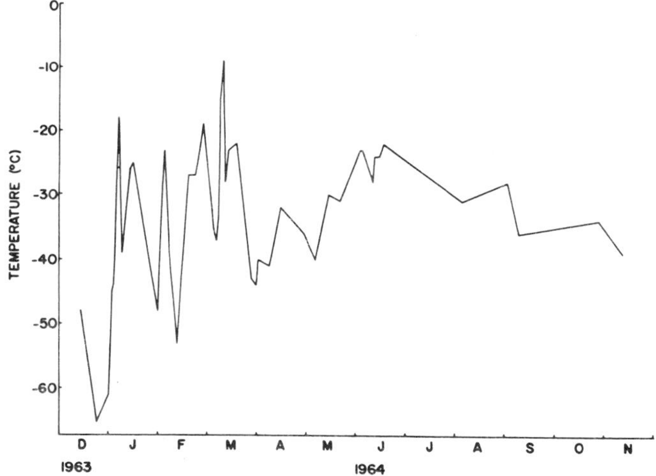

Fig. 4.63. Annual range of stratospheric temperature at the 35-km level over Fort Greely.

clear that the data presented here for the winter period occupy a very special place in time in that the stratospheric air over Fort Greely at these levels was occasionally significantly warmer than the general case for this period although it illustrates a large amount of variability; that is, temperatures at 40-km altitude in early January are illustrated by these data to have been warmer than —10°, which is greater than the observed difference in temperature at 40 km between White Sands Missile Range and Fort Greely on 13 January. Since temperatures in the region of the order of —50° are the rule during the early winter season, it is clear that conditions in the stratosphere are highly variable with the amplitudes of the swing of the order of 40°C occurring in cycles of a month or less. Thus in late January and mid-February we find temperatures over Fort Greely to be down to nearly —50°C.

A most striking aspect of the data presented in Fig. 4.63 is the very significant difference in short-period variability that is evidenced between the winter and summer seasons. Gross variations of temperature over a few days' time is the rule, at least from the beginning of the winter storm period until initiation of the spring reversal, after which the high-latitude temperature stabilizes and exhibits only minor variations during the summer season. These data point toward an annual cycle in temperature data at 35-km altitude in high latitudes which ranges from the minus fifties in winter to the order of —20°C in the summer. This is the variation induced by the seasonal aspect of the sun's incident radiation.

Warm periods in winter at the 35-km level over Fort Greely must result from circulation effects during the turbulent winter season. In part these temperature changes may be advective, since Fort Greely is located between the cold low near the pole and the ridge of warm air usually situated in middle latitudes. In addition, longitudinal variations produced by the lack of symmetry of the winter polar low would undoubtedly contribute to the variability observed in Fig. 4.63. Daily 10-mb charts provide evidence of rotation of the major axis of the low system back and forth across North America at much the same rate as these changes are observed to occur. Even so, it appears unlikely that these factors can produce the results presented on occasion in these data. Temperatures in the negative teens probably cannot be advected into the Fort Greely region at 35 km as a result of the lack of a source.

A remaining source of heating to produce the observed temperatures is downward motion. Many researchers attribute the temperature rises in the middle stratosphere during sudden warmings (Fig. 4.25) to descending motions as that phenomenon apparently moves poleward and downward. The strong meridional flows which have been measured by the MRN at all stations during the winter season indicate a considerable convergence of stratospheric mass at high latitudes. There is very little reason to expect this inflowing air to rise in polar regions in the absence of a significant heat source. A most likely circumstance is that it will descend, provided that the MRN data thus far available are representative of the hemispheric distribution of meridional wind speeds.

If descending motion does indeed represent the heat source, the data now available provide information on the mode of occurrence of these important circulations. Inspection of Fig. 4.63 indicates that temperature changes of as much as 10° per day must be accounted for. Assuming adiabatic conditions as a first approximation, this means that the sinking motions must be of the order of 1 km per day or more, since inaccuracies in our assumptions will require a greater heat source. This reduces to a local negative vertical velocity of approximately 1 cm/sec, which does not seem at all excessive. Heat loss processes will tend to reduce the temperature gain effected by any compressional heating, so in order that there be maintained a mean temperature above the radiationally established equilibrium values it is necessary that a continuous mean negative vertical velocity be a general characteristic of stratospheric high latitudes. The data of Fig. 4.63 indicate that the mean temperature of the latter part of the winter storm period is approximately 20° higher than is to be expected from radiational control. Several investigators have estimated radiational cooling of the polar winter stratosphere at several degrees per day, so combining these values we obtain a mean descending motion of the central polar stratosphere of the order of 3 cm/sec.

That is a highly significant value if the above approximations prove valid. While the MRN data are inadequate at this point to resolve the matter owing to gross data skips in the longitudinal coordinate, it is interesting to consider the fact that, since the adiabatic process is relatively invariant with height in the stratosphere, the influx of mass into the polar stratospheric region must have a peculiar vertical profile in order to produce the observed temperature profile changes. For instance, if all of the inflow were produced by meridional winds in the mesosphere which then descended through the polar stratosphere there would be an exponential decrease in this descending vertical wind speed with height. To obtain the observed heating in the lower stratosphere such a circulation system would require descending currents in the stratopause region in the meters per second range, which we can conclude do not exist even from the meager data now available.

The stratopause is a thermally defined surface which occupies a position similar to that of the earth's surface in that it forms the top of a stable layer and is overlaid by an atmosphere of at least conditional instability. The stratopause is a more diffuse boundary than the earth's surface, but when the increased scale of most atmospheric parameters is considered the analogy with the earth's surface may not be too crude. In any case, the stratopause represents a very stable feature of the thermal structure of the upper atmosphere. Only in winter polar regions does the definition become a bit slippery, with instances illustrated in Figs. 4.61 and 4.62 providing cases where the high-latitude stratopause is high (above 50 km) or low (37 km) according to the circulation system in effect.

MRN data are as yet insufficient to map the physical configuration of the stratopause on a global scale. By approximating the tropopause temperature with the temperature at the 50-km level we can get an impression of the variability in that region. The curves of Fig. 4.64 illustrate the annual course of 50-km level temperature at White Sands Missile Range (solid) and Fort Greely (dashed) for 1964. It must be remembered that the variations in these curves include both the temperature changes in a stable layer and advective changes of the vertical temperature structure. The latter are most pronounced at the winter pole, but may well be significant at all locations.

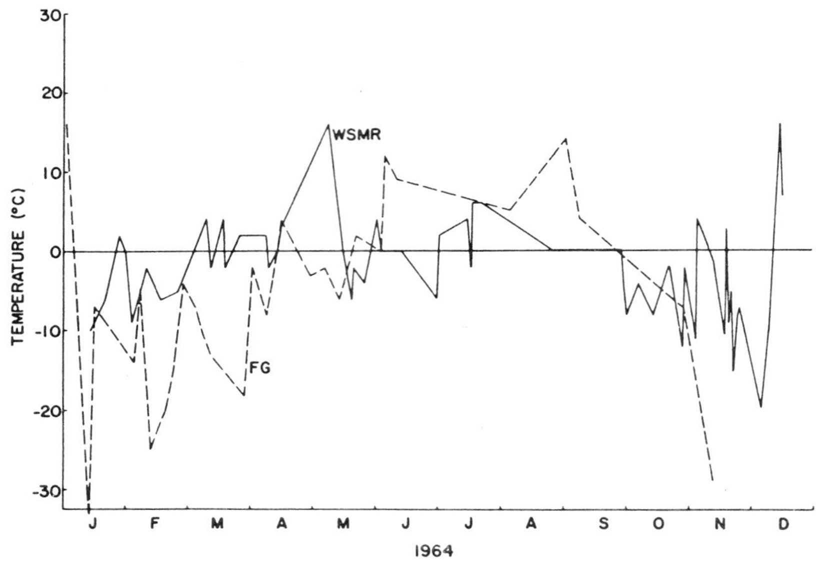

Fig. 4.64. Annual temperature variation at 50-km altitude over White Sands Missile Range and Fort Greely.

The data of Fig. 4.64 point out that there is summer warming and winter cooling of the 50-km level which is small at low latitudes and increases with increasing latitude. Peak-to-peak differences in mean values of the curves appear to be a few degrees at White Sands Missile Range and of the order of 20° at Fort Greely. It will be remembered that the circulation data of Section 4.1 imply a positive meridional height gradient in subtropical regions of the winter hemisphere during the winter storm period to support the easterly circulation invasion from the summer hemisphere. The data of Fig. 4.64 indicate an early recovery of the White Sands Missile Range temperature to nominal summer values after an early winter low.

A principal result of the temperature structure and variations discussed above is to restructure the vertical density distribution. Under static conditions the vertical pressure distribution is related to the vertical temperature structure by Eq. 1.2, which is a form of the hydrostatic relationship. Using the equation of state and an assumed composition it is possible to transform a temperature profile into a density profile. According to Thiele (1963) the density data thus obtained should be accurate to within 5% at all levels and to within 2% under average conditions in the stratosphere.

As could be predicted from the above relations, the cold winter stratosphere and its underlying troposphere shrink and produce a reduced density at each level. This reduced density has its maximum at high latitudes when the winter vortex has its maximum intensity. Warm temperatures resulting from continuous solar irradiation in the polar summer conversely expand the polar atmosphere so that constant-density surfaces are lifted, and the density at a particular level is thus increased. As was indicated in discussion of the temperature data, a maximum seasonal change occurs in the stratospheric region between the very stable tropopause and stratopause boundary surfaces. The data thus far indicate the level of maximum temperature change to be in the 30- to 40-km altitude range, with significant variations at all levels above the stratonull. Expansion of an atmospheric environment resulting from heating is a cumulative affair, with contributions from each lower elemental layer being additive to that at a particular level. Thus, the change in level of a particular pressure surface during a warming event will increase with height throughout the layer and evidence a maximum at the top. The amount of expansion contribution will show a maximum at the level of maximum temperature increase, of course.

These considerations point to the likelihood of minimum densities occurring in the winter season and maximum densities occurring in summer. A maximum in annual variation should also be observed in the density data for high latitudes and at high altitudes. These points are illustrated by Thiele's (1963) mean data for the fall and summer seasons at White Sands Missile Range and Fort Churchill in Fig. 4.65. In this case the seasonal breakdown was in the standard quarter-year periods, with the summer defined as June, July, and August. Maximum depression of the winter polar low occurs early in winter, and thus falls in the fall season of these data.

Annual variations in density of the stratosphere are demonstrated by these data to be most significant at high latitudes, with principal deviations in the negative direction during the winter seasons. At the stratopause altitude of 50 km the Fort Churchill range is approximately 0.5 gm per cubic meter and that for White Sands Missile Range is approximately 0.1 gm per cubic meter. The annual variation of mean density

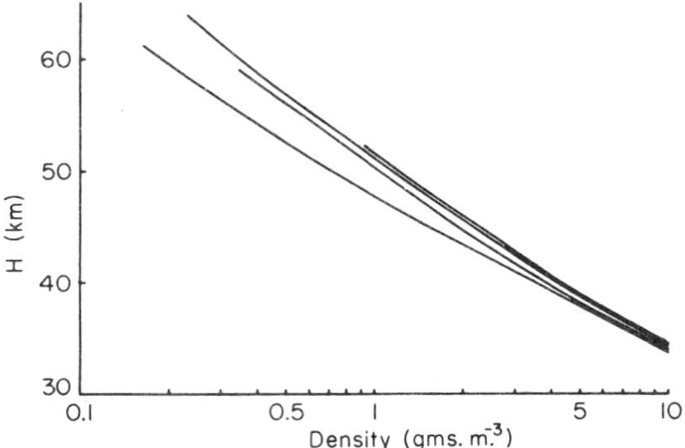

Fig. 4.65. Mean density distribution in the vertical over White Sands Missile Range (middle curves) and Fort Churchill (outer curves) for the fall (lower curves) and summer (upper curves) seasons. (Courtesy Journal of Applied Meteorology)

then increases from roughly 10% at 30 degrees latitude to roughly 50% at 60 degrees latitude. It should be remembered that these are mean values obtained from data taken over a three-month period, so rather large excursions beyond these conservative values must be expected. In view of the fact that ambient density has a strong bearing or effect on several atmospheric processes, it is obviously important to have a global picture of the stratospheric density structure. Quiroz (1961a) has shown

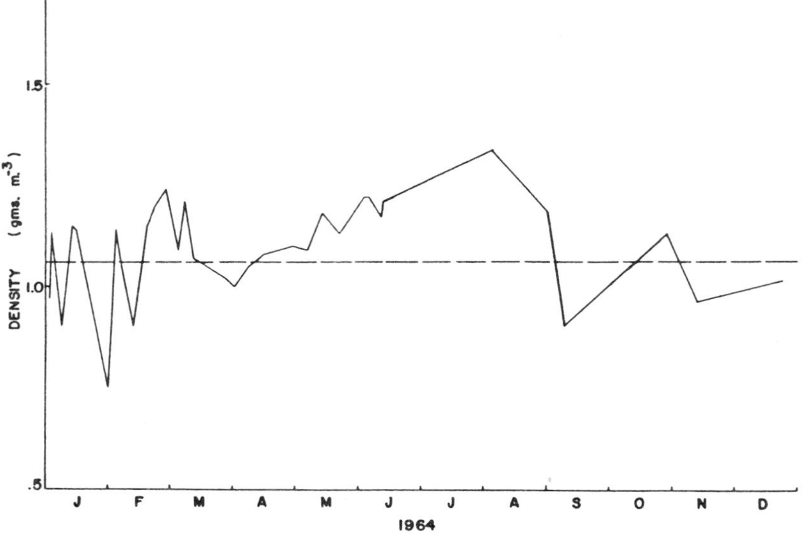

Fig. 4.66. Annual variation in atmospheric density over Fort Greely at 50-km altitude during 1964 as determined from MRN temperature-height soundings.

that an isopycnic layer appears in available stratospheric and mesospheric data at an altitude of approximately 80 km. Maximum amplitude of the annual oscillation apparently occurs immediately above the stratopause. MRN data on temperature and density are optimum at the approximate level of 50 km and thus the current data availability makes the stratopause an ideal place for inspection of density variations. Data on the annual variation for 1964 at 50-km altitude at Fort Greely are presented in Fig. 4.66. The density value prescribed by the 1962 U. S. Standard Atmosphere for the 50-km level (approximately 1.03 gm/meter^3) is illustrated by the dashed line.

As was indicated by the mean data of Fig. 4.65 for Fort Churchill, low densities are characteristic of the winter season, with lowest values during the early winter season when the circumpolar low is at its maximum strength. As can be noted from the 10-mb charts for both winter seasons, Fort Greely is in a protected position north of the persistent Pacific high-pressure region and thus is not generally representative of the density field in the polar low; that is, considerably lower values of density at 50 km are to be expected if observations should be taken within the low system. As could be predicted from the known temperature distribution at this level most of the short-term variability is observed during the winter season.

4.7 Detail Structure

The observational window which is thus far available to us for inspection of upper stratospheric circulation and structure is limited by applicable experimental techniques. The hemispheric circulation and very low Rossby number perturbations of that flow can be explored, over North America at least, by the current MRN distribution of stations and observational program of Monday, Wednesday, and Friday noon firings. Special features of the general monsoonal circulation can be analyzed by intensified observational schedules at certain stations if the phenomenon lends itself to such probing techniques. The middle range of atmospheric variability, characterized by tropospheric disturbances such as thunderstorms, standing waves, and hurricanes, falls into the latter category, which proves difficult to sample properly with the limited resources of the MRN. A third scale of variability in the upper stratosphere which lends itself to sampling with current MRN techniques is small-scale detailed structure with vertical dimensions of the order of a few tens of meters to a few hundreds of meters. In general we can probe directly only the vertical profile of these small-scale features, and thus there are many questions for which answers are not readily available in these data.

A most important result of the exploration of the upper atmosphere with small rocket vehicles and their special sensors has been the discernment that there is a large amount of detailed structure in the vertical profiles of all of the atmospheric parameters which have thus far been sampled with adequate sensitivity. There is also evidence that these small-scale variations exist in dimensions below the threshold of detection of the sensors at all upper stratospheric levels. It is evident, therefore, that these phenomena fall in size ranges which tax the sensitivities of the sensing systems. There is some reason to question the possibility that these detailed features are the result of errors in the observational and data processing systems since it is possible to demonstrate experimentally that such information can be introduced into the data by the sensors, or tracking, telemetry, and data reduction techniques. On the other hand, there is much evidence that these sources of error only contaminate the information contained in the profiles, and that they do not represent a major obstacle to analysis of the upper atmosphere for certain ranges of these small scales of variability.

An example of the type of problem encountered here is to be found in the evolution of parachute sounding of the stratosphere. Very minor problems were encountered in producing an effective wind sensor for routine use in the MRN which had characteristics entirely suitable for analysis of the general flow in the upper atmosphere. When the sensors

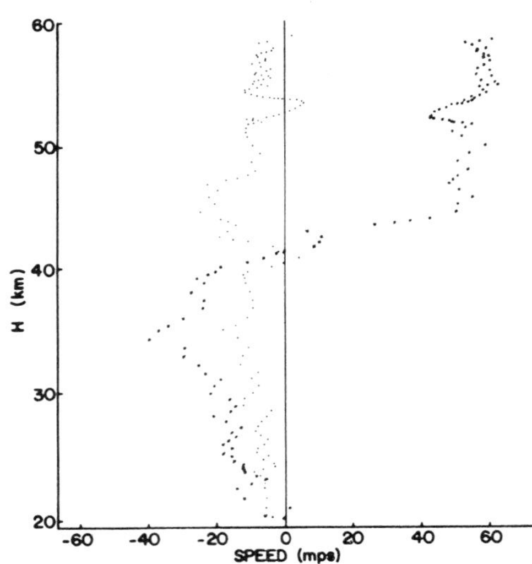

Fig. 4.67. An example of component wind profile data from the Arcas parachute wind sensor and sophisticated data acquisition and processing techniques. The observation was obtained at White Sands Missile Range on 7 February 1964 at 2200 MST. Dots represent meridional and crosses represent zonal data points.

were tracked with the most precise of available radars and the data subjected to the most applicable reduction techniques, wind profiles of the type illustrated in Fig. 4.67 were obtained. Input perturbations of such a parachute system are the spin of the rocket, which is known to be several cycles per second, and the coning motion of the rocket at the time of parachute ejection. The latter motions probably fall into the range of the observed oscillations, and elementary analysis of the motions of a simple pendulum of the system's dimensions point toward this input appearing in the data. However, once free of input sources the oscillations should decay in a systematic manner, and it is the failure to do this which points most forcefully toward the detailed structure indicated by the system being a characteristic of the environment.

It has been satisfactorily demonstrated that only a small fraction of the observed variability has cyclic inputs which can be expected to result from ejection events. Thus, the motions observed must largely be the result of some forcing function other than those characteristic of ejection. The possibility still remains that variable aerodynamic interactions between the parachute and the atmosphere cause torques on the system. This could happen through the partial collapse of the leading or downwind edge of the parachute as the system enters a new stratum of flow, which could then result in induced pendulum motions and impairment of the system's ability to sense the wind field accurately. These possible sources of error are being investigated with on-board camera systems and other measurements of the motions of the parachute as it falls. While the details of the parachute's wind sensing function are still not clear, it is relatively well-established that the observed variations in the wind profiles represent actual variations in the wind field.

A second analysis of this aspect of the upper atmosphere's actual detail structure came with the development of a sensitive temperature-sensing capability. The 10-mil-diameter ceramic bead thermistor employed in this effort is somewhat inadequate in response at the mesospheric levels and is surely smoothing the data there, but in the stratospheric region where the sensitivity of the bead and the reduced fall rate of the parachute permit short-period response a complex-detail vertical structure is clearly apparent. This is illustrated in the example of Fig. 4.68. The possibility that the variations were produced by alternate exposure to the sun and the shade of the parachute was evaluated and ruled out, as were several other possible sources of error.

A third upper-atmospheric measuring system of considerable sensitivity which has measured much detail structure in vertical profiles is the Regener ozonesonde. Data obtained by this instrument on balloon platforms in the lower stratosphere illustrate a large amount of small-scale

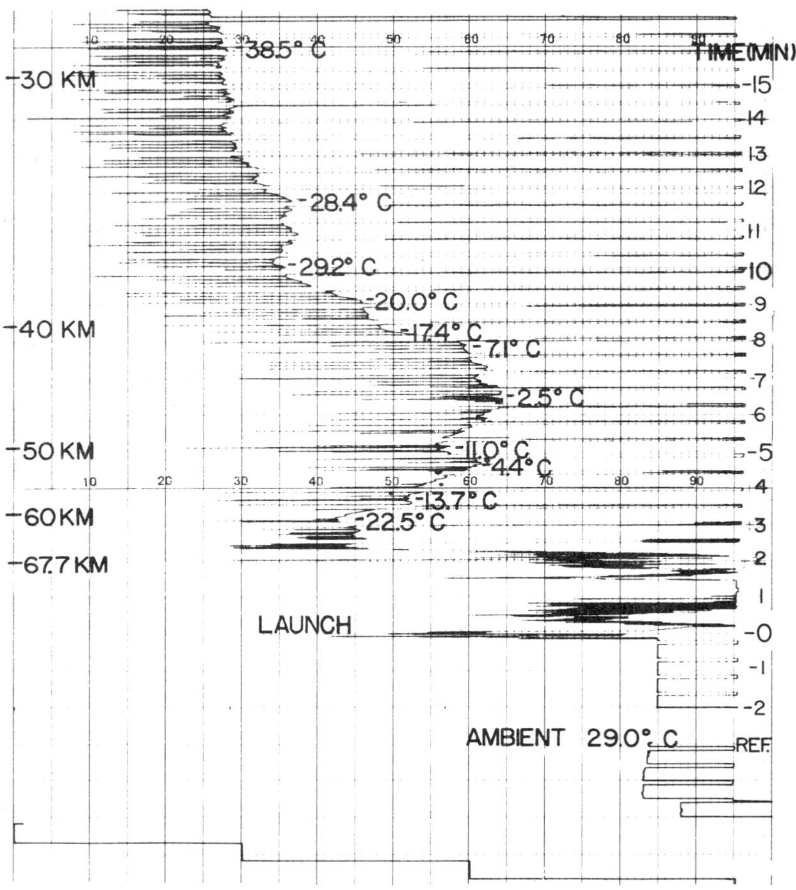

Fig. 4.68. A relatively typical recording of vertical temperature structure obtained with the 10-mil-diameter bead thermistor as it is supported by the Arcas parachute, telemetered by an Arcasonde, and recorded by the GMD-1 meteorological system. This sounding was obtained at 0300 GTM, 30 June 1965, at White Sands Missile Range.

detail on the larger features of variability which can be attributed to advective motions. These detail structures fade away at the highest altitudes reached by the balloon systems, but it is probable that sensitivity considerations enter in here also. An example of these data is presented in Fig. 1.3.

One of the upper-atmospheric sensors which can be expected to have a most nearly symmetrical response relative to any interaction with the atmosphere is the Robin sphere, which has been developed by the

Air Force Cambridge Research Laboratories. The Robin is a pressurized rigid balloon with an enclosed radar reflector which is designed to measure density through radar observation of the fall rate, at the same time providing for observation of the wind field. Except for errors in the tracking system and in the data reduction techniques, the data obtained with this sensor should provide a true representation of the wind profile, except possibly to smooth the stronger gradients owing to lack of sensitivity. Two special series of Robin soundings were accomplished at Eglin Missile Test Range, resulting in the acquisition of 11 profiles on 10 May 1961 and 16 profiles on 12 October 1962. An example of the type of data contained in these profiles is illustrated in Fig. 4.69.

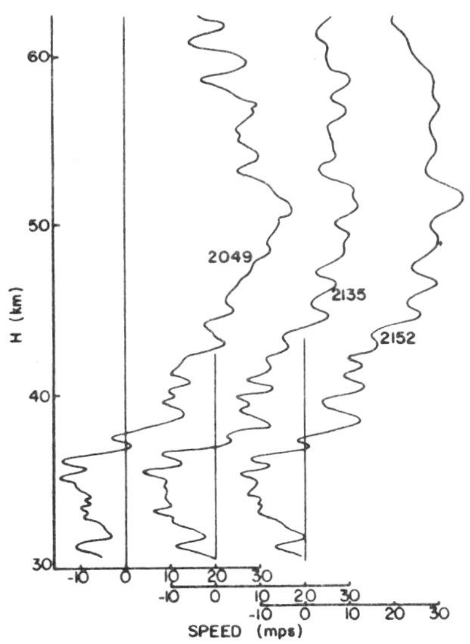

Fig. 4.69. Example of zonal wind profiles obtained by the United States Air Force at Eglin Missile Test Range on 12 October 1962 at the times indicated with the Robin sphere, AN/FPS-16 radars, and automatic data recording and processing systems.

There are certain similarities to be noted in the profiles presented here. In general, however, the detail features change from one profile to the next. Time spacing between the observations in Fig. 4.69 were 46 and 17 minutes, respectively. These data indicate that the time variability of the generating phenomenon must be measured in minutes or less, or that the horizontal scale of these features must be measured in hundreds of meters or less. The former description is applicable if the generator is a transient phenomenon such as a gravity wave, and in this case the field of these detail structures can be expected to be very fluid and variable. If these detailed structures should be static characteristics of the environment, then the variable observational data are simply the result of varying sample aspect as these inhomogeneities drift past in the general flow. Resolution of these questions is of basic importance in understanding the dynamics of the upper atmosphere.

These data were analyzed for the characteristics of the detail structure by simply measuring the vertical distance between peaks in the zonal wind profile and categorizing these data by the mean altitude at which they occurred (Webb, 1965). A total of 244 data points were obtained from the May series for an average of 22 per sounding. In October there were a total of 412 points for an average of 26 per sounding. These data and a linear least-squares fit are illustrated in

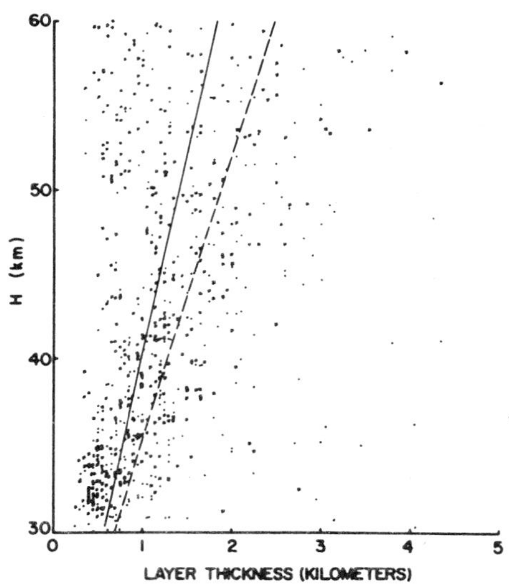

Fig. 4.70. Vertical scale of stratospheric detail structure as a function of altitude. Data are represented by dots and a dashed line in the summer case of 10 May 1961 and by crosses and a solid line in the winter case of 12 October 1962.

Fig. 4.70 for each of the sets of data, the dots and dashed line indicating the May case and the crosses and solid line indicating the October case.

These data indicate that the scale of the detail structure increases with height in the upper stratosphere during both the summer and winter seasons. In the summer data the vertical dimensions of these features average about 700 meters at an altitude of 30 km, and increase relatively uniformly to more than 2 km at an altitude of 60 km. The detail structure of the winter season appears smaller, ranging from approximately 600 meters at 30 km to about 1.8 km at 60-km altitude. A most important difference in the two sets of data concerns the small-scale features that appear in the upper portions of the October data. These data imply that there is a source for such features in the upper stratosphere or in the mesosphere which is peculiar to the winter season. The winter westerly winds which are strong in October have their maximum speeds near the stratopause, or at approximately the 55-km level, and above that level the wind speed decreases with height, with shears of considerable magnitude the rule in the lower mesospheric region. The temperature also decreases with height, and the combination of these two positive lapse rates could well be the factor which, through a turbulent exchange mechanism, converts energy from the general circulation into local heating and associated effects with a resulting production of detail features in the vertical profiles.

In view of the considerable stability of the upper stratosphere there is a distinct possibility that the detailed structures of the vertical profiles do not represent turbulent motions of a tropospheric character at all, but are manifestations in the wind field of internal motions in the medium which are wave motions controlled principally by gravitational effects. Such waves are common in fluids when the flow is forced past an obstruction and have their best known tropospheric analogy in the standing lee waves observed when strong wind systems operate in the vicinity of mountain ranges. At the upper end of the atmospheric profile we have information on the fact that such rather stable wave motions exist in the upper atmosphere in the form of well-defined waves in noctilucent clouds and in the deformations in meteor trails in the lower ionosphere which appear to result from the wind field. All of these data suggest a wide range of sizes of these detail features at all levels, although the trend toward larger sizes with increasing height is evident in all of the data. If these features of the vertical wind profiles should be the result of propagating wave energy through the atmosphere, it is well known that the amplitude of such features should increase rapidly with height owing to the rapid decrease in atmospheric density. In fact, the variation should be roughly exponential with height. To test this factor, data points have been taken from the least-squares fits of Fig. 4.70 at 35- and 55-km altitudes and plotted on a semilogarithmic scale. A straight line has been drawn through the data and extended to the surface and into the ionosphere in Fig. 4.71.

A mean value of vertical scale of small-scale features was obtained by Greenhow and Neufeld (1959) as is indicated in Fig. 4.71 through

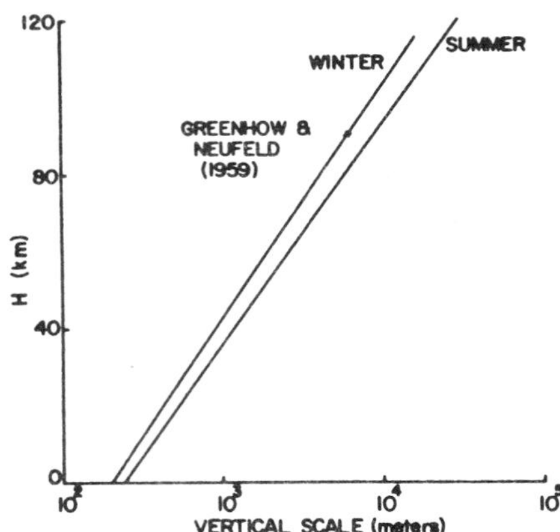

Fig. 4.71. Variation of vertical dimensions of detail structure with altitude. Data at 35 and 55 km were taken from the least-squares fit of Fig. 4.70 to establish these exponential relations.

radar observation of the meteor trails generated in the lower ionosphere of 6 km, with a mean altitude of occurrence at 90 km. Their data are in agreement with winter data obtained from the stratospheric Robin soundings. On extrapolation to the surface we obtain values of approximately 250 meters for the scale of vertical details in the wind profiles. The summer data in the stratosphere, when extrapolated into the lower ionosphere, show a steady increase over the mean winter value with increasing height. The surface value for the summer case appears to be about 20% greater than that indicated for the winter season.

It is quite possible that these data offer a measure of turbulence in the tropospheric sense and thus that the increase noted above the stratopause in the small-scale detail structure concerns the increased instability which is expected to exist in the mesosphere. If so, it is only coincidental that the winter stratospheric data did fit with the ionospheric data. In that case the winter curve of detail structure would not be a straight line on the semilogarithmic plot, and increased resolution would show minimums between 20- and 30-km altitude and again between 60 and 70 km, where the zonal wind profiles exhibit minimums.

It seems far more likely that both processes are going on at the same time. Thus, body waves set up by turbulent motions over rough terrain, about local circulations such as clouds, and as a result of local instabilities can be expected to make their influence felt at remote points in the atmosphere, and if they propagate upward their amplitude will necessarily experience the growth illustrated in Fig. 4.71. On the other hand, it is unrealistic to assume that turbulent overturning of the free atmosphere does not occur when the possibility of dynamic instability exists. The mesosphere is a region of the atmosphere where the likelihood of instability is obvious, and thus it is no real surprise to observe such a result.

It is pertinent to ask why, if waves postulated in earlier paragraphs are so prolific, they are not more obvious to us in the troposphere. From the meteorological point of view, this is principally because available sensors lack the sensitivity in those regions where such motions might be identified. The lee waves downwind from mountain ranges are relatively well understood, but the occurrence of lenticular-type clouds under other conditions and on frequent occasions points to the likelihood that the motions which produce these clouds are quite common in the troposphere. The scale of the motions which we must look for in the troposphere is such that the information is very difficult to extract from radiosonde data which fail to resolve details in the 100-meter range. We have in recent years inadvertently acquired a new sensor which is better designed for the purpose: that sensor is the high-speed aircraft.

Thus far we have spoken only of the vertical scale of the small-scale features of atmospheric profiles. Experience, along with other considerations, tells us to expect the horizontal scales of these phenomena to be larger, generally by an order of magnitude or more. The horizontal extent of these perturbations will then be of the order of a few kilometers in the troposphere, and the traversal of several cycles of such a phenomenon at rates which cause the input forces on the vehicle to resonate can have significant effects on the flight path and the structure of the aircraft. In order for the effect to be significant, there must be just the right combination of factors; that is, the aircraft must traverse the region occupied by the waves at a particular combination of speed, azimuth, and elevation angle before results of the interaction are noticeable. One can expect that there will be marked variability in the reports of such phenomena as a result of the above factors as well as the variability with time. We have described above the atmospheric interaction with high-speed aircraft which has come to be commonly known as "clear air turbulence" (CAT) (Reiter, 1962, 1963).

The turbulence which is well known to pilots of relatively low-speed aircraft operating in the troposphere results from variations in the wind vector which, through aerodynamic interaction, cause the aircraft to be accelerated in an uncontrolled fashion. If the input perturbations are random the probability of structural failure is very small, even with the strongest gusts which are encountered. If the inputs to the airframe should become cyclic at the resonant frequency of the structure, the amplitude of the perturbations required to endanger the aircraft becomes smaller as the number of repetitive cycles encountered increases. It fast becomes impossible to economically engineer a reasonable safety factor into the vehicle if one has to allow for a large number of cycles of this input.

4.8 Diurnal Variations

Because the earth's rotation introduces a diurnal variation in many atmospheric variables it is to be expected that a region tied so closely to the solar radiant flux as is the stratosphere will exhibit pronounced symptoms with a 24-hour period. About the only possibility that this would not occur is if the time constants of response are very long with respect to the diurnal period. On theoretical grounds there is every reason to expect reaction rates to be measured in seconds or less at the stratopause, so the lag in an equilibrium adjustment should be very small. On the other hand, the lifetime of an ozone molecule may be measured in years at the stratonull level, so surely a diurnal variation resulting from ozone absorption of solar radiant energy at that level would be quite negligible. Since ozone is at least a major contributor to the heat exchange processes which result in the negative lapse rate of the stratosphere, it is probable that its response to the solar radiant flux will be a good indicator of the atmosphere's response in this region.

These considerations point toward an increasing response of the stratosphere to a diurnal heat input with increasing altitude. The very slow response of ozone generating and destroying processes at low levels in the stratosphere is a direct measure of the available radiant flux for heating of the local region, and since that source is negligible the only likely alternate source of heat is conduction from adjacent regions. The convective transport of heat in the very stable stratosphere should be very small, so that there would be small likelihood of significant transport from this mechanism which is so very efficient in the troposphere. Diffusion processes, both molecular and eddy, will result in a heat flow from the warm stratopause downward into the lower portions of the stratosphere, but the time constants for significant relaxation of the general gradient appear large relative to a diurnal period and are thus probably of small significance to this problem.

A most important factor in the establishment of a thermal equilibrium profile in the stratosphere is the gross density variation with height. The heat capacity of a unit volume of air varies directly with the density, and thus becomes rapidly less with increasing altitude. A uniform heat input would result in a negative lapse rate under these conditions, other factors being negligible. In addition, the quantity of solar ultraviolet energy available is steadily depleted with decreasing altitude so that little or none is available at low altitudes for interaction with the atmosphere. Rather than a uniform heat input with height, there is actually a strong positive gradient in this parameter with height in the upper stratosphere. All factors combine then to indicate that if there is a diurnal variation to be observed in the stratosphere it will have a greater amplitude in the upper stratosphere than in the lower stratosphere.

While the conduction of heat is very probably negligible over the gross dimensions of the stratosphere, it is a fundamental process that will have a certain efficiency in transporting heat from the warm upper stratosphere into the cold lower stratosphere.

Our knowledge of the vertical *Austausch* coefficient in the stratosphere is inadequate at this time, but we can consider the flow by molecular diffusion, which for the specific heat flow through a unit area is given by

$$\frac{dh}{dt} = Ds\gamma, \tag{4.1}$$

where D is the diffusion coefficient, s the specific heat, and γ the lapse rate. The diffusion coefficient is directly related to the mean free path of the medium, and the mean free path increases rapidly with height in the stratosphere. The diffusion coefficient also varies with the temperature, increasing as the temperature increases, with the result that the gradient in magnitude of the molecular diffusion coefficient should be pronounced in the upper stratosphere.

This factor will reduce the degree to which the adiabatic assumption is applicable. It will dampen the diurnal oscillation by conducting heat away from the region of maximum temperature, a process which takes time and will thus introduce lag into the characteristics of the diurnal oscillation that results from combined direct ultraviolet absorption of this transported heat. There is, then, reason to expect a decrease in amplitude with decreasing altitude of any diurnal variation and for an increasing lag of the cycle with decreasing altitude. Variations in the phase of the diurnal oscillation resulting from heat input into the stratosphere should provide a measure of the vertical distribution of this heat transfer mechanism.

Above the stratopause a decidedly different mechanism will control the transfer of heat in the vertical. The mesosphere is thermally conditionally stable in general, but it is very likely that the vertical motions which are assumed to produce detail structure in the stable stratosphere must precipitate considerable vertical motion in the mesosphere. Convective overturning is probably a characteristic feature of the mesosphere, and the resultant mixing will cause a heat flux upward in the same way such a flow is generated in the troposphere. This mechanism for adjusting the asymmetrical vertical heat input distribution is probably highly efficient, with a large eddy diffusion coefficient resulting in rapid adjustment of the upper portion of the heat sinks which lie above and beneath the mesosphere. It is obvious from the previous discussion of molecular diffusion that this factor will also exhibit a high efficiency in the mesosphere. Eddy diffusion can be expected to control the flow upward during the period of heat gain at the stratopause when local gradients are likely to produce instabilities, while molecular diffusion will control the heat transport during other portions of the diurnal variation.

This picture of the vertical structure of heat transport mechanism has been drawn under rather static assumptions; that is, we have discussed the heat exchange processes in much the same way which one applies to the troposphere, where the earth's surface affords a rather stable boundary. This is apt to be a poor approximation in the stratosphere, where mobility is a principal characteristic. One must expect the heated air to expand, and again using static relationships, order-of-magnitude estimates of the vertical expansion and contractions between pressure surfaces associated with the diurnal heat oscillation can be obtained through the relation (Panofsky, 1956)

$$\Delta h = \frac{R\Delta \bar{T}_v}{980} \ln \frac{p}{p'}, \tag{4.2}$$

where R is the universal gas constant and ΔT_v the change in mean virtual temperature. In the case where a 10-km-thick layer centered at 50 km experiences a virtual temperature rise of 10°C, an expansion in the vertical of approximately 360 meters will occur unless compensating effects change the situation. Part of the input heat will thus go into the work of expanding the gas rather than into raising the temperature. Since this occurs in a strong positive density lapse rate, a major portion of this energy flux will be directed upward.

Again, it must be remembered that these effects do not occur in a static medium. Winds are normally quite strong at the stratopause, and it is obvious that any expansion of the environment will necessarily result in an adjustment in the circulation. These adjustments will have a profound effect on the final distribution of all dynamic and thermodynamic parameters. They are boundary conditions, and their exact configuration is probably the least known of the several relevant factors. We can, however, describe the manner in which these variations will occur relative to the diurnal pattern. For instance, the disturbance will propagate westward around the earth near the stratopause at velocities ranging from zero in polar regions to almost one-half a kilometer per second in equatorial regions. In the simple case cited above, using mean parametric values, vertical flows of at least 1.7 cm/sec are indicated.

This ridge of high pressure will provide an acceleration of the general circulation. Since the ridge stretches from high latitudes in the Southern Hemisphere to high latitudes in the Northern Hemisphere, the zonal winds will be diverted in the poleward direction as the ridge approaches in the morning and will flow from high latitude toward the equator in the evening as the diurnal pressure wave recedes over the horizon. Our above estimate of the mean vertical wind resulting from the heat expansion was based on a uniform gradient over a six-hour period, and is undoubtedly lower than the maximum to be observed. For instance, if a major portion of the heating occurred in a two-hour period, vertical velocities would be of the order of 5 cm/sec. In this case accelerations could easily be above 10^{-2} cm/sec^2, possibly in the 0.1 order of magnitude.

In addition to the nongeostrophic aspects of the circulation about this fast moving pressure wave, interesting effects may develop as a result of the special features of the vertical wind structure. A rather general wind profile in the winter season is characterized by a strong negative wind speed lapse rate in the upper stratosphere, a maximum wind speed in the 100- to 200-meter/sec range at approximately the height of the stratopause, and a positive wind speed lapse rate in the lower mesosphere. If the lifting due to heating is confined to a relatively thin layer of the ozonosphere around the level of the stratopause, inspection of the effect of the expansion on these profiles before and after would indicate that the profile would be bodily lifted, in addition to other changes. Thus, in the lower portion a lifting of the wind profile would cause the wind speed to become lower at a particular level, while in the upper portion of the scene the reverse is true, and the wind speed should be observed to increase during the lifting process. These considerations provide us with certain check points which may be emphasized in our analysis of the diurnal oscillations which solar heating may introduce into the stratospheric circulation.

Studies of fluid motions which are related to the earth's rotational and orbital program were initiated in analysis of sea level responses to the several disturbing factors. The moon's gravitational effect proved to be the dominant celestial factor in the complex diurnal and semidiurnal variations of the ocean surfaces. These periodic variations of diurnal order became known as tides in the common language, and time has clouded the origin of this term until today it is frequently used to designate any long period motion, be it in earth, ocean, or atmosphere. There is really no objection to this, since the term is actually addressed to the motions themselves, and the origin of the perturbations is quite another subject. As is currently the usage in the literature, we will speak of tidal motions of the atmosphere.

While there is still some confusion as to what we do mean by the expression "atmospheric tides," it is fairly clear that gravitational effects are negligible, at least insofar as an outside body is concerned. The air is heated and cooled alternately, and the resulting expansion and contraction involves gravitational energy exchange in the geopotential field. Control of this function does not rest with gravitational considerations, however, but is guided by the aspect of the radiant flux, which varies with the earth's rotation. As is common with transport of energy through radiation transfer processes, there is a lag in the occurrence of peak amplitude in the response which is not unlike the gravitationally induced response which is lagged by inertial effects. One of the most obvious differences between these two mechanisms is the fact that a semidiurnal oscillation is a principal result of the gravitational mode of tidal impulse, while the diurnal can be expected to dominate the radiationally induced oscillations. It is clear, however, from the discussion in the early part of this section that tidal oscillations induced by heat disposition also are not simple.

Numerous attempts have been made to observe and analyze atmospheric variations that are analogous to the oceanic tidal motions. Conflicting results have been obtained, principally because the data have been obtained in the lower atmosphere, where many complicating factors cloud the picture. Radio studies of ionospheric meteor trail drifts and structural motions first presented data on the diurnal oscillations characteristic of that region of the atmosphere. The advent of meteorological rocket vehicles prepared the way for a systematic analysis of short-period fluctuations in the wind field and also made possible a parallel observation of variations in the temperature structure. A very excellent series of soundings was obtained at White Sands Missile Range on 7–9 February 1964 when wind profiles were obtained at two-hour intervals for a consecutive series of 13 observations. The wind measurements were made by AN/FPS-16 radar tracking of the 4-meter-diameter Arcas parachute system described in Chapter 2. Temperatures were obtained from a 10-mil-diameter bead thermistor with the Delta flight unit which telemeters the data into the standard AN/GMD-1 meteorological sounding system ground station. A second series of wind profiles, using both Arcas and Judi systems, was obtained on a similar schedule on 21–22 November 1964.

An example of the type of data obtained during the February series of observations is illustrated in Fig. 4.72. Examination of Fig. 4.4 indicates

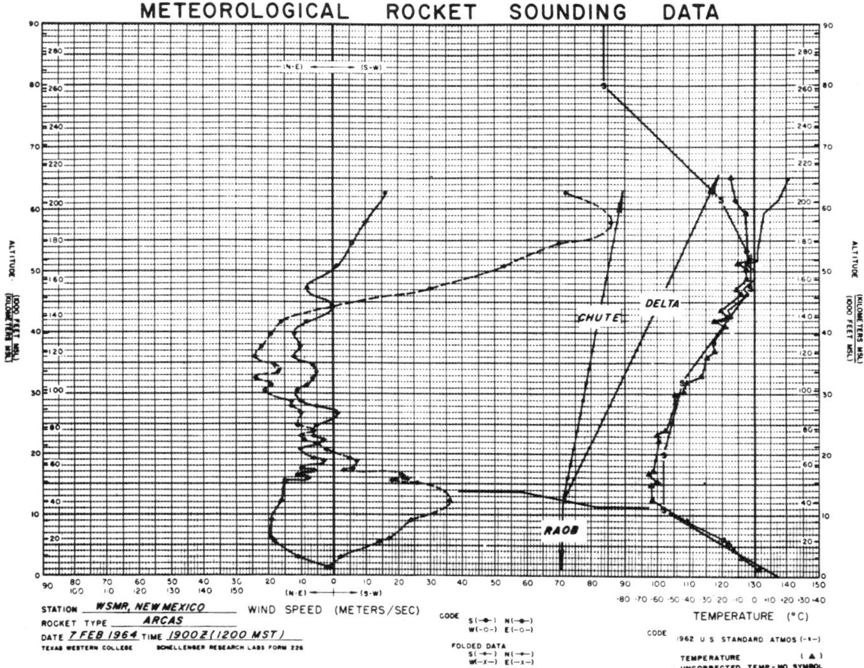

Fig. 4.72. A typical example of the component wind profiles obtained during the 7–9 February 1964 diurnal sample.

that the stratospheric circulation was in a relatively typical pattern of the late winter season. Miers (1965) has analyzed these data for diurnal-type variations and has concluded that the winds veer clockwise for the period of the observations, completing a cycle in 24 hours and exhibiting a diurnal period.

Miers (1965) obtained the diurnal distribution of zonal wind speeds presented in Fig. 4.73 and the meridional variations illustrated in Fig. 4.74 through statistical analysis of the experimental data for 7–9 February 1964. (See Table 4.3.) There is a strong diurnal cycle in the data, with a maximum of 26 meters/sec at an altitude of 45 km and a secondary maximum of 24 meters/sec at an altitude of 58 km. It is striking that these two layers of gross variation in the zonal wind should be so effectively separated, and even more so that they should be so thoroughly out of phase. Peak winds in the tidally perturbed layer above the stratopause occur at noon local time. Beneath the stratopause the maximum is at midnight, or just a half cycle out of phase. The division would appear to be just above the 50-km level according to the data of Fig. 4.73, and thus is at the immediate level of the stratopause.

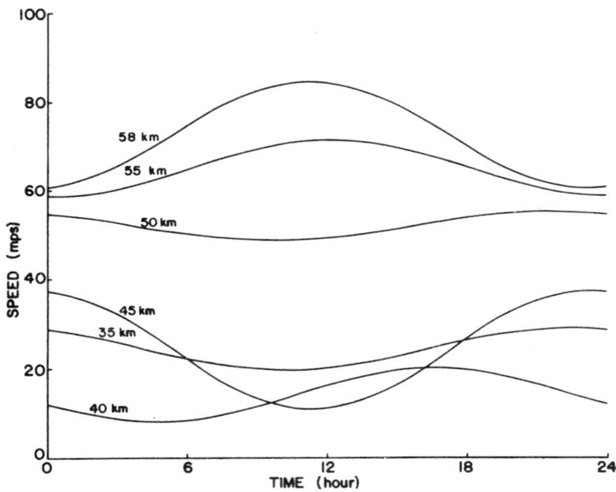

Fig. 4.73. Diurnal variation in the zonal wind speed over White Sands Missile Range at various levels in the stratospheric circulation on 7–8 February 1964.

TABLE 4.3

Coefficients of Equations of Component Wind Variability[a]

Altitude (km)	A_u	B_u	C_u	A_v	B_v	C_v
	Diurnal: $u = A_u + B_u \sin \omega_0 t + C_u \cos \omega_0 t$					
35	−24.29	4.65	−0.45	−13.77	1.84	1.63
40	−14.11	1.08	5.86	−11.07	−2.26	0.61
45	24.22	−12.31	4.79	−12.05	4.34	−9.03
50	51.93	−3.11	−0.15	−5.51	1.94	−0.06
55	64.97	5.34	−3.37	−1.30	5.47	−0.39
58	72.45	11.32	−3.96	5.76	0.48	0.78
	Semidiurnal: $u = A_u + B_u \sin 2\omega t + C_u \cos 2\omega t$					
35	−24.29	−0.24	2.51	−13.77	−0.78	+0.23
40	−14.11	1.32	1.95	−11.07	0.83	0.48
45	24.22	1.13	4.92	−12.05	−0.38	2.05
50	51.93	0.77	1.34	−5.51	−2.58	2.01
55	64.97	3.05	0.33	−1.30	−0.88	1.57
58	72.45	3.88	0.82	5.76	−5.90	1.70

[a] Obtained by Miers (1965) through harmonic analysis of meteorological rocket data for White Sands Missile Range on 7–9 February 1964. Speeds are in meters per second when the equations have the form shown.

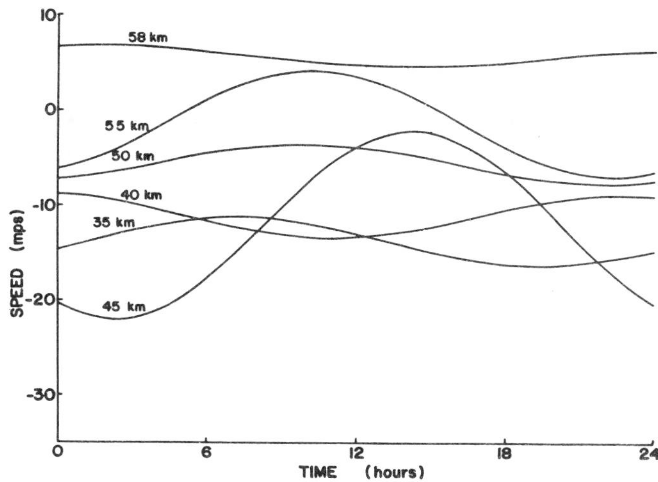

Fig. 4.74. Diurnal variation in the meridional wind speed over White Sands Missile Range at various levels in the stratospheric circulation on 7–8 February 1964.

These data then picture diurnal variations of the zonal wind to be characterized by strong daytime and weaker nighttime speeds above the stratopause and by weaker daytime and stronger nighttime speeds below the stratopause. Such a distribution is to be expected if the disturbance is a thermally incited vertical expansion of the upper stratosphere which is confined to a thin stratum located just below the stratopause. The vertical divergence associated with this increase in volume will result in a decrease in wind speed in the expanded layer. If higher levels of the atmosphere should resist this upward expansion, convergence will result and the zonal flow will be accelerated as is indicated by the data for 55 and 58 km.

The data used here are limited in vertical extent, so details of the diurnal wind structure above the stratopause are incomplete, and are absent at higher altitudes. The data are relatively complete at the lower boundary of the diurnal cycle, however, and it is clear that there is a very sharp gradient in the diurnal oscillation. As was pointed out by Miers, the phase of the diurnal oscillation shifts rapidly and the amplitude decreases rapidly with decreasing altitude. At 35-km altitude the diurnal amplitude is less than 10 meters/sec in these data and becomes indistinguishable in this sequence of soundings at lower altitudes. It seems likely that the region of enhanced speeds above the stratopause extends throughout the mesosphere and is capped by the stable negative lapse

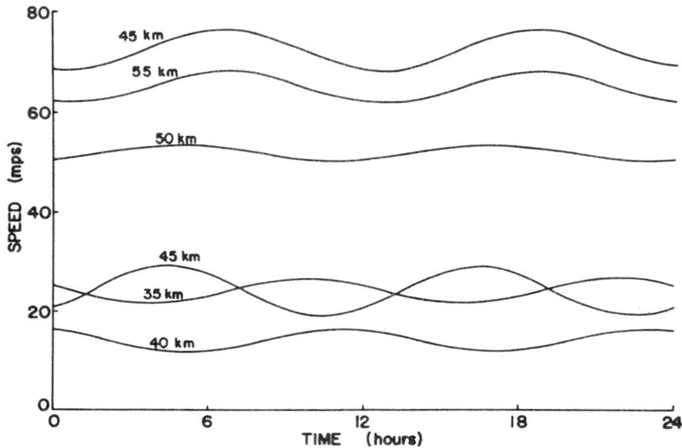

Fig. 4.75. Semidiurnal variation in the zonal wind speed over White Sands Missile Range at various levels in the stratospheric circulation on 7–8 February 1964.

rate of the ionosphere. If the above assumptions are correct, the amplitude should increase with altitude up to near the mesopause level.

Inspection of meridional aspects of diurnal oscillations also shows strong peaks at 45-km and at 55-km altitude, with only a slight difference in phase. Amplitude of the lower oscillation is double that of the upper one, and curiously enough it exhibits a phase lag of approximately 60 degrees. The separating surface at the stratopause is well-defined, as are the lower boundary shifts in phase and decreases in amplitude. A notable difference exists in the 58-km data, however, where little or no diurnal oscillation is evidenced, in contrast to the maximum which was observed in the zonal data. The meridional winds associated with this expansion due to solar heating are more southerly both above and below the stratopause, which, for this location, implies the approach of the enhanced thickness region advancing from the east.

Various estimates of atmospheric tidal motions have emphasized the importance of semidiurnal oscillations. Miers' analysis points toward the diurnal tidal oscillation being definitely stronger than the semidiurnal component, at least in this sample. For instance, at 45 km the semidiurnal component has an amplitude of 10 meters/sec while the diurnal oscillation at the same level is more than twice as great. The same phenomenon of a minimum in amplitude at the stratopause is apparent in the semidiurnal analysis illustrated in Figs. 4.75 and 4.76. A decrease in amplitude and marked shifts in phase are also apparent in the 35- and

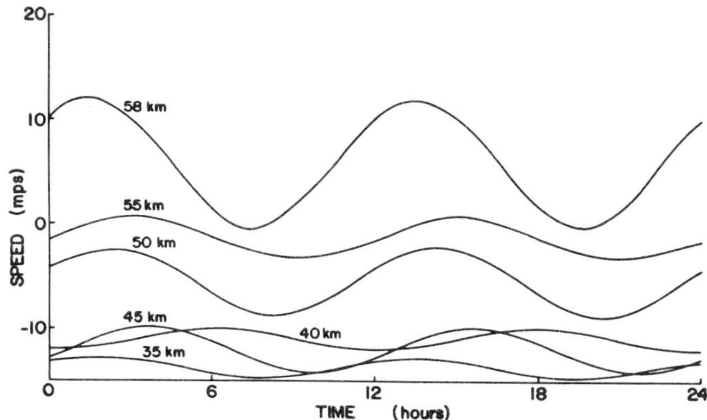

Fig. 4.76. Semidiurnal variation in the meridional wind speed over White Sands Missile Range at various levels in the stratospheric circulation on 7–8 February 1964.

40-kilometer region, which forms a lower boundary of the region in which tidal motions dominate.

At all levels, peaks in the zonal winds are found near 6 AM and 6 PM, except in the lower boundary layer. The higher levels evidence diurnal oscillations with amplitudes a few per cent smaller than at the 45-km level. In the meridional data of Fig. 4.76, however, amplitude of the semidiurnal data shows an increase with height, exceeding 12 meters/sec at the 58-km level. The amplitude of the meridional semidiurnal oscillation is thus in excess of the similar diurnal component at that level.

Thirteen sets of temperature data obtained during the 7–9 February 1964 series of firings have been analyzed by Beyers and Miers (1965) to produce the solid curve in Fig. 4.77. The temperature difference indicated here represents the difference at each level between a mean daytime and a mean nighttime curve, which were obtained by averaging the 1400 and 1600 temperature profiles and the 0400 and 0600 LST profiles. Very little difference is noted in this diurnal temperature range at the 35-km level, but a steady increase is noted above that level, with a maximum of almost 20° difference in the 50- to 55-km altitude layer. The data are not adequate to discern the variation at higher levels, so the diurnal difference may have been even greater at the 60-km level.

During the second series of wind observations on 21–22 November 1964, temperature profiles were obtained at 1400 and 1600 in the afternoon and at 0400 and 0600 in the morning. The difference in the diurnal average curves is illustrated by the dashed curve of Fig. 4.77. The situation in this case is more complex, with an almost complete absence

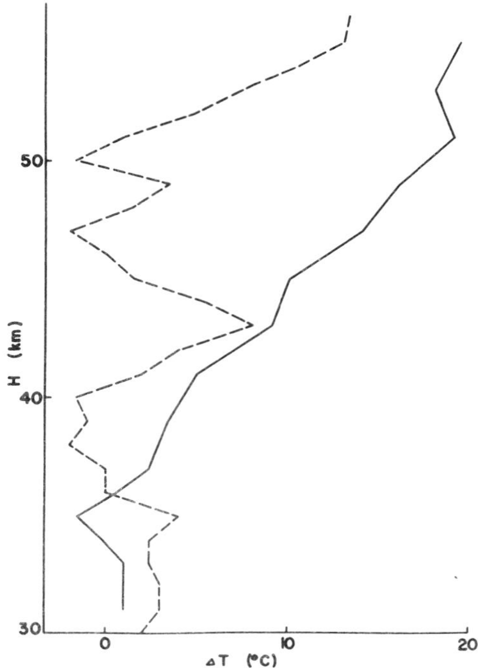

Fig. 4.77. Diurnal temperature range obtained by taking the difference between averages of 1400 and 1600 LST and 0400 and 0600 LST soundings for the data of 7 February 1964 (solid) and 21–22 November 1964 (dashed).

of the diurnal wave in the 45- to 50-km region, where it was quite strong in the 7–9 February data. A fairly strong diurnal variation is noted in the 40- to 45-km layer, and above 50 km the difference increases rapidly to a diurnal temperature variation of almost 14° in the 55- to 60-km altitude range.

These data would indicate that diurnal heating of the upper stratosphere is not a simple phenomenon. There are no obvious reasons for the lack of a diurnal variation in lower levels of the upper stratosphere during the November case, and it seems unlikely that errors of that magnitude could be present. It would appear that some special phenomenon occurred during the November sampling period. Inspection of Fig. 4.4 does not indicate a special case for that observational sample, although the rather small vertical motions which could have destroyed the radiationally generated diurnal temperature change could have easily gone undetected owing to the very limited amount of data available. Individual data points obtained at two-hour intervals in the February series show significant short-term changes from one sounding to the next. Since it is very difficult to accept the data obtained in November as characteristic of the static case, it would appear to indicate the presence of local perturbations in the vicinity of the stratopause. The comparatively unstable region above and the high wind speeds that are representative of the early winter season provide possibilities for vertical motions and a resulting corrugation of the stratopause surface from the nice smooth surface that we fancy.

It is also possible that the unusual structure illustrated here is the result of local variations in composition. Contaminants such as water vapor or dust could have a pronounced effect on the radiational equilibrium, either by changing the absorption or re-radiation or by modifying the concentration of ozone. The latter is a rather inviting possibility, since it implies strong horizontal gradients in composition which can be advected across a station such as White Sands Missile Range. Our discussion above relative to interaction between the hemispheric circulations is precisely of this nature. The interaction occurs near the stratopause, starting in subtropical regions and working into middle latitudes. It is clear from Figs. 4.9 through 4.20 and 4.21 that this interaction does not get into full swing until mid-December, so the event captured in this November sample must be a very special case. Little is known about the structure of this interaction, and a determination of the vertical and horizontal scales of the vortices would be of great value. Unfortunately the amount of data obtained in the November case is inadequate for this purpose.

The data samples obtained in the samples presented above relative to diurnal oscillations in the stratospheric circulation over White Sands Missile Range are in reasonable agreement with the picture of stratospheric response presented in the early portion of this section. Since these early data were confined to low latitudes and there might well be different effects at high latitudes, a similar experiment was conducted at Fort Greely, Alaska on 21–23 March 1965. A temperature variation of between 15° and 20°C was observed in the stratopause region, and equivalent wind variations were observed. These data would then indicate that the input perturbation in the atmosphere is relatively evenly distributed in the latitudinal direction over the entire sunlit side of the earth.

In addition to a more detailed look at the polar ends of the heated zone, the tropical regions provide a most interesting point for inspection of the diurnal oscillation. In advance of the heat wave meridional winds from equator toward the poles are expected, and the available data tend to confirm this configuration. If so, a general divergence is indicated in tropical regions ahead of the pressure wave. Behind the pressure wave an opposite meridional flow should form as the thickness contours move to the west.

The data presented above indicate that diurnal temperature variations compose a significant portion of the total meridional temperature

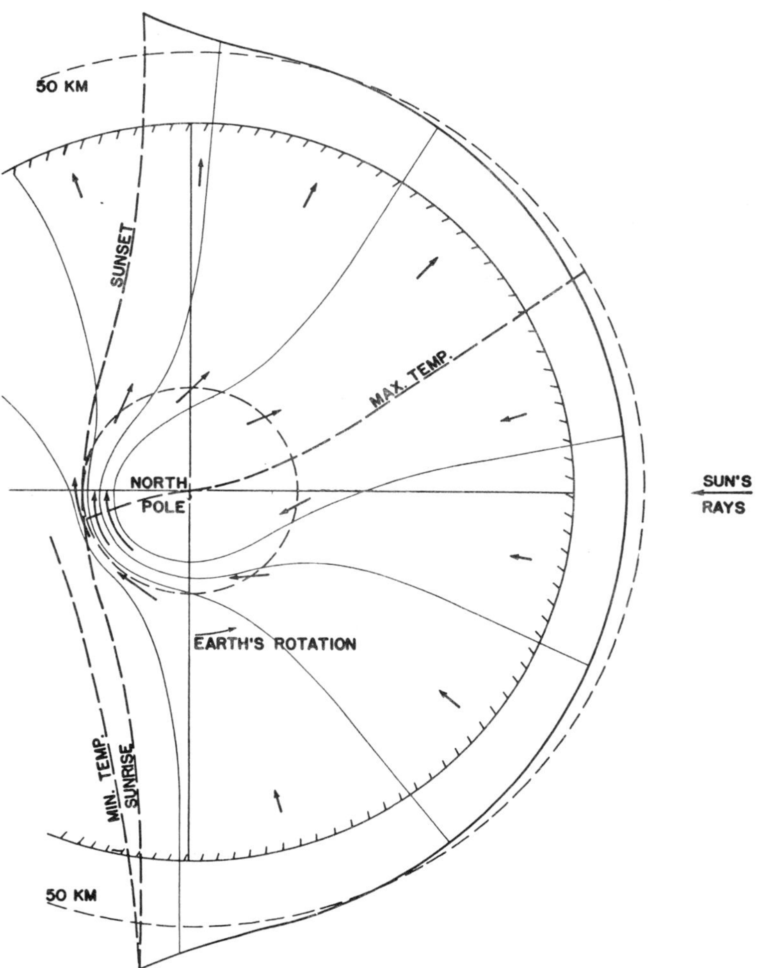

Fig. 4.78. Temperature and wind field of the stratospheric tidal circulation projected on the equatorial plane for the Northern Hemisphere a few weeks after the spring equinox. The thin lines represent temperature or thickness contours and the arrows illustrate the wind field. This diurnal circulation system is superimposed on the general structure of the early summer circulation.

differences which drive the monsoonal circulations. It is clear that as the subsolar point moves into the summer hemisphere there will be significantly different hemispheric responses to this perturbation. The situation in the summer hemisphere a few weeks after the equinox is illustrated in Fig. 4.78. A ridge of heated expanded air caps the pole and extends equatorward, lagging the subsolar point by a couple of hours. Diurnal perturbations of the wind field induced by the daily rotation of this heat wave around the earth will surely exhibit different characteristics at high latitudes from the picture presented by the data discussed above.

At low latitudes the diurnal wind oscillation can be considered, with a reasonable degree of approximation, to exhibit a simple meridional oscillation. This is represented in Fig. 4.78 by the equator-to-pole arrows in the morning hours between sunrise and the crest of the diurnal heat wave, and by the pole-to-equator arrows during the afternoon and night. With increasing latitude the diurnal component of the flow will turn toward the west in the morning sector and return from the northeast in the afternoon and nighttime region. In the polar region the diurnal component of the stratospheric circulation is largely zonal, with the acceleration occurring principally in the nighttime mesosphere where the shadow of the earth results in maximum temperature gradients between the cool night sky and the warm pool of polar air. This circulation of air about the polar region has been termed the "stratospheric tidal jet."

Consideration of the level of penetration of solar radiant energy as a function of latitude as is illustrated in Fig. 4.78 indicates that the diurnal circulation will rise in level with increasing latitude. At lower latitudes diurnal oscillations of the stratospheric circulation are centered at the stratopause level. This is illustrated by the mean data on meridional vertical wind profiles presented in Fig. 4.79, which indicates that the amplitude of the meridional component of the diurnal motion increases with latitude as was predicted by Stolov (1955). In addition, these data also support the concept of a rise in altitude of the tidal motion with latitude. If these data are accurate in that the tidal motions are located above 60 km at 60 degrees latitude, then the highest altitude reached by the stratospheric tidal jet should fall in the upper mesosphere in the 70- to 80-km range.

The perturbation imposed on the stratospheric circulation by this diurnal heat wave thus distorts the hemispheric circulation, particularly in the case of the summer easterlies. An average flow is produced by the average meridional temperature difference between equatorial and polar regions, and this is the picture generally presented to describe the stratospheric circulation. In the case of data from the MRN, mean profiles

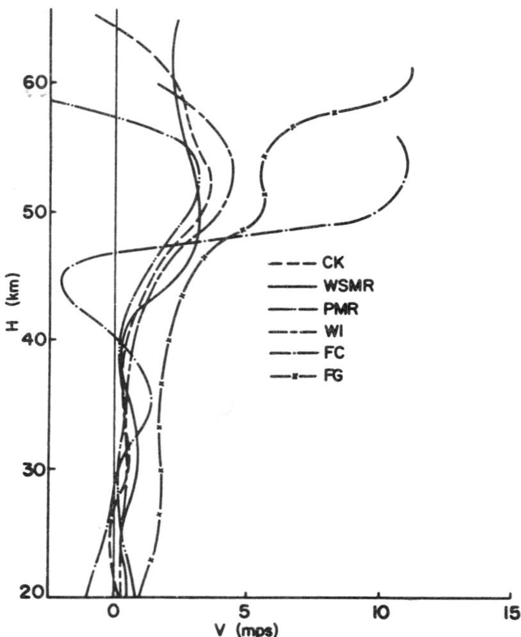

Fig. 4.79. Mean meridional wind profiles for the summer season at selected MRN stations. The data were processed at 5-km intervals, and a wind from the south is positive.

(Fig. 4.79) and SCI data (Figs. 4.4, 4.5, 4.23, and 4.24) are representative for that portion of the summer easterly circulation along the ridge of high temperature and pressure (that is, just after noon local time). Since the temperature of this region is elevated at that time it is clear that the equator-to-pole meridional temperature gradient will be reduced. The zonal flow observed in the MRN data should therefore be less than the average summer stratospheric zonal flow, and should be less than the flow that could be observed in the nighttime portion of the hemisphere by almost the full diurnal variation in the zonal wind.

The opposite effect will appear in the winter hemisphere, where effects of the diurnal variation extend only into upper middle latitudes. In this case, the maximum over-all temperature difference between equator and pole is to be found along the ridge of heated air, and the MRN sample is of that strongest portion of the winter westerly zonal circulation. The mean westerly winds of the winter hemisphere should, therefore, be lower than that indicated by direct summaries of the

MRN data. In addition, the lowest values of the zonal circulation should appear in the nighttime sector, where diurnal cooling of low latitudes will serve to reduce the temperature difference from there to the continuously cold polar night. The presence of this large diurnal perturbation results in a bias in the MRN data concerning the stratospheric zonal circulation which varies from zero at the equinox times to a maximum during the solstice periods, with opposite signs and differing values characteristic of the two seasons. Available data indicate that the magnitude of this zonal variation in lower latitudes falls in the 20-meter/sec range. Since the diurnal circulation should circle the poleward end of the heated ridge in upper middle latitudes from a westerly direction as the winds flow poleward ahead of the heat wave and equatorward behind it, it appears likely that maximum westerly winds would be produced there at altitudes well within MRN observational capability. As has been pointed out before, in the summer case the principal axis of the diurnal circulation should be found at mesopause altitudes in the high-latitude sky, and special provisions will have to be made for observing this flow.

Some similarity exists in meridional flows between the hemispheres, principally in that in both cases the flow is from the equator toward the pole ahead of the heat wave and the reverse after the wave passes. At low latitudes the amplitude of this oscillation appears to be of the order of 40 meters/sec. During the solstice periods the situation is far different in the two hemispheres. In summer the meridional flow becomes zonal as it attempts to execute a circumpolar trajectory about the pool of hot air attached to the high-latitude end of the ridge of high pressure. The diurnal circulation effectively turns away from the centerline of the heat wave to do this. In winter the diurnal circulation vector turns toward the approaching heat wave and becomes zonal as the wave passes by at lower latitudes. The diurnal circulation then continues to turn clockwise in the lee of the heat wave to become meridional from the north.

It is clear from the points discussed above that the tidal motions of the stratospheric circulation constitute a significant portion of the dynamic processes that occur in that region. The data at this time are inadequate for construction of a complete picture of the phenomenon, but are sufficient to clarify the importance of this mechanism in establishing dynamic equilibrium in the stratospheric circulation as well as to provide information on circulation adjustments of composition parameters derived from static considerations. Only with global coverage by an observational system such as the current MRN can these inferred circulation effects be properly evaluated

4.9 Composition

Diffusion effects will tend to separate the gases in the atmospheric mixture so that each element will present a particular vertical distribution in the earth's gravitational field. If the other variables such as temperature should remain constant with height, the density of each molecular constituent would assume an exponential vertical profile in the equilibrium case. Such a distribution is not found in the atmosphere, principally owing to eddy mixing and a variable heat content resulting from inhomogeneous absorption of solar energy. The actual distribution of atmospheric gaseous material is complex, and thus far is inadequately observed. Difficulties in accomplishing such observations have limited the amount of data acquired and contribute to considerable uncertainty, so that today it is common practice to assume that turbulent mixing maintains a thoroughly stirred gaseous mixture throughout the stratosphere.

This assumption is based on a few data points obtained principally in midlatitudes and a general belief that eddy diffusion will dominate molecular diffusion up to mesopause altitudes. With the acquisition of synoptic-type data by the MRN it is possible to obtain an estimate of the global intensity of large-scale advection in specific instances and to compare these data with the intensity of molecular diffusive transport. Diffusion through molecular motion is strongly sensitive to the mean free path of environmental molecules, which increases rapidly with height in the stratosphere. The diffusion coefficient (D) varies directly with the 3/2 power of the absolute temperature (T), and inversely with the mean free path (λ). These factors are summed up in the following relation:

$$D \propto \frac{T^{3/2}}{\lambda}. \tag{4.3}$$

The characteristic value of the diffusion coefficient of water vapor in air is 0.237 cm^2/sec at sea level, increasing to approximately 3 at the base of the upper stratosphere and to over 1500 cm^2/sec at the stratopause. On the basis of an estimated rms displacement (ξ) with time (t) of the molecules given by the relation

$$\overline{\xi^2} = 2Dt \tag{4.4}$$

we obtain small-scale molecular diffusion mixing rates of 2½ cm/sec at the stratonull and 60 cm/sec at the stratopause. Molecular diffusion is roughly symmetrical and thus may be compared with the vertical component of the wind advective transfer. In view of the great thermal stability of the upper stratosphere, it is most likely that molecular diffusion will first achieve ascendency in the vertical component, and thus it appears likely that the vertical profile of stratospheric moisture will be controlled by the diffusion process.

Mean seasonal meridional upper stratospheric wind profiles at least indicate the existence of convergence and vertical motions in higher latitudes. These vertical motions are very likely to be upward in the summer and downward in winter. The summer upward motions are likely to be at higher latitudes while the winter subsidence is more probably in middle latitudes near the maximum intensity of the strato-spheric and tropospheric circulations. During the summer season upper stratospheric meridional winds are very light, with the only significant flow in the diurnally driven stratopause current sheet. Using measured values of the outflow from tropical stratospheric regions it is possible to show that the replacement current across the tropopause is a vertical wind of less than 0.1 cm/sec in the mean and thus is smaller than the diffusive transport term.

To illustrate the results of this diffusion control of equatorial strato-spheric structure we can very simply consider the equilibrium condition which would prevail in a still isothermal atmosphere, neglecting variation of gravity (g). Each molecular species (m) will assume an exponential vertical distribution of concentration (n) as defined by the hydrostatic relation

$$n = n_0 e^{-h/H}, \tag{4.5}$$

where h is the altitude for which n is to be evaluated and the scale height (H) is defined as (KT/mg). Using molecular weight values of 18 for water vapor and 29 for air, 300°K for the atmospheric temperature, and 2.5×10^{23} and 2.5×10^{25} for the sea level number concentrations of water vapor and air, respectively (corresponding to specific humidity of 10 gm/kg, which is characteristic of surface values in tropical regions), we obtain $H_{\text{air}} = 8.43$ km and $H_{H_2O} = 13.58$ km, which gives the profiles illustrated in Fig. 4.80. These curves point out the rather obvious fact that under diffusion control the mixing ratio of water vapor to air will increase with height. The lighter water vapor will establish a smaller lapse rate and thus increase in relative numbers as the total number density decreases with height. This factor would result in an order-of-magnitude increase in mixing ratio between the surface and the stratopause.

Now the earth's equatorial atmosphere does not have an isothermal layer and it is not still. Low temperatures of the equatorial stratopause will reduce the water vapor present in that region, as they also reduce the concentration of air. Advective and convective motions of the

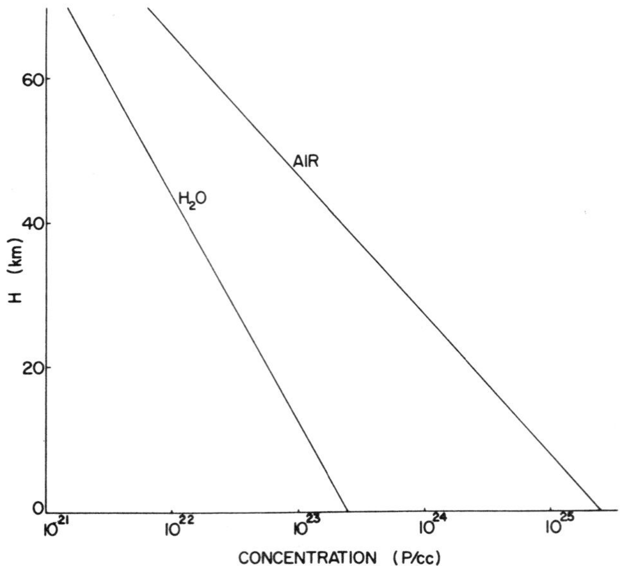

Fig. 4.80. Vertical distribution of number concentration of water vapor and air molecules in a hypothetical isothermal earth atmosphere. The constants used are typical of equatorial regions.

atmosphere produce the gross deviations of observed vertical distribution of water vapor in the lower troposphere from the roughly exponential one to be expected from diffusion effects. Upward and downward flows in the troposphere are asymmetric, with upward motions strong and generally confined to relatively small cross sections, while downward motions are generally widespread and thus of low speed. The diffusion process, which is generally symmetrical, is overpowered by these wind systems so that the structure of the water vapor profile in the troposphere is controlled by air motions and the associated precipitation processes rather than diffusion.

The stratosphere in equatorial regions is remarkably still relative to vertical motions in the summer periods. Lateral motions have been observed to be negligible below 40 km from MRN observation, and there is adequate reason to expect vertical motions to be small in view of the considerable thermal stability of the stratosphere. Available data indicate that the residence time of air in the stratospheric region between the tropopause and 40-km altitude is of the order of several months, and thus is adequate for diffusion processes to establish control of the vertical water vapor distribution. In view of the several complications discussed above, it is difficult to assess the stratospheric profile which will result from the gravitationally established diffusion profile.

A second factor which will exert an influence on the vertical water vapor profile of the stratosphere is the strong positive thermal gradient of the region. As predicted by Chapman (1918), water vapor will diffuse from the cold tropopause toward the warm stratopause because the stratopause is effectively dry. In addition, the nonlinear (exponential) relationship expressed by Eq. (4.5) requires that a particular temperature increase with height more efficiently contain the water vapor, and this in turn requires that the mixing ratio increase with height as a result of the observed negative temperature lapse rate of the upper stratosphere. The amount of the increase in water vapor mixing ratio, if equilibrium is achieved under this thermally guided diffusion process, turns out to be of the order of 1 gm/kg between the tropopause and the stratopause. A postulated stratospheric water vapor distribution based on the above considerations and available experimental data is illustrated in Fig. 4.81.

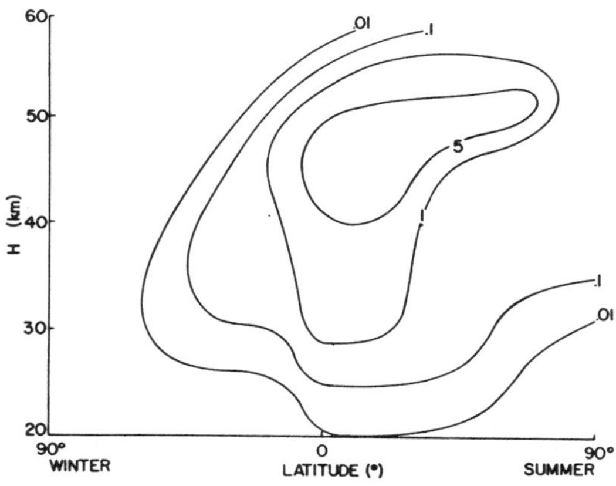

Fig. 4.81. Postulated stratospheric water vapor mixing ratio in grams per kilogram for 1 December in the Northern Hemisphere.

The above considerations are at variance with other estimates of the water vapor content of the stratosphere where the principal mechanics had been assumed to be convective transport from the ocean surface to the stratopause. If that mechanism is the principal mode of establishing the stratospheric moisture distribution, then the excess water vapor would be condensed out of the ascending air and removed through precipitation processes. This method of water vapor introduction into the stratosphere sets firm limits on the amount of this constituent to be found there by the value of the saturation vapor pressure at the low temperatures of the tropical tropopause. It has been common to assume that the mixing ratio remained constant above that level, which assumes a dominance of convective-type interchange, since otherwise diffusive separation would occur between the water vapor and air molecules and the mixing ratio would once more increase with height. It should be remembered that the stratospheric residence times are only adequate for diffusive equilibrium in tropical regions, and advective transfers will be the rule in the stratosphere at middle and high latitudes. The result is a moist tropical and dry high-latitude stratosphere.

The question of a water layer at high altitudes immediately arises. Photochemical destruction of water molecules is to be expected in the presence of a strong ultraviolet flux in the 1000-angstrom region of the solar spectrum. Energy from the sun is increasingly available with height in the mesosphere and ionosphere, so that, in effect, there is a lid on the stratospheric cloud of water vapor, with the maximum mixing ratio to be found near the stratopause.

It should be noted that this water vapor destruction will have a strong diurnal characteristic. Water vapor will be rapidly elevated above the stratopause by convection in the turbulent mesosphere, so that during nighttime hours there will be a significant upward extension of the moist cell in the tropical upper atmosphere. With sunrise the uppermost water molecules will be exposed to dissociating radiation, and the altitude of a particular water vapor concentration will retreat toward the stratopause. These diurnal variations will appear as small perturbations in the water vapor profile throughout the region where diffusive equilibrium prevails.

The question of changes in state resulting in significant changes in composition and also exerting a strong thermodynamic influence on circulation processes is of special interest. Lack of visual evidence of condensation products is often cited as proof of their absence, but the quantities involved clearly indicate that such a phenomenon would not lend itself to detection by surface visual observation. It appears likely that condensation occurs in the upper atmosphere, and that a most effective moisture-depleting mechanism is the gravity-powered fallout of meteorite particles upon which stratospheric water vapor has condensed. Any consideration of dynamic physical processes in the stratosphere must then include allowance for heat exchange through changes in state of water, as well as for radiational heating and cooling of the stratosphere resulting from the presence of significant amounts of water vapor. For these reasons it is important that the actual water vapor structure of the stratosphere be determined. More effective experimental techniques will be required than have previously been available. It is clear, however, that the most opportune place for conclusive analysis is in tropical regions. Here the nearest to a steady state must exist, and the large variability that will characterize middle- and high-latitude data can be avoided in the preliminary studies.

The upper stratosphere is subjected to a downward flux of meteoroid debris which provides a population of particulate material for this remote region of the earth's atmosphere. No observational data of the natural concentration or distribution of these particles are available, so it will be necessary to construct a hypothetical population for this region without even the rough guidelines provided by observational data in other layers of the upper atmosphere. In attempting to do this, it will be assumed that the large thermal stability of the upper stratosphere will preclude convective overturning and thus will not favor the introduction of particulate material into the upper atmosphere from beneath. This assumption may well be unjustified for the submicron particles owing to the extended residence times which their very low fall rate specifies, particularly in the lower layers of the upper stratosphere. Very small vertical wind components in the upward direction could be exceedingly efficient in introducing new particles into this region. Further attention to this important possible source of upper stratospheric pollution must await the acquisition of more adequate data on transport processes in that region.

Sedimentation of particulate material downward across the stratopause boundary into the upper stratosphere is quite likely to be variable in space and time. A positive meridional gradient is to be expected in the general case, with a certain amount of local variability in the higher latitudes. Upon this spatial and short-term temporal distribution will be superimposed an annual variation with an amplitude factor of approximately 3. The peak of this annual variation occurs in the Northern Hemisphere summer relative to its input into the earth system, and thus on the average should appear at the stratopause approximately 15 days later, with the minimum spaced approximately six months later. These comments are based on extrapolation of data obtained from incoming particles in higher momentum ranges than are of prime interest here and thus could be erroneous in the smaller-size range. Satellite observations of these smaller particles are inadequate for resolution of this variable, so for our purposes here the orbital space distribution will be assumed to be governed by a simple logarithmic extrapolation for all sizes of meteoroids.

A best estimate of stratospheric particulate concentration against altitude if no destructive processes are present is presented in Table 4.4.

TABLE 4.4

Estimated Seasonal Concentration of Meteoroid Particles in the Stratosphere[a]

Altitude (km)	Northern Hemisphere		Southern Hemisphere	
	Summer	Winter	Summer	Winter
50	46	16	16	5
45	86	30	30	10
40	168	58	58	20
35	336	116	116	39
30	720	249	249	84
25	1560	540	540	162

[a] Assumed to have average radius 10^{-5} cm and 2.5 gm/cm^3 density. Applicable if no destructive processes occur. Units are particles per cubic centimeter.

This table was constructed on the assumptions that the particulate matter is composed of smooth spheres of 10^{-5}-cm radius and a density of 2.5 gm/cm^3. This size is probably representative of the majority of extraterrestrial particles which are of concern relative to atmospheric contamination, both in size and number. As is discussed in Chapter 5, these particles are decelerated from escape velocity to terminal fall speed by the atmosphere in the altitude region of 100 km. The input of these particles is not symmetrical about any of the earth axes, and the stratospheric circulation can be expected to complicate the ditsribution through transport during the two weeks which these particles spend in the mesosphere and during their stay in the stratosphere. The considerable enhancement indicated in the Northern Hemisphere summer is the result of variations in meteoroid density with position in the orbit.

Fall rates of the more important particles expected in the upper stratosphere are presented in Fig. 4.82. A particulate density of 2.5 gm/cm^3 is assumed. Diffusive growth through condensation could be important during the cool nighttime, but is unlikely to be important in daytime, leaving principally coagulation processes to alter the size distribution and concentration that are nominally to be expected from differential fall rate considerations. This process should result in a shift of the size distribution toward larger particles, and the effect should become increasingly important at lower altitudes throughout the upper stratosphere owing to the efficiency of Brownian motions in inducing collisions.

Inspection of Fig. 4.82 indicates a residence time per kilometer of 500 hours for a 0.01-micron radius particle, 50 hours for a 0.1-micron particle, and 5 hours for a 1.0-micron particle at the stratopause. Residence times of 40, 2½ and ½ months are estimated for the same particles

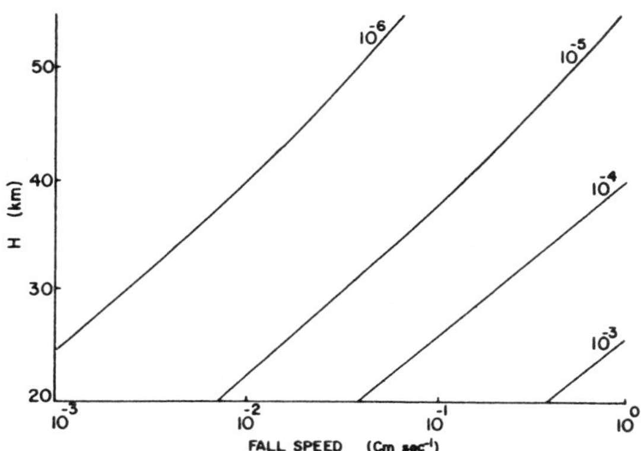

Fig. 4.82. Stratospheric fall rates of small spherical particles of 2.5-gm/cm³ density. Radii in centimeters are indicated on the curves.

in the vicinity of the stratonull surface at the base of the upper stratosphere. Integrations of the transit times for these particles yield total upper stratospheric residence times of 50, 4, and ½ years for 0.01-, 0.1-, and 1.0-micron radius particles, respectively. These residence times are very long compared with transport mechanism time constants, so the spatial distribution of these particles will be profoundly modified by any systematic circulation systems. The particles, if they maintain the same size, can be considered to be carried with the winds. Circulation data obtained by the Meteorological Rocket Network show a net equator-to-pole flow during the winter season at all levels in low and middle latitudes over the North American sector of the Northern Hemisphere, while during the winter storm period a strong circulation away from the pole is observed. A convergence in particulate concentration will result in lower latitudes owing to this horizontal flow, and some divergence will occur in the high latitudes. More importantly, the vertical flows which will necessarily be associated with this convergence can have a strong influence in altering the particulate concentration.

There is little to support the concept of upward motions in the upper midlatitude stratosphere in winter. Present evidence points strongly to a zone of descending upper stratospheric air in a circumpolar belt between 50 and 55 degrees latitude. This vertical transport will result in compression of the stratospheric air into a smaller volume and thus enhance the particulate concentration. This process can only be expected to increase the concentration by a small factor, and is far less efficient than the mesospheric case, where the concentrating mechanism hinges on altering the Eulerian fall rate of the particulates.

Our hypothetical model of particulate population of the upper stratosphere must be significantly different for the summer and winter seasons. In summer the upper layers of the region beneath the stratopause will be denied input of the smaller particles owing to vertical velocities in the mesosphere, and there will be a significant shift of concentration poleward (at approximately 65 degrees latitude) due to reduced particulate fall rates in that mesospheric region. The winter season, on the other hand, will evidence a band of enhanced particulate concentration in the 50- to 55-degree latitude region, with maximum concentration appearing at the base. Concentrations will generally increase by two orders of magnitude between top and bottom of the upper stratosphere and by an order of magnitude between equator and pole in the horizontal plane. Smaller-scale inhomogeneities described above will produce order-of-magnitude horizontal increases in concentration at the top in limited high latitudes in summer and factor-of-two increases at the base in the 50- to 55-degree latitude belt in winter.

As a surprise to meteorologists' early assumptions, the upper stratosphere is probably not free of cloudy condensation. In general, the density of the cloud formations is insufficient for observation from our less than vantage point, attempting to detect these tenuous clouds through an obscure foreground. On occasion the density of the stratospheric cloudy condensation is enhanced to the extent that it is detectable, and there is reason to believe that these events are associated with strong convective activity which owes its origin to development of lee waves above mountain ranges. In part these clouds are assumed to be a part of the lee wave phenomenon because of their particular appearance, exhibiting very little motion and a fineness of detail which is characteristic of lenticular clouds of the troposphere. These clouds have been given the name nacreous (or mother-of-pearl), and have been observed in the altitude range from 20 to 30 km during the winter season at relatively high latitudes.

Scattered sunlight from nacreous clouds is sometimes quite intense during periods of sunrise and sunset, permitting them to be detected through thin cirrus clouds and to block out starlight. They display colors, which are described by observers as being oriented relative to the clouds' mass rather than sun position, although this observation could well be due to the restricted size of the clouds. The cloud particles scatter sunlight after the fashion of water droplets, although the environmental temperature is of the order of $-45^\circ C$ or colder, and is thus well below the point at which spontaneous solidification would be expected to occur in the troposphere. The size of the nacreous cloud particles is estimated to fall in the less than one micron radius range, and it is just possible that the individual cloud particles do actually condense in this cold environment as liquid droplets and the latent heat of condensation serves to keep the droplet warm during its very short lifetime as it is transported through the cloudy region of the stratosphere. These considerations present no real problems to the observational data, since the time of occurrence is related to the time of high winds in the troposphere when the formation of lee waves is most likely, and the observed data on scattering of sunlight indicate a narrow spectrum of droplet sizes which is characteristic of rapid expansion and low moisture concentrations, restricting the activity of factors which normally serve to broaden the size distribution.

As was discussed in Chapter 3, samples of particulate concentrations in the stratosphere to altitudes of 30 km have revealed values of one or fewer particles per cubic centimeter. This does not present a real problem under the conditions that we have supposed to exist for these clouds, however, since normal condensation nuclei such as Aiken nuclei are not required when supersaturations of considerable degree are assumed, because condensation could occur on large ions which are plentiful at these levels. Such a mechanism would provide an explanation of the remarkably uniform sizes, an aspect of tropospheric cloud formation which is eliminated by the general size distribution of condensation nuclei which are always present and have a strong influence on the droplet size distribution during the formative stages.

The existence of nacreous clouds offers proof that water vapor is present in the stratonull region of the upper atmosphere. The general cleanliness of the region inhibits condensation except at considerable supersaturations, and thus the clouds are only occasionally visible. The conditions in which large upward motions of small scale are present are related to the formation of stable lee waves which require strong tropospheric winds as well as reasonably good observing conditions. These specifications are best realized in wintertime at high latitudes. It is probable that the lifetime of these cloud particles is very short, limited to the period in which they take part in the wave motions of the lee waves.

Diurnal variations of the temperature and wind fields of the stratopause region discussed in Sec. 4.8 do not agree well with early concepts of ozone structure in the upper stratosphere. Early ozonospheric structure theories, based on meager data, predicted diurnal variations of 5°C or less. Discovery that stratopause temperatures execute a much larger diurnal variation than expected has raised questions relative to previous assumptions concerning physics of ozone-solar ultraviolet interaction, with particular reference to the vertical ozone structure in the upper stratosphere and mesosphere. A common assumption regarding the vertical ozone profile is that a constant mixing ratio prevails above the approximate 30-km maximum altitude to which data are usually obtained.

An ozonesounding instrument based on the Regener chemiluminescent observational techniques has been developed by Randhawa (1966) and flown at White Sands Missile Range in the Arcas rocket system. Two soundings, obtained during the evening hours, are illustrated in Fig. 4.83.

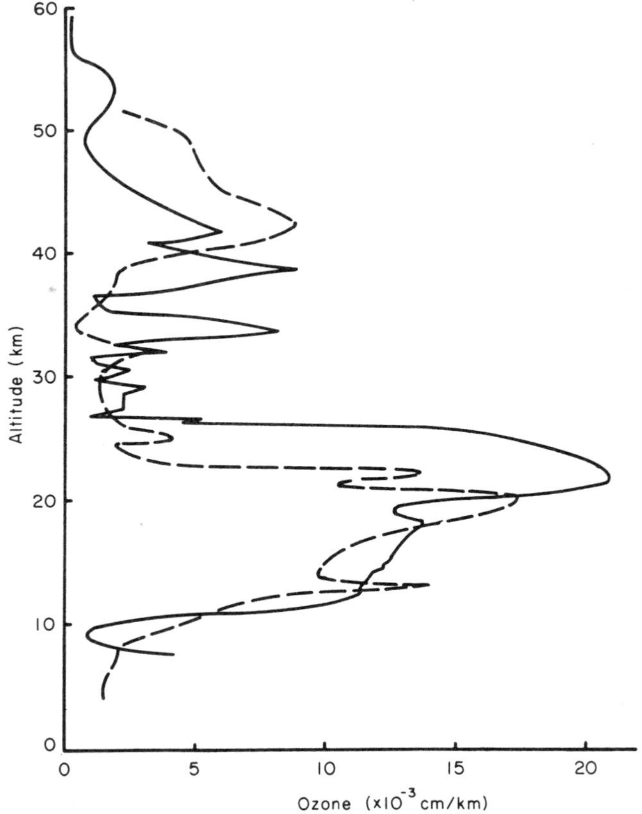

Fig. 4.83. Ozone concentration profiles obtained with chemiluminescent techniques from the Arcas rocket by Randhawa (1966) over White Sands Missile Range at 2100 MST, 25 October 1965 (dashed curve), and 2006 MST, 23 November 1965 (solid curve).

Most striking features of these data are the relatively strong secondary peaks in ozone concentration and the gross detail structure indicated in both profiles of the upper stratosphere. The 25 October observation shows a peak in the 35–40-km altitude region with a magnitude of almost one half the principal peak concentration. On 23 November, the secondary peak is located at 40–45 km altitude and is of approximately the same strength, although this sounding illustrates an appreciably smoother profile.

A nighttime secondary peak in ozone concentration had been predicted by Leovy (1964) as a result of continued formation of ozone from atomic oxygen residue after the dissociating solar flux disappears. This effect was expected at altitudes well above the stratopause, however, due to an assumed lack of atomic oxygen at lower altitudes. These data would indicate the possible existence of greater concentrations of atomic oxygen in the upper stratosphere than had previously been assumed. In addition to this possible chemical enhancement of ozone, the diurnal expansion and contraction of the stratopause region introduces a new variable which will result in dilution and concentration of ozone in the region by day and night, respectively, by an amount of the order of several percent.

Perhaps of greater importance is the possibility that the upper stratosphere is not the dry environment which is generally supposed. The assumed atomic oxygen concentration might be significantly modified if relatively large quantities of water vapor were available for dissociation, in addition to the commonly assumed atomic oxygen source through dissociation of molecular oxygen. Since other considerations would indicate that water vapor, if it is present at the stratopause in significant quantities, would result in strong latitudinal variations in the thermal and chemical equilibrium of the region, it is of utmost importance that the upper stratospheric distribution of this material be established. There is reason to expect that the stratospheric structure of ozone and water vapor will be very closely related.

4.10 Acoustic Structure

The upper stratosphere exhibits an increasing speed-of-sound structure with height as a result of the negative lapse rates of temperature and wind that are characteristic of the region. Except in polar regions during the winter there is a strong increase in temperature with height throughout the stratosphere, and except for an equatorial reversal zone, the polar regions, and a short time in the spring and fall during the monsoonal reversal the stratospheric winds present a strong negative lapse rate with height. These two gross factors combine to produce a complex environment for propagation of acoustical energy. Since the local speed of sound is of significance in many dynamic problems, the most obvious of which are the lift and drag experienced by a high-speed vehicle, a knowledge of the sonic structure is a requirement for the conduct of experiments and operations in and through the upper atmosphere.

As has been noted, the composition of the upper stratosphere is not known in detail. It has become commonplace to assume that variations in composition are small and that the Eulerian speed of sound is determined by the relation

$$v = \left(\frac{k}{\rho}\right)^{1/2}, \tag{4.6}$$

where k is the adiabatic bulk modulus and ρ is density. Incorporation of the equation of state and the fact that the adiabatic bulk modulus is related to the pressure directly through the ratio (Γ) of specific heat at constant pressure (C_p) to the specific heat at constant volume (C_v) we can obtain the expression

$$v = (\Gamma RT)^{1/2}, \tag{4.7}$$

where R is the gas constant of the specific gas and T is the temperature. Gutenberg (1951) has used values of $R = 2.87 \times 10^6$ cm^2 sec^{-2} deg^{-1} and $k = 1.403$ (for dry air) to obtain the relation

$$v = 20.06(T)^{1/2}, \tag{4.8}$$

which is commonly used in atmospheric sound propagation work, with speed-of-sound vertical profiles resulting which are quite similar to the temperature profiles of Fig. 1.1.

Now the ratio of specific heats of water vapor is smaller than that of dry air, given as $\Gamma = 1.324$ at 100°C in the Handbook of Chemistry and Physics, 42nd Edition. The gas constant is the universal gas constant divided by the molecular weight of the gas. The molecular weight of water vapor is significantly lower than that of air, so that this parameter is reduced if water vapor is present. Combination of these opposing effects results in an over-all increase in speed of sound with increased water vapor, amounting to a few meters per second enhancement of sound speed under humid conditions.

The speed of sound in still air is, within the limits of the above assumptions, a function of the atmospheric temperature structure. Thus it decreases with height in the troposphere, increases with height in the stratosphere, decreases with height in the mesosphere, and finally increases with height in the ionosphere. The speed of sound at the stratopause is roughly the same as at the earth's surface, but the strong temperature inversion of the ionosphere results in very high sound speeds in that region. This simple picture is complicated by the variations in temperature structure with latitude. At the poles the vertical variations are quite small in the winter season in the troposphere and stratosphere, although the mesodecline is strong. Slowest atmospheric sound speeds are to be found at the mesopause level in summer, where temperatures fall to their lowest observed values, apparently as low as 130°K.

Lowest speeds of sound in the troposphere and stratosphere occur in tropical tropopause regions. On occasion the polar tropopause may show a comparable value, but it is generally warmer. These considerations suggest that there is a wave guide structure in the atmosphere relative to sound propagation. Between the surface and the stratopause the positive and then negative speed-of-sound lapse rates will trap any energy which enters the layer within certain acceptance angles. The mode of operation of this wave guide phenomenon is illustrated in Fig. 4.84, in which a symmetrical point sound source located near the

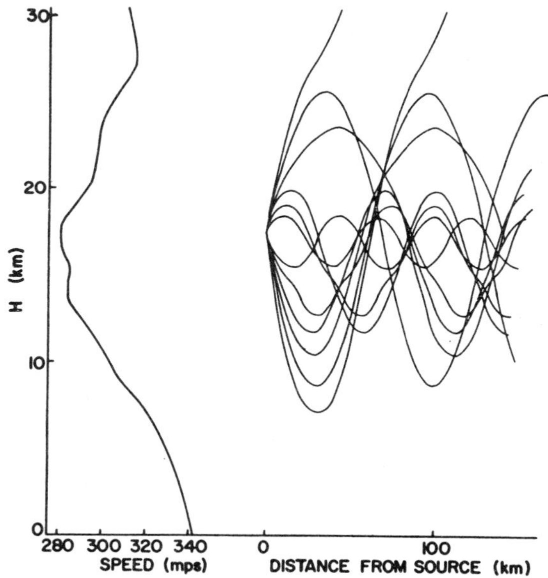

Fig. 4.84. Ray paths from a point acoustic source located near the tropopause which are trapped by the troposphere-stratosphere acoustic duct. The curves presented here are for rays with injection angles of 10, 14, 18, 22, 26, and 30 degrees above and below the horizontal.

tropopause is analyzed to determine subsequent positions of the sound front. The ray paths of only those incident elements that are trapped by refractive effects are illustrated here, and the temperature- and wind-determined speed of sound is somewhat unusual in this case in that the sonic stratopause is rather low in altitude, such as that obtained for a sound wave moving westward during the winter season.

These ray paths were obtained through simple application of Snell's law,

$$\frac{\sin i}{\sin i'} = \frac{v}{v'}, \tag{4.9}$$

using the speed-of-sound profile applicable to the particular azimuth along which the sound front was moving with the incident angles (i) of interest. In this particular case the vertical speed-of-sound structure is such that the trapped ray paths do not intersect the earth's surface after the first cycle in the wave guide. This is not necessarily characteristic, however, and if the speed of sound at the upper level is sufficiently high, a dominant mode of propagation involves reflection of the sound front at the earth's surface as a part of each cycle in the duct. It is also obvious that the sound front may traverse the general atmospheric ducting structure with the ionospheric positive speed-of-sound gradient in conjunction with surface reflection to confine the sound front to the duct.

In addition to these gross modes of sound channeling in the atmosphere similar phenomena can occur on a microscale. Small variations in the speed-of-sound structure can act to duct incident energy into these small detail features of the atmosphere. Generally the acceptance angles are very small compared with the major ducts. The most important result of the action of these "subducts" is the separation of a portion of the acoustic energy out of the expanding acoustic front. As long as a portion of the front is subducted it will travel at a speed different from that of the rest of the front, and when it is released from the subduct it will occupy a different position in time and space from that occupied in the front before the subduct was encountered. These points are illustrated in Fig. 4.85. A particular detail in the vertical speed-of-sound profile is used here to point out its effect on an incident wave front. Such detail structure is characteristic of upper stratospheric vertical profiles, which exhibit an irregular structure with a profusion of sizes in the vertical dimensions of these features. Meteorological rocket systems have been used in attempts to analyze the horizontal extent of these features, and while a few of them can be identified from one sounding to the next, the great majority cannot be traced through subsequent soundings when they are made at intervals as short as one-half hour.

The ray paths which intersect the mouth of the subduct will be trapped in this small-scale feature and confined to it over its horizontal extent. The speed of propagation of this element of the incident wave front will be lower than the speed of those rays which escape subducting

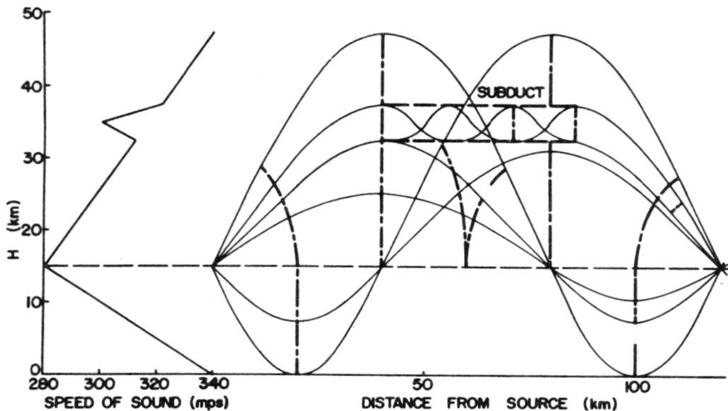

Fig. 4.85. Action of a stratospheric subduct in altering the structure of an acoustic front. Horizontal extent of the subducts is a matter of conjecture at this time, although from the data they appear to be short relative to a half cycle in the principal wave guide. The heavy dash-dot lines represent positions of the acoustic front at various times.

and swing to higher altitudes, but will then be higher when those ray paths swing back into the center of the principal duct where the speed of sound is minimum. Confinement to the subduct will result in separation of the affected frontal element from its proper place in the hemispherical front, and that separation will increase with the time during which the subduct is involved. On emergence from the subduct the frontal element will reenter the gross ducting environment but will be separated in space and time from its original position in the sound front. Certain of the ray paths generated by a sound source will not intersect the mouth of the subduct, but will encounter it from below or above. In that case the wave front will not be trapped in the subduct, but will be deviated from the trajectory specified by the over-all speed-of-sound gradient. The effect will become more pronounced as the angle of encounter becomes more acute.

The net result of interaction between a smooth sound front and an inhomogeneous environment will be to fracture and reorganize the frontal surface. The degree to which these effects will be carried out will depend on the number of subducts encountered and the vertical and horizontal sizes and intensities of the subducts. Now we have a capability of observing these features in the vertical with meteorological rocket probes, and it would appear possible, under simplified circumstances, to detect the cause and effect relating these two separate techniques of upper stratospheric observation. The result of such observations should be a measure of the horizontal scale of these phenomena.

The vertical scale of these features of the atmosphere has been analyzed by Webb (1962) using the MRN data on wind and temperature. The horizontal extent of these subducts is unknown at this time, however, so it is not possible to depict the exact effects on a sound wave. It is known from grenade experiments and other sources that the shape of an acoustic pulse is not seriously modified in the passage from high altitude to the surface. In the case of an explosion on the surface, the pressure perturbations that are observed at the first reflection point on the surface (usually 100 to 200 km meters away) are generally very complex. An example of this type of data is illustrated in Fig. 4.86. If

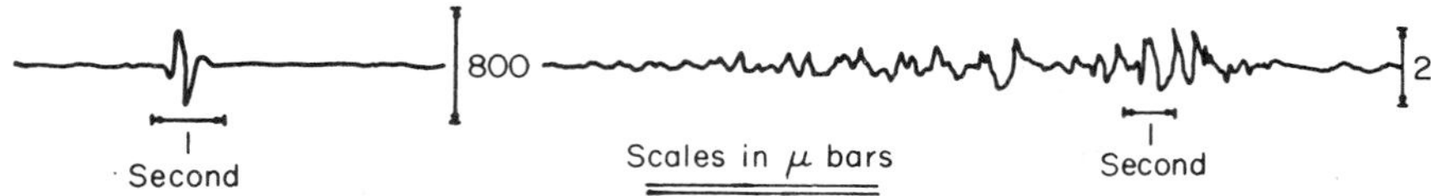

Fig. 4.86. Pressure oscillations observed on the ground at 13 miles' and 140 miles' distance, respectively, from a 1.2-ton explosive source. These curves were obtained by Jack Reed of Sandia Corporation.

subduct action is the cause of this complexity, it is clear that a number of these items must be involved in each traverse of the stratosphere. This means that the horizontal extent of the subducts must be limited to tens of kilometers at most, since the sound front is in the stratosphere less than 100 km during each loop.

Experimental data obtained at the turn of the century (F. J. W. Whipple, 1926) demonstrated that low-frequency sound waves traveled over great distances. Dean (1959) has calculated losses which would be experienced in certain assumed atmospheres according to the relation

$$\alpha = \frac{2.5 \times 10^{-7}(T^*)^{0.3}f^2}{P^*} + \frac{5.3 \times 10^{-4}(T^*)^3 f}{P^*} + \frac{10^2 f}{(T^*)^{2.5}e^{8.13/T^*}} \times \frac{2f_0 f}{f^2 + f_0^2}\left(\frac{\text{db}}{\text{mile}}\right), \tag{4.10}$$

where f is the number of cycles per second,

$$f_0 = \frac{P^*}{(T^*)^{0.8}}\left[900\left(\frac{wT^*}{P^*}\right)^2 + 500\,\frac{wT^*}{P^*} + 50\right],$$

$T^* = T/273$, T is degrees Kelvin, P^* is pressure in atmospheres, and w is absolute humidity in grams per cubic meter.

The first term in Eq. (4.10) represents absorption by viscothermal-rotational processes, the second term represents radiational effects, and the third term incorporates vibrational absorption processes. Evaluation of these various loss mechanisms by Dean (1959) has resulted in tabulated estimates of absorption of sound of various frequencies in representative atmospheres, a portion of which is reproduced in Table 4.5, which indicates that sound can indeed travel over great distances in the atmosphere. These calculations apply only to that portion of the atmosphere below the stratopause, but it is clear that low-frequency sounds enjoy a very favorable environment in the upper atmosphere. This trapping, coupled with a generous supply of low-frequency sound sources, makes the atmosphere a very noisy place, resulting in the nickname of "noisesphere" for that region of the atmosphere bounded by the earth's surface and the stratopause. High-intensity sounds have been observed to propagate through one or more oscillations in the lower (noisesphere) duct of the atmosphere with the ear serving as the sensor, while more sensitive devices tuned to very low frequencies have provided observation of more than one circuit of the globe for very long-wavelength oscillations of high initial intensity. Thus, sounds in the 10- to 20-cps range are in the audio range at distances of a few hundred kilometers, while pressure perturbations of 1-km wavelength have propagation distances in the tens of kilometers range. As has been mentioned above, and as is to be expected of any atmospheric phenomenon, these pressure oscillations are considerably modified in their transit. These modifications of the wave form, however, become less obvious as the wavelength of the disturbance approaches the width of the wave guide through which it is passing. Such is the case for waves with frequencies below 0.1 cps.

Surface observations of the sound pressure levels of ambient pressure fluctuations have provided background noise levels in the frequency range below 1 cps ranging from 0.1 to 10 dynes/cm^2. This is the total input to the detector, and thus is composed of various sources, only one of which is those pressure perturbations arriving at the sensor at a speed equivalent to sound velocity. An important source of this contamination of the desired signal is pressure fluctuations induced into the sensing system owing to local accelerations in the atmospheric flow field. Allowance for these nonacoustic outputs of the sensing systems leaves a relatively high level of noise at these low frequencies, making the earth's troposphere and stratosphere a very noisy place if one's sensitivity to these fluctuations were on a level with the audio range of a threshold of approximately 0.0001 dyne/cm^2 at 1000 cps. The latter point has been demonstrated by observations taken at high altitude on constant-altitude balloons (Webb *et al.*, 1959; Wescott, 1964), which illustrated the same high noise levels.

Where does this noise come from? Theoretical work by Lighthill

TABLE 4.5

Absorption of Sound in the Atmosphere[a]

Altitude (km)	100% Rel. humid.			50% Rel. humid.			10% Rel. humid.		
	1 cps	10 cps	100 cps	1 cps	10 cps	100 cps	1 cps	10 cps	100 cps
0	6.7×10^{-5}	7.2×10^{-4}	1.2×10^{-2}	6.8×10^{-5}	7.9×10^{-4}	1.8×10^{-2}	8.6×10^{-5}	2.6×10^{-3}	2.0×10^{-1}
5	9.4	1.4×10^{-3}	5.7	1.0×10^{-4}	2.5×10^{-3}	1.7×10^{-1}	2.8×10^{-4}	2.0×10^{-2}	1.7×10^{0}
10	2.8×10^{-4}	1.7×10^{-2}	1.1×10^{0}	4.7	3.5×10^{-2}	1.1×10^{0}	1.1×10^{-3}	8.5	5.3×10^{-1}
15	1.0×10^{-3}	4.3	1.1×10^{-1}	1.1×10^{-3}	4.2	1.1×10^{-1}	1.2	4.2	1.0
20	2.2	4.6	1.4	2.6	4.1	1.3	3.0	3.6	1.2
25	5.0	6.1	2.6	6.4	4.8	2.4	7.9	3.7	2.3
30	9.2	9.8	5.5	1.3×10^{-2}	7.0	5.2	1.4×10^{-2}	5.0	5.0
35	1.5×10^{-2}	1.8×10^{-1}	1.2×10^{0}	2.1	1.2×10^{-1}	1.1×10^{0}	2.0	8.7	1.1×10^{0}
40	2.3	3.5	2.5	3.3	2.4	2.4	3.0	1.7×10^{-1}	2.3
45	3.8	6.0	5.2	4.9	4.7	4.8	4.8	3.5	4.7

[a] Calculated by use of Eq. (4.10) (Dean, 1959). Units are decibels per mile.

(1952) and Meecham and Ford (1958) has indicated that turbulent regions in the vicinity of high-speed streams of a gas can be expected to be prolific generators of pressure perturbations which radiate energy away from the source with the speed of sound. These pressure variations are scaled according to the dimensions of the jet and the magnitude of shears involved, falling in the audio range for a source such as a jet engine. Under atmospheric conditions such as the tropospheric and stratospheric jet streams the dimensions are much larger and the shears far weaker, so the resulting frequencies of pressure oscillations are lower. If the turbulent motions about the high-speed jets of air in the atmospheric flow field should prove to be an important source of acoustical energy, there are immediately several predictable distinctive aspects of the resulting sound field. First, there should be strong seasonal variations in the noise level from this source, with a strong maximum in the winter season when both the tropospheric and stratospheric jet streams are active. The summer season should be appreciably quieter with the absence of the tropospheric jet stream and the relatively small shears of the stratospheric easterly circulation. During the winter storm period there should be appreciable day-to-day variations in noise level since both the stratospheric and tropospheric circulations exhibit gross changes, sometimes together and sometimes independently.

In addition to these long-term variations in strength of the noise sources, it is obvious that there will be a spatial distribution of this source of acoustic energy. Both stratospheric and tropospheric jets have their peak intensity in middle latitudes, although there is a seasonal progression, with the jets working toward lower latitudes in the fall and back toward the pole in spring in the tropospheric case with a general trend toward advancing from middle latitudes toward high latitudes in the stratospheric case. A considerable amount of meandering is the general rule for the tropospheric jet, and is sometimes a characteristic of the stratospheric jet during the winter storm period. With respect to altitude, strong concentrations of the generated energy are to be expected in the strong gradients below and above the jet stream wind maximums. These layers of maximum turbulent exchange should fall in the vicinity of 10- and 20-km altitude for the tropospheric jet case and in the 30- to 40- and 60- to 70-km altitude ranges for the stratospheric circulation case. Evidence for the existence of these turbulent zones in the stratospheric circulation has already been obtained from the MRN data (Webb, 1962, 1963), while the presence of a turbulent layer in the base of the tropospheric jet is now well documented by the clear air turbulence experience of modern aircraft.

Additional sources of low-frequency acoustical energy concern those natural and man-made events that involve pressure oscillations of large dimensions in the atmosphere. These include ocean waves, lightning discharges, landslides, explosions, earthquakes, meteoroid falls, and supersonic flight, to point out a few. Cook and Young (1962) have pointed out other probable sources such as tornadoes and auroral activity. When the superior propagation characteristics of the atmosphere are considered it is clearly to be expected that the troposphere and stratosphere will be noisy places indeed. The problem, as the experimenter quickly finds, is one of filtering the desired information from a heterogeneous background of relatively white noise.

Thus far we have considered the propagation problem in the atmosphere from a vantage point of drifting with the medium. In practice, the speed-of-sound structure of the noisesphere is formed by the collective contributions of thermal, wind, and composition structure, which are individually variable and thus contribute to a complex wave guide. Initial assumptions that the upper atmosphere would be a uniform medium have proved to be unfounded, and instead a large amount of detail structure is observed in the vertical profiles of all parameters that are measured with adequate sensitivity. Frequent soundings show changes with time which allow identification of specific features from one observation to the next in only a few cases. Superimposed on this small-amplitude structure is seasonal variation generated by the wind component, which on occasion completely eliminates the sound channel. A final factor about which we know the least, results from the possible presence of water vapor and other minor constituents in the upper atmosphere which could have a pronounced effect on the propagation of sound.

Winds provide the most variable aspect of stratospheric structure, with variations from easterly at approximately 50 meters/sec in the summer to westerly at 100 to 200 meters/sec in the winter. Since the winds stay relatively light in the region between the two circulation systems, this annual variation in the stratospheric winds can either strongly intensify the ducting characteristics of the noisesphere or completely eliminate the duct, according to the direction in which the sound wave is traveling. This vector character of the atmospheric speed-of-sound structure when the wind component is included is usually accounted for through use of zonal and meridional components of speed-of-sound structure as is illustrated in Fig. 4.87. The curves presented here represent winter mean sonic profiles for a sound wave traveling from the east, above, and west, respectively, for the curves from left to right. Westerly winds of the winter season in midlatitudes exhibit a strong negative lapse rate in the stratospheric circulation, which adds

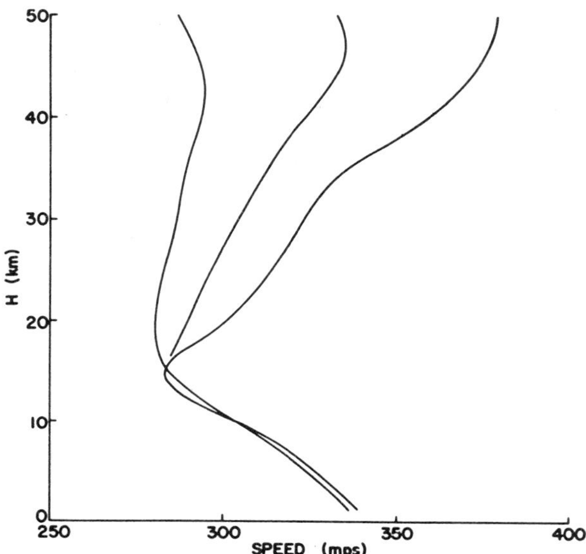

Fig. 4.87. Mean winter seasonal speed of sound profiles over White Sands Missile Range for a sound wave traveling horizontally from the west (right curve), from the east (left curve), and vertically (center curve).

directly to the usual negative thermal lapse rate in that region to produce a ducting configuration which is much stronger than normal in the upper portion, causing greater incidence angle acceptance than usual. The sound ray paths which travel into the upper portions of this duct where the speed of sound is greater than that at the earth's surface will necessarily strike the earth's surface during the next half cycle of propagation in the duct. By using both refractive and reflective modes of propagation considerably greater amounts of energy can be carried by the duct illustrated here if the sound wave is traveling from west to east. If the sound wave is coming from the east, however, the wind shear will tend to prevent the usual concave rotation of the wave front by the thermal field, and the sound energy may in large part be lost from the duct. On occasion the wind shear is greater than the thermal gradient in terms of effect on the speed of sound and the stratospheric negative acoustic gradient is completely eliminated. Under these conditions a leak in the noisesphere develops, and the ambient noise level should decrease.

Meridional components of the speed of sound have not been presented here, principally because these components deviate only slightly from the vertical speed-of-sound distribution in the general case. There are several notable exceptions to this rule, however, the most evident one being during the winter storm period, when strong meridional winds occur at all latitudes for short periods. Strong meridional wind components are a standard feature of high-latitude circulation in the stratosphere, with the monsoonal characteristic in the meridional flow of almost equal significance with the zonal component. These winds are from the south in summer and north in winter, and thus intensify the duct for a sound front traveling toward the pole in summer and away from the pole in winter. Of course this situation would be altered in the immediate vicinity of the pole, and in view of our very small sample of data in the zonal direction, there may well be significant variations with longitude. Although the meridional components of high-latitude flows are comparable with the zonal components, they are both generally weaker than those evidenced by the strong circulation of middle latitudes. The thermal structure is also generally weaker in high-latitude locations, becoming near isothermal at times during the winter seasons, and thus the ducting features of the atmosphere are easily altered by the wind structure.

A maximum stability of the atmospheric acoustic duct is to be found in tropical regions where the winds do not reach large values and the temperature structure has its greatest amplitude in the vertical. The lowest sound speeds to be found in the noisesphere are located here, with speeds as low as approximately 250 meters/sec a possibility during the periods when peak winds of the biennial cycle combine with the cold tropical tropopause. This means that the efficiency of duct trapping will be a maximum on successive years for alternately east and west propagating sound fronts in tropical latitudes. The climatological aspects of the global speed of sound distribution have been treated in greater detail (Webb and Jenkins, 1962; Williamson, 1965), but it is clear from the data available that a full understanding of the earth's sonic environment cannot be obtained without global MRN coverage, as well as a more detailed analysis of the composition structure of the stratosphere.

An additional important aspect of the acoustic structure of the atmosphere concerns the impact of small-scale inhomogeneities on high-speed flight. "Turbulence" has a special meaning to the aviation industry which is at variance with the traditional meaning assumed by meteorologists and fluid dynamicists. To the airman, any atmospheric interaction with his aircraft which causes the vehicle to move uncontrollably will be interpreted as "turbulence," without regard to the source or nature of the atmospheric factors which produced the undesirable motion. When flight speeds were limited to a few hundred miles per hour the variation in atmospheric motions, which the meteorologist calls turbulence, was the obvious cause of atmospheric-induced fluctuations in the flight path. As the speed of an aircraft is increased, however, the relative effectiveness of the wind in causing deviations in the flight path is diminished, and if this were the sole cause of turbulence input flight should become more smooth. Experience has shown that this is not the case, but that very severe perturbing factors may be encountered in high-altitude subsonic and supersonic flight.

The experiences that cause a pilot to report turbulence are actually variations in the lift and drag of the aerodynamic surfaces of the aircraft. This change is accomplished in the case of wind variations by changing the angle of attack of the airstream on the surfaces and by changing the relative speed of the aircraft through the environment. The latter effect results from the inertia of the aircraft, and either an acceleration or a deceleration must be accomplished before the aircraft is in equilibrium with the new wind field. As has been pointed out (Webb, 1963), it is possible to alter the lift and drag of a high-speed vehicle by modes other than wind, so that a readjustment in velocity is required before equilibrium is again established. This is so because the lift and drag of a high-speed vehicle are functions of the Mach number, and the Mach number is directly related to the local speed of sound. The relevant relations are

$$D = C_D A V^2, \tag{4.11}$$

where C_D is a drag coefficient, A is the characteristic area, and V is the velocity of the vehicle relative to the environment, and

$$M = V/v, \tag{4.12}$$

where v is the local speed of sound. Thus, without any change in the relative speed between the aircraft and the environment, the equilibrium may be upset by a change in the local speed of sound which will cause the vehicle to speed up or slow down or to rise or fall until a new equilibrium is established. Such changes in the local speed of sound may be implemented by changes in the temperature or composition of the environment as is indicated by Eq. (4.6).

In general the lift and drag of a vehicle are related in a complex fashion to the Mach number which related its speed to the speed of sound of the environment in which it is operating. For most vehicles this function is a constant at low speeds, usually beginning to increase at a Mach number of approximately 0.75. A maximum in the function is usually obtained at about Mach one, where the coefficient may be more than double the low-speed value. Above Mach one the coefficients usually decrease with increasing Mach number, but owing to the velocity squared term in Eq. (4.11), the total drag or lift generally increases steadily with increasing speed, in most cases rather sharply. It may be shown that variations in the speed of sound in a gas resulting from changes in temperature of the gas occur according to the relation

$$\frac{2v}{2T} = \frac{v}{2T}. \tag{4.13}$$

Other thermodynamic parameters such as the molecular weight and ratio of specific heats are similarly related.

It is well known that the temperature and composition structure of the atmosphere are inhomogeneous, principally as a result of advective transport in the process of eddy mixing between adjacent differing strata. When these eddy motions involve vertical motions they will in turn induce additional thermodynamic variations. It is probable, then, that the variations in local speed-of-sound structure will not be unrelated to the detail wind structure, but that both effects will be in operation at the same time. A particular structure would be resolved principally into a wind effect in the case of a low-speed vehicle, but would be principally due to sound speed variation in the case of a very high-speed traverse. Intermediate flight speeds which are characteristic of present day aircraft operations probably experience a mixture of these turbulence-inducing effects, but at the higher speeds attained by missiles and reentry vehicles much of the atmospheric variability which makes for a rough ride should be thermodynamic in origin.

The atmospheric variations which produce bumpiness in high-speed flight have rather large spatial dimensions compared with those which meteorologists normally assess as turbulence. The vehicle traverses these features at sufficiently high speed that the vehicle reaction is in that range which cannot be controlled out by the pilot, but which allows sufficient time for the vehicle to respond. As the input time is decreased the vehicle becomes less able to respond to the input forces, and these forces then impose maximum stress on the structure of the airframe. It is in the latter category that these atmospheric inputs become of most consequence to the safety of the vehicle. A single perturbation from atmospheric inhomogeneity sources is usually of small effect, but when a series of such inputs is imposed on a vehicle at the resonant frequency of the airframe structure, very large stresses may be generated.

Internal waves in the atmosphere have just the characteristics that would represent a maximum threat to high-speed flight. Such wave motions are perhaps best represented by the familiar lee waves on the downwind side of a mountain ridge when subjected to high-speed winds. These waves produce large deviations in the wind field as well as the thermodynamic structure, and the dimensions of these variations are systematic with a scale which depends on the approach azimuth; that is, if the line of flight is along the ridge line of the mountain range the variations experienced may well be easily controlled out of the flight path with no noticeable effect on the vehicle. On the other hand, if the flight path crosses the wave formation perpendicular to the ridge line a number of synchronous inputs may be experienced. At some in-between angle of approach the phase relations may be exactly right to effect resonance and produce a maximum turbulent input to the vehicle. Now the mountain wave is a very special case which is intense enough to be very obvious, but there is reason to believe that the atmosphere is permeated by such phenomena, which are not generally obvious owing to our lack of sensitivity in detecting such structure. Most of our radiosonde data of upper atmospheric structure are isolated nearly vertical soundings with data points obtained at intervals of approximately 300 meters. Wave formations of considerable importance could well be hidden in such data and only become apparent when the sensors become more adequate, such as a high-speed vehicle, or when they propagate into a less dense region of the atmosphere where energy conservation will require increased amplitudes of motions which can then be detected.

The large amount of detail structure of the stratosphere which has come to light in the MRN data could then be ascribed to the presence of systematic wave motions. It is important to realize that the output of the MRN sensing systems probably does not truly represent the character of the originating waves. These data simply represent the aspect viewed by that particular sampler, and the actual structure of the wave must be deduced from these data. We do have some indication of the nature of these waves by the photographic evidence presented by observations of noctilucent clouds which are illustrated in Figs. 5.14–5.17. These photographs give the impression of surface-type waves, much like ocean waves when viewed in the inverted position (Fig. 5.17), which would indicate that the formation mechanisms have much the same characteristics observed in wave motions such as lee waves; that is, the waves which are such an obvious part of noctilucent cloud formations appear to have their principal axis in near horizontal orientation. The wide spectrum of sizes which appear in the noctilucent cloud observations would indicate the probable existence of a similar wide spectrum in the lower atmosphere, except for a reduction in amplitude of the oscillation.

REFERENCES

Appleman, H. S. (1962). A comparison of climatological and persistence wind forecasts at 45 km. *J. Geophys. Res.* **67**, 767–772.

Appleman, H. S. (1963). The climatological wind and wind variability between 45 and 60 km. *J. Geophys. Res.* **68**, 3611–3617.

Appleman, H. S. (1964). A preliminary analysis of mean winds to 67 kilometers. *J. Geophys. Res.* **69,** 1027–1031.

Armstrong, C. L., and R. D. Garrett. (1960). High altitude wind data from meteorological rockets. *Monthly Weather Rev.* **88,** 187–190.

Attmannspacher, W. (1960). Über die Existanz hochstratosphärischen Nullschicht. *Meteorol. Rundschau* **13,** 38.

Attmannspacher, W. (1961). Windgeschwindigkeiten in der hochstratosphärischen Nullschicht. *Meteorol. Rundschau* **14,** 125.

Attmannspacher, W., J. Börstinger, and J. Wiehler. (1964). Preliminary results on rocketsonde measurements of wind and temperature in the upper stratosphere and in the mesosphere at Cape San Lorenzo during October/November, 1963. German Weather Service, Central Office, Offenbach A. M., Germany, Sci. Report No. 3.

Barrett, W. E., L. R. Herndon, Jr., and H. J. Carter. (1949). A preliminary note on the measurement of water-vapor content in the middle stratosphere. *J. Meteorol.* **6,** 367–368.

Bates, D. R. (1951). The temperature of the upper atmosphere. *Proc. Phys. Soc. (London)* **B64,** 805–821.

Batten, E. S. (1961). Wind systems in the mesosphere and lower ionosphere. *J. Meteorol.* **18,** 283–291.

Batten, E. S. (1963). A model of the annual temperature variations at 30° N between 30 and 60 km. Rand Corp., Santa Monica, California, RM-3564-PR.

Baynton, H. W. (1961a). Rocketsonde temperature profiles over White Sands, New Mexico. *Bull. Am. Meteorol. Soc.* **42,** 11–16.

Baynton, H. W. (1961b). AN/AMQ-15 rocketsonde tests at White Sands Proving Ground, New Mexico. *Bull. Am. Meteorol. Soc.* **42,** 34–41.

Bellamy, J. C. (1961). Requirements for high-altitude meteorology, 1965–1975. *Navigation* **18,** 156–163.

Belmont, A. D. (1964). The reversal of stratospheric winds over North America. General Mills, Inc., Minneapolis, Reports 1–4, AF19 (604)-6618.

Beyers, N. J., and B. T. Miers. (1965). Diurnal temperature in the atmosphere between 30 and 60 km over White Sands Missile Range, New Mexico. *J. Atmospheric Sci.* **22,** 3.

Brasefield, C. J. (1954a). Measurement of atmospheric humidity up to 35 km. *J. Meteorol.* **11,** 412–416.

Brasefield, C. J. (1954b). Winds at altitudes up to 80 km. *J. Geophys. Res.* **59,** 233–237.

Brekhovskikh, L. M. (1956). Focusing of sound waves by nonuniform media. *Akust. Zh.* **2,** 124–132.

Brekhovskikh, L. M. (1960). "Waves in Layered Media." Academic Press, New York.

Brown, J. A. (1960). Arcas high-altitude meteorological rocketsonde system. *Trans. N. Y. Acad. Sci.* [2] **22,** 175–183.

Chapman, S., and T. G. Cowling. (1952). "The Mathematical Theory of Non-uniform Gases," 150 pp. Cambridge Univ. Press, London and New York.

Chapman, S. (1930). A theory of upper atmospheric ozone. *Mem. Roy. Meteorol. Soc.,* **3,** 103–125.

Charney, J. G., and P. G. Drazin. (1961). Propagation of planetary-scale disturbances from the lower into the upper atmosphere. *J. Geophys. Res.* **66,** 83–109.

Charney, J. G., and J. Pedlosky. (1963). On the trapping of unstable planetary waves in the atmosphere. *J. Geophys. Res.* **68,** 6441–6442.

Charney, J. G., and M. E. Stern. (1962). On the stability of internal baroclinic jets in a rotating atmosphere. *J. Atmospheric Sci.* **19,** 159–172.

Clark, J. E. (1913). Air currents at a height of 50 miles as indicated by the bolide, Feb. 22, 1909. *Quart. J. Roy. Meteorol. Soc.* **39,** 115–126.

Conover, W. C., and C. J. Wentzien. (1955). Winds and temperatures to forty kilometers. *J. Meteorol.* **12,** 160–164.

Cook, R. K. (1962). Strange sounds in the atmosphere, Part I. *Sound* **1,** 12–16.

Cook, R. K., and J. M. Young. (1962). Strange sounds in the atmosphere, Part II. *Sound* **1,** 25–33.

Court, A., A. J. Kantor, and A. E. Cole. (1962). Supplemental atmospheres, U. S. Air Force, Cambridge Research Laboratories, L. G. Hanscom Field, Massachusetts, Project 8624, AFCRL-62-899.

Cox, E. (1948a). Upper atmosphere temperatures from remote sound measurements. *Am. J. Phys.* **16,** 465–474.

Cox, E. (1948b). Upper atmosphere temperatures from the "Helgoland big bang." *Bull. Am. Meteorol. Soc.* **29,** 78–80.

Craig, R. A. (1950). The observations and photochemistry of atmospheric ozone and their meteorological significance. *Meteorol. Monographs* **1,** 1–50.

Craig, R. A. (1951). Radiative temperature changes in the ozone layer. *Compendium Meteorol.* pp. 292–302.

Craig, R. A. (1960). Dynamic analysis of a stratospheric warming. *J. Geophys. Res.* **65,** 2485.

Craig, R. A., and G. Ohring. (1958). The temperature dependence of ozone radiational heating in the vicinity of the mesopeak. *J. Meteorol.* **15,** 59–62.

Craig, R. A. (1965). "The Upper Atmosphere: Meteorology and Physics," 509 pp. Academic Press, New York and London.

Crary, A. P. (1950). Stratosphere wind and temperatures from acoustical propagation studies. *J. Meteorol.* **7,** 233–242.

Crary, A. P. (1952). Stratosphere winds and temperatures in low latitudes from acoustical propagation studies. *J. Meteorol.* **9,** 93–109.

Crary, A. P. (1953). Annual variations of upper air winds and temperatures in Alaska from acoustical measurements. *J. Meteorol.* **10,** 380–389.

Crary, A. P., W. B. Kennedy, and V. C. Bushnell. (1954a). Atmospheric winds and temperatures at heights up to 50 km as determined by acoustical propagation studies. *Proc. Toronto Meteorol. Conf., 1953* pp. 9–13. Roy. Meteorol. Soc., London.

Crary, A. P. (1954b). Revision of high level wind temperatures in Bermuda. *J. Meteorol.* **11,** 257.

Crary, A. P., and V. C. Bushnell. (1955). Determination of high-altitude winds and temperatures in the Rocky Mountain area by acoustic soundings, October, 1951. *J. Meteorol.* **12,** 463–371.

Curtis, A. R., and R. M. Goody. (1956). Thermal radiation in the upper atmosphere. *Proc. Roy. Soc.* **A236,** 193–206.

Danielsen, E. F. (1964). Radioactivity transport from stratosphere to troposphere. *Mineral Ind. Penn. State Univ.* **33,** 1–7.

Dean, E. A. (1959). Absorption of low frequency sound in a homogeneous atmosphere. Texas Western College, El Paso, Schellenger Research Laboratory, Contract DA-29-040-ORD-1237, 86 pp.

De Quervain, A. (1916). Das Meteorbeben vom 29 Juli 1915. *Jahresber. Schweiz. Erdbebendienstes, Zurich* pp. 1–13.

Diamond, M. (1963). Sound channels in the atmosphere. *J. Geophys. Res.* **68,** 3459–3464.

Diamond, M., and O. M. Essenwanger. (1963). Statistical data on atmospheric design properties to 30 km. *Astronaut. Aerospace Eng.* **1,** 68–69.

Ditchburn, R. W. (1956). Absorption of ultraviolet radiation by the atmospheric gases. *Proc. Roy. Soc.* **A236,** 216–226.

Dobson, G. M. B., A. W. Brewer, and J. T. Houghton. (1962). The humidity of the stratosphere. *J. Geophys. Res.* **67,** 902–903.

Donn, W. L., and M. Ewing. (1962a). Atmospheric waves from nuclear explosions. *J. Geophys. Res.* **67,** 1855–1866.

Donn, W. L., and M. Ewing. (1962b). Atmospheric waves from nuclear explosions —Part II: The Soviet test of 30 October 1961. *J. Atmospheric Sci.* **19,** 264–273.

Elford, W. G. (1959). Winds in the upper atmosphere. *J. Atmospheric Terrest. Phys.* **15,** 132–136.

Elterman, L. (1953). A series of stratospheric temperature profiles obtained with the searchlight-probing technique. *J. Geophys. Res.* **58,** 519.

Elterman, L. (1954). Season trends of temperature, density and pressure to 67.7 obtained with the searchlight-probing technique. *J. Geophys. Res.* **59,** 351–358.

Elterman, L. (1964). Altitude variation of Rayleigh, aerosol and ozone attenuating components in the ultraviolet region. U. S. Air Force Cambridge Research Laboratories, L. G. Hanscom Field, Massachusetts, Environmental Research Papers No. 20.

Embrey, L. A. (1957). Bibliography for the International Geophysical Year. Publication of U. S. National Science Foundation, Washington, D. C., 51 pp.

Faust, H. (1960). Untergradientische Winde in der winterlichen Nullschicht 2. *Meteorol. Rundschau* **15,** 1–22.

Fensenkov, V. G. (1948). Obstoistel'stra Padeniia Sikhote Sikhote-Alinskogo Meteorita. *Astron. Zh.* **15,** 190–200.

Finger, F. G., S. Teweles, and R. B. Mason. (1963). Synoptic analysis based on meteorological rocketsonde data. *J. Geophys. Res.*, **68,** 1377–1399.

Fleagle, R. G. (1958). Inferences concerning the dynamics of the mesosphere. *J. Geophys. Res.* **63,** 137–146.

Fleagle, R. G., and J. A. Businger. (1963). "An Introduction to Atmospheric Physics," 346 pp. Academic Press, New York and London.

Fujiwhara, S. (1912). On the abnormal propagation of sound waves in the atmosphere. *Japan Central Meteorol. Obs. Bull.* **2,** 1–45.

Gazaryan, Yu. L. (1956). Waveguide propagation of sound in nonuniform media. *Akust. Zh.* **2,** 133–136.

George, J. L., and W. H. Peake. (1960). Survey of the literature of temperature determination at altitudes above 120,000 ft. Ohio State University, Antenna Laboratory, Report No. 973-1 DA-36-039-sc-84516, February, AD 235 897.

Gerson, N. C. (1956). General circulation of the high atmosphere. *J. Geophys. Res.* **61,** 351–353.

Goldberg, L. (1954). Absorption spectrum of the atmosphere. *In* "The Earth as a Planet" (G. P. Kuiper, ed.), pp. 434–490. Univ. of Chicago Press, Chicago, Illinois.

Goody, R. M. (1954). "The Physics of the Stratosphere," 187 pp. Cambridge Univ. Press, London and New York.

Gowan, E. H. (1947a). Ozonosphere temperature under radiative equilibrium. *Proc. Roy. Soc.* **A190,** 219–226.

Gowan, E. H. (1947b). Night cooling of the ozonosphere. *Proc. Roy. Soc.* **A190,** 227–231.

Greenhow, J. S., and E. L. Neufeld. (1959). Measurements of turbulence in the 80 to 100 km region from the radio echo observations of meteors. *J. Geophys. Res.* **64,** 2129–2133.

Groves, G. V. (1963). U. K. meteorological rocket grenade studies. *In* "First International Symposium on Rocket and Satellite Meteorology" (H. Wexler and J. E. Caskey, Jr., eds.), pp. 42–59. Wiley, New York.

Gutenberg, B. (1929). Die Entstehung der normalen Schallzonen bei Explosionen. *Z. Geophys.* **2,** 260–266.

Gutenberg, B. (1951). Sound propagation in the atmosphere. *Compendium Meteorol.* pp. 366–375.

Gutnick, M. (1961). How dry is the sky? *J. Geophys. Res.* **66,** 2867–2871.

Hanel, R. A., W. R. Bandeen, and B. J. Conrath. (1963). The infrared horizon of the planet earth. *J. Atmospheric Sci.* **20,** 73–86.

Haurwitz, B. (1954). The zonal wind field in the upper atmosphere. New York University, Dept. Meteorol. and Ocean., Sci. Report 7.

Haurwitz, B. (1957). Solar activity and atmospheric tides. *J. Geophys. Res.* **62,** 489–491.

Haurwitz, B. (1961). Frictional effects and the meridional circulation in the mesosphere. *J. Geophys. Res.* **66,** 2381–2391.

Haurwitz, B. (1962). Wind and pressure oscillation in the upper atmosphere. *Arch. Meteorol. Geophys. Bioklimatol.* **A13,** 144–166.

Haurwitz, B. (1964). Tidal phenomena in the upper atmosphere. *World Meteorol. Organ. Tech. Note* **58,** 1–27.

Havens, R. J., R. T. Koll, and H. E. LaGow. (1952). The pressure, density and temperature of the earth's atmosphere to 160 km. *J. Geophys. Res.* **57,** 59–72.

Hesstvedt, E. (1963). On the water vapor content in the high atmosphere. *Geophysica Norvegica* **25,** 1–18.

Hubert, W. E. (1961). Rocket observations of high level meridional flow over North America during 1960 and 1961. *Monthly Weather Rev.* **90,** 259–262.

Hulbert, E. O. (1957). Physics of the upper atmosphere. *Meteorol. Monographs* **3,** 160–181.

Jenkins, K. R., and W. L. Webb. (1959). High-altitude wind measurements. *J. Meteorol.* **16,** 511–515.

Jenkins, K. R., W. L. Webb, and G. Q. Clark. (1960). Rocket soundings of high-atmospheric meteorological parameters. *IRE, Trans. Military Electron.* **4,** 238–243.

Johnson, F. S. (1953). High-altitude diurnal temperature changes due to ozone absorption. *Bull. Am. Meteorol. Soc.* **34,** 106–110.

Johnson, F. S. (1954). Rocket observations of atmospheric ozone. *Proc. Toronto Meteorol. Conf., 1953* pp. 17–26. Roy. Meteorol. Soc., London.

Johnson, F. S., and W. L. Webb. (1963). The atmosphere and near space. *Astronaut. Aerospace Eng.* **1,** 81–86.

Johnson, F. S., J. D. Purcell, R. Tousey, and K. Watanabe. (1952). Direct measurement of the vertical distribution of atmospheric ozone to 70 km altitude. *J. Geophys. Res.* **57,** 157–177.

Johnson, F. S., J. D. Purcell, and R. Tousey. (1954). Studies of the ozone layer above New Mexico. *J. Atmospheric Terrest. Phys.* **1**, Spec. Suppl., 189–199.

Johnson, N. K. (1946). Wind measurements at 30 km. *Nature* **157**, 24.

Joint Scientific Advisory Group. (1961). The Meteorological Rocket Network and analysis of the first year in operation. *J. Geophys. Res.* **66**, 2821–2842.

Jones, L. M., J. W. Peterson, E. J. Schaefer, and H. F. Schulte. (1959). Upper air density and temperature: Some variations and an abrupt warming of the mesosphere. *J. Geophys. Res.* **64**, 2331–2340.

Julian, P. R. (1965). Some aspects of tropospheric circulation during midwinter stratospheric warming events. *J. Geophys. Res.* **70**, 757–768.

Junge, C. E. (1963). "Air Chemistry and Radioactivity," 382 pp. Academic Press, New York and London.

Keegan, T. J. (1961a). Observed variations of winds and circulations in the mesosphere. *Bull. Am. Meteorol. Soc.* **42**, 126.

Keegan, T. J. (1961b). Winds and circulations in the mesosphere. *J. Am. Rocket Soc.* [N.S.] **31**, 1060–1066.

Keegan, T. J. (1962a). Synoptic patterns at 100,000 to 200,000 feet. Proceedings of the National Symposium on Winds for Aerospace Vehicle Design, Vol. II. U. S. Air Force, Cambridge Research Laboratories, Surveys in Geophysics, No. 140, pp. 195–210.

Keegan, T. J. (1962b). Large-scale disturbances of atmospheric circulation between 30 and 70 km in winter. *J. Geophys. Res.* **67**, 1831–1838.

Kellogg, W. W. (1952). Temperatures and motions of the upper atmosphere. "Physics and Medicine of the Upper Atmosphere, Albuquerque," pp. 54–74.

Kellogg, W. W. (1960). Upper atmosphere studies. *Trans. Am. Geophys. Union* **41**, 179–183.

Kellogg, W. W. (1960). The dynamics of the polar mesosphere in winter. *Trans. Am. Geophys. Union* **41**, 620.

Kellogg, W. W. (1961). Chemical heating above the polar mesopause in winter. *J. Meteorol.* **18**, 373–381.

Kellogg, W. W. (1961). Warming of the polar mesosphere and lower ionosphere in winter. *J. Meteorol.* **18**, 373–381.

Kellogg, W. W., and G. F. Schilling. (1951). A proposed model of the circulation in the upper stratosphere. *J. Meteorol.* **8**, 222–230.

Kennedy, W. B. (1955). Further acoustical studies of atmospheric winds and temperatures at elevations of 30 to 60 km. *J. Meteorol.* **12**, 519–532.

Kennedy, W. B., and L. Brogan. (1954). Determination of atmospheric winds and temperature in the 30–60 km region by acoustic means. Univ. of Denver, Denver Res. Inst. Final Rept., Contract AF19(122)-252, AD-36 812.

Khvostikov, I. A., M. N. Izakov, G. A. Kokin, Yu. V. Kurilova, and N. C. Livshitz. (1963). Investigation of the stratosphere by means of meteorological rockets in the USSR. *In* "First International Symposium on Rocket and Satellite Meteorology" (H. Wexler and J. E. Caskey, Jr., eds.), pp. 34–41. Wiley, New York.

Kiss, E. (1961). Annotated bibliography on upper atmosphere structure. *Meteorol. & Geoastrophys. Abstr.* **12**, 776–827.

Kochanski, A. (1955). Cross sections of the mean zonal flow and temperature along 80°W. *J. Meteorol.* **12**, 95–106.

Kriester, B., K. Labitzke, R. Scherhag, and R. Stuhrmann. (1963). Daily and Monthly Northern Hemisphere 10-Millibar Synoptic Weather Maps of the Year

1963. *Meteorol. Abhandl. Inst. Meteorol. Geophys. Freien Univ. Berlin* **40,** 1–100.

Kuiper, G. P., ed. (1949). "The Atmosphere of the Earth and Planets," 749 pp. Univ. of Chicago Press, Chicago, Illinois.

Labitzke, K. (1962). Beiträge zur Synoptik der Hochstratosphäre. *Meteorol. Abhandl. Inst. Meteorol. Geophys. Freien Univ. Berlin* **28,** 1–93.

Labitzke, K. (1965). On the mutual relation between stratosphere and troposphere during periods of stratospheric warmings in winter. *J. Appl. Meteorol.* **14,** 91–99.

Lenhard, R. W., Jr. (1962). Hourly wind variability at 35 to 65 kilometers over Eglin Air Force Base, Florida. *Bull. Am. Meteorol. Soc.* **43,** 94.

Lenhard, R. W., Jr. (1963). Variations of hourly winds at 35 to 65 kilometers during one day at Eglin Air Force Base, Florida. *J. Geophys. Res.* **68,** 227–234.

Lenhard, R. W., Jr., and J. B. Wright. (1963). Mesospheric winds from 23 successive hourly soundings. U. S. Air Force, Cambridge Research Laboratories, L. G. Hanscom Field, Massachusetts, Report No. 63-836.

Leovy, C. (1964). Radiative equilibrium of the mesosphere. *J. Atmos. Sci.,* **21,** 238–248.

Lettau, H. (1951). Diffusion in the upper atmosphere. *Compendium Meteorol.* pp. 320–330.

Ley, W. (1951). Upper atmosphere, its exploration and exploitation. *Aeron. Eng. Rev.* **10,** 20–24.

Libby, W. F. (1963). Moratorium fallout and stratospheric storage. *J. Geophys. Res.* **68,** 2933–2937.

Lighthill, M. J. (1954). On sound generated aerodynamically, II. Turbulence as a source of sound. *Proc. Roy. Soc.* (*London*) **A222,** 1–32.

London, J. (1959). Dynamics of the mesosphere. New York University, Final Report AF19(604)-1738, AD 232 548.

McDonald, J. E. (1963). Cloud-ring in the upper stratosphere. *Weatherwise* **16,** 99, 148.

Mantis, H. T. (1960). On a diurnal variation of stratospheric winds. *J. Meteorol.* **17,** 465–468.

Massey, H. S. W., and R. L. F. Boyd. (1958). "The Upper Atmosphere," 333 pp. Hutchinson, London.

Mastenbrook, H. J., and J. E. Giner. (1961). Distribution of water-vapour in the stratosphere. *J. Geophys. Res.* **66,** 1437–1444.

Masterson, J. E. (1959). Review of meteorological sounding rockets. *Proc. 1st Intern. Symp. Rockets Astronautics, Tokyo,* 1958, pp. 216–223.

Masterson, J. E., W. E. Hubert, and T. R. Carr. (1961). Wind and temperature measurements in the mesosphere by meteorological rockets. *J. Geophys. Res.* **66,** 2141–2151.

Mathur, L. S. (1950). Reflection of sound waves from the stratosphere over India in different seasons of the year. *Indian J. Meteorol. Geophys.* **1,** 24–34.

Meecham, W. C., and G. W. Ford. (1958). Acoustic radiation from isotropic turbulence. *J. Acoust. Soc. Am.* **30,** 318–322.

Miers, B. T. (1963). Zonal wind reversal between 30 and 80 km over the southwestern United States. *J. Atmospheric Sci.* **20,** 87–93.

Miers, B. T. (1965). Wind oscillations between 30 and 60 km over White Sands Missile Range, New Mexico. *J. Atmospheric Sci.* **22,** 4.

Miers, B. T., and N. J. Beyers. (1964). Rocketsonde wind and temperature measurements between 30 and 70 km for selected stations. *J. Appl. Meteorol.* **3,** 16–26.

Mitra, S. K. (1952). "The Upper Atmosphere," 2nd ed., 713 pp. Asiatic Society, Calcutta.

Moore, R. G. (1959). Project Ozarc. *Bull. Am. Meteorol. Soc.* **40,** 375.

Morris, J. E., and B. T. Miers. (1964). Circulation disturbances between 25 and 70 km associated with the sudden warming of 1963. *J. Geophys. Res.* **69,** 201–214.

Mukherjee, S. M. (1952). Landslide and sound due to earthquakes in relation to the upper atmosphere. *Indian J. Meteorol. Geophys.* **3,** 240–257.

Murcray, D. G., F. H. Murcray, W. J. Williams, and F. E. Leslie. (1960). Water vapor distribution above 90,000 ft. *J. Geophys. Res.* **65,** 3641–3649.

Murcray, D. G., F. H. Murcray, and W. J. Williams. (1962). Distribution of water vapor in the stratosphere as determined from infrared absorption measurements. *J. Geophys. Res.* **67,** 759–766.

Murgatroyd, R. (1955). Wind and temperature to 50 km over England. Anomalous sound propagation experiments 1944/45. Meteorological Office, Great Britain, Geophysical Memoirs, No. 95, p. 33.

Murgatroyd, R. J. (1957). Winds and temperatures between 20 km and 100 km. *Quart. J. Roy. Meteorol. Soc.* **83,** 417–458.

Murgatroyd, R. J., and R. M. Goody. (1958). Sources and sinks of radiative energy from 30 to 70 km. *Quart. J. Roy. Meteorol Soc.* **84,** 225–234.

Murgatroyd, R. J., and R. Singleton. (1961). Possible meridional circulation in the stratosphere and mesosphere. *Quart. J. Roy. Meteorol. Soc.* **87,** 125–135.

Murray, F. W. (1960). Dynamic stability in the stratosphere. *J. Geophys. Res.* **65,** 3273–3305.

Nazarek, A. (1950). The temperature distribution of the upper atmosphere over New Mexico. *Bull. Am. Meteorol. Soc.* **31,** 44–50.

Newell, H. E., Jr. (1951). Temperatures and pressures in the upper atmosphere. *Compendium Meteorol.* pp. 303–310.

Newell, H. E., Jr. (1955). Rocket data on atmospheric pressure, temperature, density and winds. *Ann. Geophys.* **11,** 115–144.

Newell, R. E. (1961). The transport of trace substances in the atmosphere and their implications for the general structure of the stratosphere. *Geofis. Pura Appl.* **49,** 137–158.

Newell, R. E. (1963). Preliminary study of quasi-horizontal eddy fluxes from Meteorological Rocket Network data. *J. Atmospheric Sci.* **20,** 213–225.

Nicolet, M. (1960). The properties and constitution of the upper atmosphere. *In* "Physics of the Upper Atmosphere" (J. A. Ratcliffe, ed.), pp. 17–71. Academic Press, New York and London.

Officer, C. B. (1958). "Introduction to the Theory of Sound Transmission." McGraw-Hill, New York.

Ogden, D. E., and D. B. Swinton. (1960). Arcas temperature data in the mesosphere. *Monthly Weather Rev.* **88,** 191–192.

Ohring, G. (1958). The radiation budget of the stratosphere. *J. Meteorol.* **15,** 440–451.

Panofsky, H. (1956). "Introduction to Dynamic Meteorology," 243 pp. Pennsylvania State Univ., University Park, Pennsylvania.

Pant, P. S. (1956). Circulation in the upper atmosphere. *J. Geophys. Res.* **61**, 459–474.

Pekeris, C. L. (1937). Atmospheric oscillations. *Proc. Roy. Soc.* **A158**, 650–671.

Pfeffer, R. L. (1962). A multi-layer model for the study of acoustic-gravity wave propagation in the earth's atmosphere. *J. Atmospheric Sci.* **19**, 251–255.

Pfeffer, R. L., and J. Zarichry. (1962). Acoustic-gravity wave propagation from nuclear explosions in the earth's atmosphere. *J. Atmospheric Sci.* **19**, 256–263.

Pressman, J. (1954). The latitudinal and seasonal variations of the absorption of solar radiation by ozone. *J. Geophys. Res.* **59**, 485–489.

Pressman, J. (1955a). Diurnal temperature variations in the middle atmosphere. *Bull. Am. Meteorol. Soc.* **36**, 220–223.

Pressman, J. (1955b). Seasonal and latitudinal temperature changes in the ozonosphere. *J. Meteorol.* **12**, 87–89.

Priester, W., H. A. Martin, and K. Kramp. (1960). Diurnal and seasonal density variations in the upper atmosphere. *Nature* **188**, 202–204.

Quiroz, R. S. (1961a). Seasonal and latitudinal variations of air density in the mesosphere. *J. Geophys. Res.* **66**, 2129–2139.

Quiroz, R. S. (1961b). Air density profiles for the atmosphere between 30 and 80 km. U. S. Air Force, Air Weather Service, TR 150, AD 254 659.

Randhawa, J. S. (1966). Ozone measurements with rocket-borne ozonesondes. *J. Geophys. Res.* In Press.

Ratcliffe, J. A., ed. (1960). "Physics of the Upper Atmosphere," 586 pp. Academic Press, New York.

Reed, J. W. (1956). Ozonosphere winds and temperatures from acoustic observations of 1955 atomic tests. Sandia Corp., Albuquerque, Tech. Memo., 14 pp.

Regula, H. (1949a). Erforschung der Hochstratosphäre durch Schallwellen. *Meteorol. Rundschau* **2**, 263–267.

Regula, W. (1949b). Temperaturen und Winde in der oberen Stratosphäre. *Meteorol. Rundschau* **2**, 267–270.

Reisig, G. (1956). Instantaneous and continuous wind measurements up to the higher stratosphere. *J. Meteorol.* **13**, 448–455.

Reisig, G. (1958). Dynamic wind measurements into the stratosphere and indications on the structure of the wind field. *Bull. Am. Meteorol. Soc.* **39**, 436.

Reiter, E. R. (1962). Nature and observation of high-level turbulence especially in clear air. Colorado State University, Fort Collins, Colorado, Contract N189(188)55120A, 28 pp.

Reiter, E. R. (1963). Occurrence and causes of high-level turbulence—final report. Colorado State University, Fort Collins, Colorado, Contract N189(188)55 120A.

Richardson, J. M., and W. B. Kennedy. (1952). Atmospheric winds and temperatures to 50 km altitude as determined by acoustical propagation studies. *J. Acoust. Soc Am.* **24**, 731–741.

Roberts, W. O. (1963). Does variable solar activity affect stratospheric circulation? *Proc. Intern. Symp. Stratospheric Mesospheric Circulation,* pp. 341–352, Inst. Meteorol. Geophys. Freien Univ., Berlin.

Rocket Panel, Harvard College Observatory. (1952). Pressure, densities, and temperatures in the upper atmosphere. *Phys. Rev.* **88**, 1027–1032.

Rotolante, R. A., and A. M. Parra. (1965). Meteorological rocket data profile of the stratosphere, McMurdo, Antarctica. *J. Geophys. Res.* **70**, 749–756.

Sawada, R. (1956). The atmospheric lunar tides and the temperature profile in the upper atmosphere. *Geophys. Mag. (Tokyo)* **27**, 213–236.

Sawyer, J. S. (1958). Report of discussion on "Dynamical state of the upper atmosphere." *Weather* **13**, 281.

Scherhag, R. (1952). Die explosionsartigen Stratosphärenerwärmungen des Spätwinters 1951/52. *Ber. Deut. Wetterdienstes U. S. Zone* **38**, 51–63.

Scherhag, R. (1960). Über die Luftdruck-, Temperatur-und Windschwankungen in der Stratosphäre. *Abhandl. Math.- Naturw. Kl. Akad. Wiss., Mainz* pp. 87–93.

Scherhag, R. (1961). Eine erste synoptische Höhenwetterkarte der 0.5 mb Fläche. *Inst. Meteorol. Geophys. Freien Univ. Berlin. Wetterkarte* **33**, 1–20.

Sheppard, P. A. (1954). The meteorological point of view on observational data in the mesosphere. *Proc. Intern. Assoc. Meteorol., Rome,* 1953, pp. 509–513.

Sheppard, P. A. (1959). Dynamics of the upper atmosphere. *J. Geophys. Res.* **64**, 2116–2121.

Shvidkovskii, E. G. (1958). Nekotorye rezul'taty izmerenii termodinamicheskikh parametrov stratosfery pri pomoshchi meteorologicheskikh raket. *Iskusstv. Sputniki Zemli, Akad. Nauk SSSR* **2**, 10–16.

Shvidkovskii, E. G. (1959). Measurements of thermodynamic parameters of the stratosphere with the aid of meteorological rockets. *J. Am. Rocket Soc.* [N.S.] **29**, 733–736.

Sicinsky, H. S., N. W. Spencer, and G. W. Dow. (1954). Rocket measurements of upper atmosphere ambient pressure and temperature in the 30 to 75 km region. *J. Appl. Phys.* **25**, 161–168.

Siebert, M. (1961). Atmospheric tides. *Advan. Geophys.* **7**, 105–187.

Singer, S. F. (1953). Synoptic rocket observations of the upper atmosphere. *Nature* **171**, 1108.

Smith, L. B. (1960). The measurement of winds between 100,000 and 300,000 ft. by use of chaff rockets. *J. Meteorol.* **17**, 296–310.

Spencer, N. W., H. F. Schulte, and H. S. Sicinski. (1954). Rocket instrumentation for reliable upper atmosphere temperature determination. *Proc. IRE* **42**, 1104–1108.

Spitzer, L., Jr. (1952). The terrestrial atmosphere above 30 km. *In* "The Atmospheres of the Earth and Planets" (G. P. Kuiper, ed.), pp. 326–351. Univ. of Chicago Press, Chicago, Illinois.

Starr, V. P. (1959). Questions concerning the energy of stratospheric motions. U. S. Air Force, Cambridge Research Center, TN 59–665, December, AD 233 568.

Stolov, H. L. (1955). Tidal wind fields in the atmosphere. *J. Meteorol.* **12**, 117–140.

Sutton, O. G. (1960). High atmosphere research in the meteorological office. *Meteorol. Mag.* **89**, 97–98.

Teweles, S. (1959). Structure and circulation of the stratosphere. *Trans. Am. Geophys. Union* **40**, 84–88.

Teweles, S. (1961). Time section and hodographic analysis of Churchill rocket and radiosonde winds and temperature. *Monthly Weather Rev.* **89**, 125–136.

Teweles, S., and F. G. Finger. (1963). Synoptic studies based on rocketsonde data. *In* "First International Symposium on Rocket and Satellite Meteorology" (*H.* Wexler and J. E. Caskey, Jr., eds.), pp. 135–153. Wiley, New York.

Thiele, O. W. (1961). Density and pressure profiles derived from meteorological rocket measurements. U. S. Army Signal Missile Support Agency, White Sands Missile Range, New Mexico, Tech. Report 108.

Thiele, O. W. (1963). Mesospheric density variability based on recent meteorological rocket measurements. *J. Appl. Meteorol.* **2**, 649–654.

Toth, J. (1963). Über die Temperatur in der Stratosphäre und Mesosphäre. *Meteorol. Rundschau* **16**, 16–19.

Van Allen, J. A., and J. J. Hopfield. (1952). Preliminary report on atmospheric ozone measurements from rockets. *Mem. Soc. Roy. Sci. Liege* [4] **12**, 179–183.

van Mieghem, J. (1963). New aspects of the general circulation of the stratosphere and mesosphere. *Proc. Intern. Symp. Stratospheric Mesospheric Circulation* pp. 5–62. Inst. Meteorol. Geophys. Freien Univ. Berlin.

Volz, F. E., and R. M. Goody. (1962). The intensity of the twilight and upper atmospheric dust. *J. Atmospheric Sci.* **19**, 385–406.

Warnecke, G. (1961). Eine synoptische Höhenwetterkarte für das 0,5 mbar Niveau vom Sommer 1960. *Berlin. Wetterkarte* **37**, 62.

Watanabe, K. (1958). Ultraviolet absorption processes in the upper atmosphere. *Advan. Geophys.* **5**, 153–221.

Webb, W. L. (1962). Detailed acoustic structure above the tropopause. *J. Appl. Meteorol.* **1**, 229–236.

Webb, W. L., W. I. Christensen, E. P. Varner, and J. F. Spurling. (1962). Inter-Range Instrumentation Group Participation in the Meteorological Rocket Network. *Bull. Amer. Meteorol. Soc.* **43**, 640–649.

Webb, W. L. (1963). Acoustic component of turbulence. *J. Appl. Meteorol.* **2**, 286–291.

Webb, W. L. (1964). The dynamic stratosphere. *Astronaut. Aerospace Eng.* **2**, 62–68.

Webb, W. L. (1965). Scale of stratospheric detail structure. "Space Research V" (D. G. King-Hele, P. Muller, and G. Righini, eds.), pp. 997–1007. North-Holland Publ., Amsterdam.

Webb, W. L., and K. R. Jenkins. (1962). Sonic structure of the mesosphere. *J. Acoust. Soc. Am.* **34**, 193–211.

Webb, W. L., J. W. Coffman, and G. Q. Clark. (1959). A high altitude acoustic sensing system. U. S. Army Signal Missile Support Agency, White Sands Missile Range, New Mexico.

Wescott, J. W. (1964). Acoustic detection of high-altitude turbulence. University of Michigan, Contract DA-20-018-ORD-22840.

Wexler, H. (1950a). Possible effects of ozonosphere heating on sea-level pressure. *J. Meteorol.* **7**, 370–381.

Wexler, H. (1950b). Annual and diurnal temperature variations in the upper atmosphere. *Tellus* **2**, 262–274.

Wexler, H. (1961). Some aspects of stratospheric and mesospheric temperature and wind patterns. "Space Research I" (H. Kallmann Bijl, ed.), pp. 1083–1093. North-Holland Publ., Amsterdam.

Whipple, F. J. W. (1926). The detonating meteor of October 2, 1926. *Meteorol. Mag.* **61**, 253–258.

Whipple, F. L. (1954). Density, pressure and temperature data above 30 kilometers. *In* "The Earth as a Planet" (G. P. Kuiper, ed.), pp. 491–513. Univ. of Chicago Press, Chicago, Illinois.

Wilckens, F. (1961a). Meteorologische Raketensysteme, *Flugwiss.* **12**, 52.

Wilckens, F. (1961b). Bibliographi über raketenmess Verfahren in der Stratosphäre und Mesosphäre. *Flugwiss.* **11**, 99.

Wilckens, F. (1962). Bemerkungen zur der zeitigen Kenntnis der meridionalen Zirkulation in der Stratosphäre und Mesosphäre. *Meteorol. Rundschau* **15**, 23–27.

Williamson, L. E. (1963). The subpolar atmospheric acoustic structure in the autumn. *J. Geophys. Res.* **68**, 6267–6272

Williamson, L. E. (1965). Seasonal and regional characteristics of acoustic atmospheres. *J. Geophys. Res.* **70**, 249–255.

Yamamoto, G. (1962). Direct absorption of solar radiation by atmospheric water vapor, carbon dioxide and molecular oxygen. *J. Atmospheric Sci.* **19**, 182–188.

5

The Mesosphere

Introduction

The mesosphere has been defined as that region of the upper atmosphere above the stratopause in which the temperature decreases with height. Experimental data have shown its vertical dimension to be approximately 30 km, with the top established by the mesopause at roughly 80-km altitude. Above the mesopause the temperature again increases with height as ionospheric photoelectric effects become the predominant physical process. A nominal temperature decrease of 100°C is the rule, so the average mesospheric lapse rate of slightly over 3° per kilometer is somewhat weaker than the 6½° per kilometer that is usually observed in the troposphere. This structure is principally the result of turbulent mixing of this conditionally stable air and the positive lapse rate of ozone concentration which is characteristic of the mesosphere.

The lower half of the mesosphere is the scene of a light easterly thermal wind component throughout the year, resulting in a decrease with height of the winter westerlies from their peak value at stratopause altitudes and a continuation of the summer easterlies' increase with height to peak values in the upper mesosphere. This is caused by a positive mean meridional temperature gradient in both the summer and winter seasons or in both hemispheres. A different regime is found in the upper portion of the mesosphere (70–80 km) in the summer hemisphere where extremely cold temperatures occur and are reflected in a westerly thermal wind. In fact, this negative meridional temperature gradient is of global extent, ranging from the warm winter mesopause across the cool equator to a cold summer mesopause. This summer polar mesopause is the site of noctilucent cloud occurrence where the lowest temperatures of the earth's atmosphere have been observed.

A major difference in the kinetic structure of the upper atmosphere occurs near the stratopause and is evident throughout the mesosphere. The wind systems exhibit much greater variability in the mesosphere in comparison with the gross symmetry of the stratospheric circulation. Meridional components of the flow are large in the mesosphere, with values equaling and sometimes exceeding the zonal components in the mesopause region. The most probable reason for this state of affairs relates to frictional effects which remove energy from the long wavelengths of the general circulation and transform it into molecular energy (heat) through straightforward eddy viscosity effects. The principal question is the cause of the seemingly excessive circulation activity that is observed.

The physical situation in the mesosphere is similar to that of the troposphere except that the winds are strong at the base in the case of the mesosphere and the density is some three orders of magnitude lower. There is, in both cases, an underlying layer of considerable stability that is not smooth. Irregularities which result from terrain features of the earth's surface are matched at the stratopause by wave motions that provide the impetus for a vertical motion which, in the mesosphere, can grow and transform the kinetic energy into smaller eddy components. Rapid growth of these nonlinear viscous effects is probably due in part to the small reservoir of potential energy available in the thin air of the mesosphere, which is incapable of damping incident perturbations as is the case in the troposphere.

The mesopause was found experimentally to have a most unusual temperature distribution in the meridional plane. As was indicated above, the coldest point is to be found in the summer polar region, which is subjected to continuous solar radiation. The warmest point is located in the darkness of the winter pole. These facts point to it being very unlikely that thermal equilibrium is established by radiational processes in either of these cases. This is a little surprising since most of the upper atmosphere appears to experience radiational control of its temperature structure. There are, however, numerous other ways in which the temperature structure of the mesosphere can be controlled.

Several investigators have considered heating effects of particles which are guided into the polar night region of the atmosphere by the particular geometry of the earth's magnetic field. In general, the incoming particles are not sufficiently energetic to penetrate down to mesospheric altitudes in significant quantities to affect the temperature structure. It is hypothesized (Kellogg, 1961), however, that dissociation of oxygen molecules at auroral heights may be important in stratospheric and mesospheric heat balances as a result of the latent heat released by the recombination process. Subsidence is usually assumed to occur in the upper atmosphere during the winter season, and thus the density increases to produce enhanced re-association rates of the oxygenic elements. The heat thus made available to stratospheric and mesospheric regions is simply transported by convection (negative) from its ionospheric source. The process is similar to that in which water vapor effects the release of latent heat in the condensation phase of an updraft that produces saturation. Estimates of stratospheric and mesospheric heating from the influx of charged particles through the above process have generally indicated it to be inadequate as a primary source of heat.

All of the calculations of heat transfer processes are suspect at this time, however, since the basic environmental data are generally inadequate. Information on the more gross measures of the environment such as wind, temperature, and density is just now becoming available in sufficient quantity (and quality) to provide a first comprehensive view of the region. Composition of the mesosphere is only slightly known, and the rate at which physical processes such as mixing or recombination occur remains largely a matter of conjecture. Results obtained thus far are subject to considerable revision as better information becomes available, so it is premature to conclude that reactive heating processes have been eliminated from a dominant role in stratospheric and mesospheric heating in the polar night.

The gross thermal and circulation changes associated with continuous presence and absence of the heating by solar ultraviolet radiation produce a third prominent structural characteristic of the mesosphere. Shrinking associated with the cold, low-pressure system of the stratospheric circulation of the winter night causes the pressure and density to be reduced at a given altitude in the polar vortex. Summer heating, on the other hand, causes expansion and upwelling to lift the constant-pressure and -density surfaces to greater heights. This variation in density finds its maximum in the middle mesosphere, where the seasonal mean has been reported to vary from 0.103 to 0.170 gm/meter^3 from winter to summer, with extreme values of 0.071 and 0.190 gm/meter^3. These motions are accomplished isothermally in large part in mesospheric regions, since most of the work against the geopotential field is expended by expansion or contraction of stratospheric air. The vertical motions associated with this general reorganization of mesospheric structure do not, then, have the usual damping effects associated with convective motions of small elements of the fluid. These motions, however, produce horizontal pressure and density gradients in the mesosphere that will result in development of an adjusting circulation.

Assuming that the vertical pressure and density profiles are properly described by Eq. (4.5) and Fig. 4.80, the pressure gradient force will be directed from the summer pole toward the winter pole. A general derivation of the geostrophic wind equation and the thermal wind equation involves a similar orientation of the pressure gradient force and the thickness gradient force; that is, heating of an air mass will result in expansion of that air and lifting of the pressure surfaces so that they slope away from the heated region, and the air flows away from the heated region. This same expansion causes the pressure surfaces to be further separated in the heated region than they are in neighboring regions. Under these conditions the thermal wind relation (Eq. (3.1)) calls for a particular change in the horizontal wind with height. The nature of this change is such that the zonal wind of this hemispheric circulation is increased with height around a warm high-pressure or a cold low-pressure center. These are the situations to be found in the Northern Hemisphere's stratospheric circulation.

Now this new situation in the mesosphere is different in that we have derived the high-pressure region by bodily lifting all of the air above the expanding region, and the heat exchange involved is of small consequence, and is possibly even isothermal. In the mesospheric case, pressure surfaces would be closer together at the center of high pressure and spaced further apart in surrounding regions which have not been lifted, a result which is very apparent owing to the exponential density function which provides a steep gradient in the vertical that is greater with increasing pressure. The above picture of mesospheric density and pressure thickness gradient distribution in the mesosphere presumably would exist if the situation were static, and the atmosphere above the stratopause heating layer simply lifted and lowered with the seasonal heat cycle of the ozonosphere. In fact, we know from experimental data that the mesopause region is warm in winter and cold in summer. This means that the exponential density profile of the winter mesosphere falls off less rapidly with height (that is, the exponent in Eq. (4.5) is smaller), while the density gradient of the mesosphere over the summer pole is increased. This is precisely the change that is required to correct the pressure imbalance generated by the original stratospheric expansion at the summer pole and contraction at the winter pole.

We can deduce from the above discussion that a new circulation system will become operative in mesospheric altitudes and above, which will have the distinctive characteristics of vertical flows in the polar regions, upward in the summer mesopause region and downward in the winter mesopause region. Such a flow necessitates a meridional transfer of mass from summer high latitudes to winter high latitudes in the atmosphere above the mesopause, and a return flow from winter high latitudes to summer high latitudes in the atmosphere below the mesopause level. The strength of this circulation will generally depend on the strength of the stratospheric temperature difference at the two poles, although there probably will be complications due to the existence of the third heating layer of the atmosphere, the ionosphere. For purposes of this discussion, we will commit the usual error and assume that we can ignore the influence of that latter item, possibly using the familiar excuse that ionospheric effects should be small owing to the much reduced density and energy content. Actually, it is another story.

Such a flow has been observed in the 80- to 100-km region, as is described by Kochanski (1963) using meteor trail data on ionospheric winds. The results of his analysis for the meridional case are illustrated in Fig. 5.1, showing a well-defined monsoonal circulation with mean winds flowing away from the summer pole. Reversal dates in March

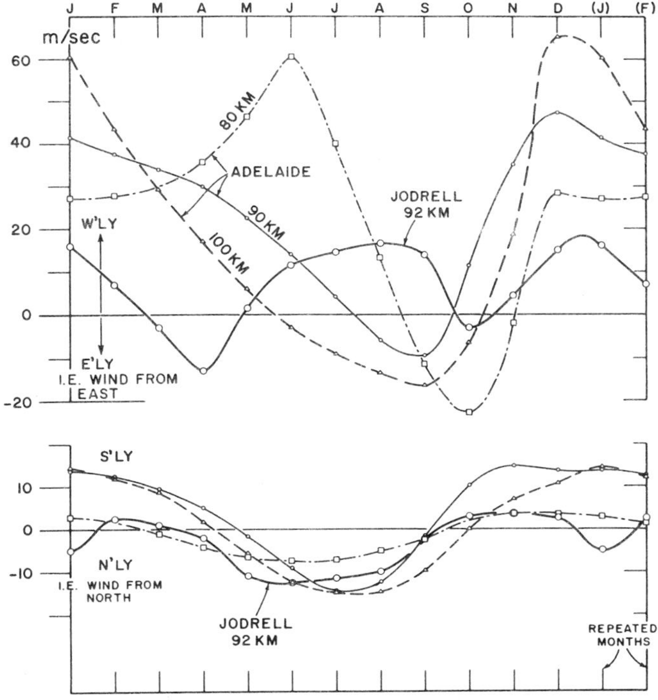

Fig. 5.1. Annual progress of zonal and meridional wind components obtained by Kochanski (1963) through analysis of meteor trail wind measurements. The data are for Adelaide, located at 35° S and 138° E, and for Jodrell Bank, at 53° N and 2° W. (Courtesy Journal of Geophysical Research)

and April in the spring and in September in the fall are in agreement with requirements of heating characteristics of the stratospheric circulation system, and the amplitude of the flow, reaching a maximum of approximately 15 meters/sec on about the dates of the summer and winter solstice, appears to be adequate to account for the mass which must be moved to achieve the required heat transport observed in the polar regions. It should be noted that some of the flow could be carried between hemispheres in higher portions of the ionosphere, but the reduced density precludes a major portion of the mesopause meridional circulation being carried by that region. While the data available in the ionospheric region immediately above the mesopause are quite meager, the considerable consistency of the data presented by Kochanski lends credence to the concept of this meridional circulation.

As has been discussed in Chapter 4, a large diurnal oscillation is observed in the stratospheric temperature and wind fields (Section 4.8), and the MRN data illustrate certain special features during the spring and fall circulation reversals (Sections 4.4 and 4.5). These factors can be considered further to evaluate MRN data obtained during the summer solstice period. The hemispheric picture is illustrated in Fig. 5.2, where the subsolar point is depicted at 23½ degrees latitude and the region of continuous illumination is indicated by the dashed circle about the North Pole. The ridge of heated air has its maximum temperature at the pole and a ridge of high temperature (and thus high pressure) extending equatorward with a considerable lag (see Section 4.8) where the day and night periods are more nearly equal. The diurnal temperature minimum, on the other hand, lies in a trough of low pressure immediately in front of the sunrise line and terminating in the nighttime sky at a latitude of approximately 60 degrees.

The diurnal temperature difference which has been observed thus far (Section 4.8) in low latitudes is of the order of 20°C in the lower mesosphere. This change occurs between approximately 5 AM and 2 PM. The thermal wind relationship of Eq. (3.1) prescribes a meridional flow from the equator toward the pole in that region if geostrophic conditions are met. The diurnal temperature gradient is not longitudinal at high latitudes, however, but is largely latitudinal with a maximum gradient in the high-latitude region from 60 to 90 degrees. In addition, the temperature difference between these extremes will be generally of the same order. The resultant horizontal gradient of temperature will be significantly larger in the high-latitude case, principally because the difference is applied across the much shorter distance between the summer pole and the nighttime darkness boundary. The diurnal tidal motion thus induced by solar heating will then be largely meridional at low

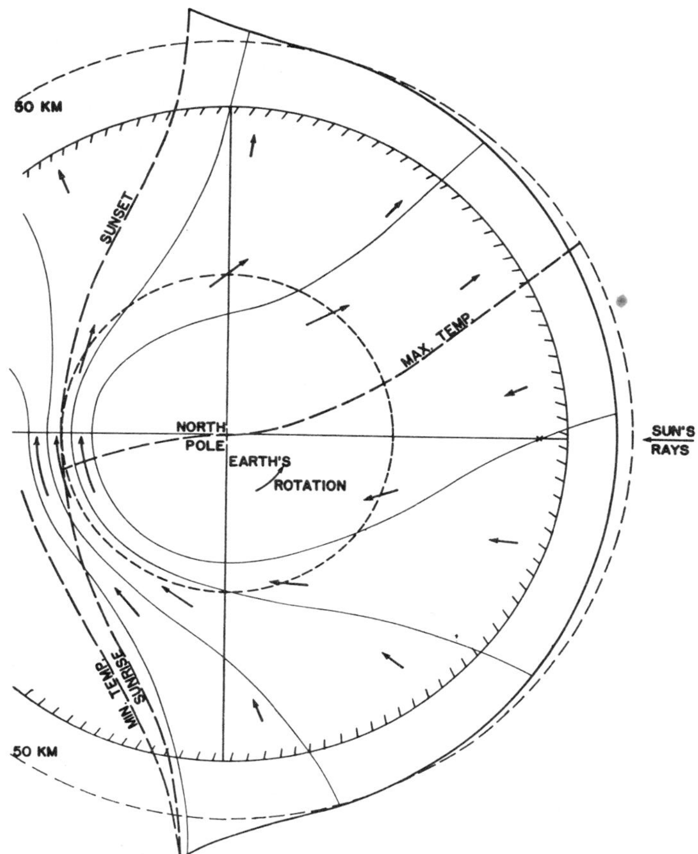

Fig. 5.2. Equatorial plane projection of the stratospheric diurnal circulation at the summer solstice. Thin solid curves are constant temperature or thickness contours, and the arrows illustrate the tidal wind field. This diurnal circulation is superimposed on the general summer easterly circulation.

latitudes, largely zonal at high latitudes, and will be strongest in high latitudes.

Inspection of Fig. 4.1 indicates that, at least under the assumed uniform conditions, this tidally induced motion will rise in altitude with increasing latitude as a result of the physical process through which the energy is transferred from the solar radiant flux into temperature of the stratospheric air through ozone interaction. This lifting of the stratum most strongly influenced by diurnal motions with increasing latitude is indicated by the mean MRN meridional data presented in Fig. 4.79.

These mean data for the summer season show an increasing southerly wind component appearing at higher altitudes for the more northerly MRN stations. This diurnal oscillation in the stratospheric wind field then has its maximum amplitude near the stratopause in middle and low latitudes and rises well into the mesosphere in polar regions. A vertical motion at high latitudes is thus a part of the tidal circulation, and the peak altitude attained must be in the upper mesosphere and must occur near local midnight in the twilight regions of the continuously illuminated polar region. This region of enhanced zonal flow has been termed the "stratospheric tidal jet" as a result of its origin in the stratospheric circulation.

All of these diurnal motions are superimposed on the general easterly circumpolar circulation of the stratospheric summer season. The high-latitude meridional temperature gradient serves to accelerate this easterly flow as it passes through the twilight zone between approximately 3 AM and 10 PM. A maximum in the easterly zonal flow at these latitudes (60–70 degrees) is then to be expected shortly after midnight and a minimum should be observed near local noon. In neither case (low or high latitude) will the diurnal oscillations be sinusoidal, and in both cases the accelerating phase (positive temperature gradient) is shorter in time than the return phase.

These comments indicate that a special circulation system will result from this solar diurnal heat input. As is pictured by the arrows of Fig. 5.2, the low-latitude motions induced by this temperature oscillation will be largely a to-and-fro meridional oscillation with little transport. At high latitudes, however, the accelerating force can be expected to produce a net circulation about the nighttime side of the continuously illuminated polar region. This diurnally induced circulation will then serve to mix the mesospheric atmosphere along meridional lines. Owing to several variables which are inadequately known and may exert important influences it is impossible to assess the efficiency of this diurnal circulation with our limited data.

As will be discussed later in this chapter, the high-latitude summer mesopause region is well known as the coldest region in the earth's atmosphere. The temperatures that appear to exist at that place and time are very difficult to explain on the basis of static and radiational equilibrium considerations. Present knowledge of mesopause structure suggests very strongly that the high-latitude features of extreme cold in summer and warm temperatures in the winter night are the result of circulation processes. If so, the physical requirements are that upward motions exist in the summer polar region and that subsidence be a characteristic of the winter polar regions.

The required upward motion which, through expansion in the reduced density environment, would supply the observed low temperatures could result from nongeostrophic motions in the tidal circulation described above. The available data indicate that the stratospheric tidal jet is located in the upper mesosphere and ascends as it circles the arctic region in the AM nighttime hours. If all or even a small part of the mass of air involved in the circumpolar tidal circulation were to be forced through momentum effects to ascend above the mesopause, that air would quickly become incorporated in the horizontal circulation of that very stable region. It is then hypothesized that the stratospheric tidal circulation is accelerated ahead of the heat wave and into the rectifying action of the Arctic circle circumpolar flow where the jet has its maximum speed. The stratospheric tidal jet then enters a divergent situation near midnight, and it is here that the maximum possibility exists that air will be transported across the mesopause and become a part of a meridional circulation.

Any air transported into the lower ionosphere will represent a loss to the diurnal component of the stratospheric circulation, and thus should be reflected in a difference in the poleward and equatorward transport. It is easy to see that such an observation is impossible with the MRN data, since if the approximate 10-meters/sec mean meridional speed obtained by Kochanski (Fig. 5.1) is assumed to be appropriate, continuity considerations concerning the transporting wind in a compensating flow at 70 km yield a value of 1 meter/sec, and at 50-km altitude the flow would only be 0.1 meter/sec. These numbers do, however, provide a clue to the capability of the stratospheric tidal jet in supplying the meridional flow of the lower ionosphere. If we assume a mean layer involved in the tidal jet of 10 km (70–80) and with known wind speeds of near 100 meters/sec, the upward motion required for this transport should not exceed a few per cent of the total mass involved in the flow. The remaining major portion of the tidal jet would return to the stratopause as the return flow of the diurnal oscillation.

A special physical characteristic concerns frictional effects, which could produce subgeostrophic winds, and thus change the principally zonal circulation of the stratosphere into a circulation in which the meridional aspect is of great importance. Haurwitz (1961a) has considered this possibility, and has concluded that frictional effects in strong wind shear regions of the upper atmosphere would result in frictional retardation or deceleration of the flow in specific layers of the upper atmosphere, but in general he concluded that the geostrophic approximation was probably valid in the stratosphere and mesosphere.

Haurwitz based his analysis of the possible effect on rather average values of vertical and horizontal gradients in the velocity field. The large amount of detail that is now known to be a general feature of stratospheric and mesospheric vertical structure, and also on a slightly larger scale in the horizontal plane, was not obvious in the data available to him for that study. The mechanism by which the energy of the general circulation is removed from the geostrophic wind through frictional acceleration would appear from the MRN data to be concentrated on a much smaller scale than that considered, and values of shears are known to be significantly higher than that assumed in these calculations. When these larger values of flow of energy out of the general circulation are considered, mean meridional winds of the order of a few meters per second at the stratopause in lower latitudes increasing to the order of 10 meters/sec at high latitudes appear to be entirely reasonable. The expected increase with height of this effect, which could be the direct result of both kinetic and eddy viscosity increases with altitude, presents no particular problem in accepting the more than 10 meters/sec meridional velocities found in the 80- to 100-km layer by Kochanski. It appears reasonable to conclude, then, that quite possibly the zonal circulation of the upper atmosphere becomes increasingly subgeostrophic with altitude from the stratopause upward, and that the energy extracted from the general circulation is largely converted into a temperature rise through viscous absorption of the energy of the general circulation.

This estimate is based not only on a most likely solution, but also the measured data, which show the upper atmosphere to contain a large amount of detailed vertical structure. This is true even in the very stable upper stratosphere where the motions that generate these features must surely be forced. The situation is more complex in the mesosphere since the possibility of instabilities is present, but even there active convective instability would not be forecasted in the absence of data on such occurrences. The result of detailed investigations of small-scale features of the upper atmosphere, in the stratosphere and ionosphere at least, has been to reveal a profusion of small-scale features in every profile that is observed with adequate sensitivity. Figure 4.69 illustrates the fact that the dimensions of these small features appear to increase exponentially with height, although the data of Fig. 4.70 indicate possible sources of additional smaller-scale structures in the mesospheric altitude range. One of the principal outstanding problems concerning our knowledge of these phenomena is information on the horizontal dimensions of the disturbances. There are very real reasons to expect the horizontal scales to vary significantly with altitude, illustrating a maximum in the stable regions of the upper stratosphere and the ionosphere. The most informative probes with which we may gain information at this time would appear to be the propagation of acoustical and electromagnetic energy through the region in the horizontal plane, or the flight of very high-speed vehicles at these levels. In each case, the principal problem will center on our acumen in interpreting the resulting data.

A start has already been made by Hines (1960) in his remarkable insight into the nature of these phenomena, relating them to the presence of internal atmospheric gravity waves. While his work was initiated in ionospheric regions using data obtained from observation of meteor trails and ionospheric drift measurements, the likelihood that the phenomenon observed in that region is simply an obvious manifestation of a general physical process has led to extension of his analysis into other regions. The curves of Fig. 4.71 are a step in this direction, and they indicate that the data are remarkably consistent. This result is probably related to the mode of data acquisition, which is, interestingly enough, limited to the stable portions of the upper atmosphere. While the mesosphere has proved to be one of the more difficult regions to inspect for these items, it may well be one of the most interesting.

We are really not without information on the nature of the wave-type motions of the mesosphere, and possibly have at our fingertips the most valuable source of data available in the entire upper atmosphere in the very special noctilucent clouds. They appear to be a veritable picture of at least one class of these body waves and illustrate for our analysis the scale in vertical and horizontal dimensions as well as orientation in a four-dimensional frame. The problems of observation are considerable, and interpretation of all features is at this time not sure, but there is a strong probability that these rare clouds today provide our most important source of information on the detail structure of the mesosphere. At least, they provide a different observational aspect of the problem, and even if they fail to provide answers, they will surely broaden our outlook so that we shall become more conscious of the physical nature of the upper atmosphere.

Haurwitz (1964) has considered the wave nature of noctilucent clouds, with special attention to whether they might be formed in the mode of surface waves on the discontinuity of the mesopause. His results were inconclusive because of the lack of adequate data, but the study indicates the data required to provide a solution. If the wave motions observed in noctilucent clouds should prove to be a surface type of phenomena, the internal effects on the fluid should decrease in an exponential fashion with distance away from the discontinuity surface, and thus their presence would be a special feature of the upper atmosphere, and could not be generalized to describe the phenomenon which produces the detail structure found in upper atmospheric profiles.

If, on the other hand, the waves in noctilucent clouds should prove to be manifestations of internal gravity waves their usefulness as a tool for analysis of upper atmospheric dynamics would be greatly enhanced. For these reasons it is essential that special observational attention be directed toward these clouds.

It is well known that water vapor molecules will be dissociated in the solar ultraviolet environment of the mesosphere. It has been postulated here that water vapor is indeed present in the upper atmosphere, having crossed the tropopause into the upper stratosphere by molecular diffusion (Chapter 4) in equatorial regions where residence times are sufficient to sustain that process, and is available for modification of a circulation system or to supply hydrogen atoms for the earth's exosphere and oxygen atoms and molecules to catalyze the several reactions of the mesosphere. Water vapor will diffuse upward in the gravitational field during the protected nighttime hours only to be eliminated by the destructive ultraviolet flux a few hours later. Since with normal incidence the solar radiation will keep the water molecules contained to near the stratopause level, there is a possibility that the mesospheric meridional circulation has a role in the transport of this constituent. A ready possibility is that the stratopause portion of the stratospheric tidal circulation would take moisture from the diffusion-enriched upper stratosphere and carry it to high latitudes in that reasonably protected stratum before lifting the water vapor into the destructive environment of the upper mesosphere. As to whether the water vapor molecules could exist until arrival at the noctilucent cloud level to take part in their formation is still very much open to question, but if their participation should prove to be an important aspect of the noctilucent problem a new light would be shed on the diurnal variation which is now suspect in the observational data. The water vapor would almost surely be dissociated before it reached the mesopause level if it were rising on the sunlight side of the summer polar region, while it would be shielded during most of its ascent on the antisolar side of the ascending current. It is possible to inspect the particles of the noctilucent clouds for appreciable water content, although not an easy observation, and the results of such study should prove most informative.

The impact of the earth's ionosphere on human activity has led to detailed inspection of that region, at times by-passing other layers of the atmosphere, even when observational techniques could obtain data. The mesosphere falls generally in the latter category, since it is somewhat electrified throughout, increasingly so with altitude. The specific conductivity varies by an order of magnitude across the mesosphere to of the order of 10^{-6} amperes per volt per meter at the mesopause. A major portion of this electrical activity results from the presence of free electrons which are far outnumbered by mesospheric ions. While the free electrons are of principal importance in the case of radio wave propagation owing to their mobility, it is clear that the ions would excel in influencing the physical environment in the presence of phenomena such as gravity waves. Stratification of electrical inhomogeneities generated by phenomena such as meteor entries could predictably be accomplished by variations in the wind field. In this case the ions would be the leaders and the free electrons would be the impurity that would tend to damp out the process.

The mesosphere is then a transition layer, separating the bulk of the atmosphere from the ionospheric condenser plate which carries the earth's fair weather electrical potential, introducing a turbulent medium between the stable upper stratosphere and ionosphere to cushion the impact of their differing diurnal variations, and providing the avenue through its polar regions for transfer of mass between these effectively separated regions of the atmosphere. The data on specific mesospheric characteristics are largely inadequate, with only enough information available to provide an elementary insight into the physical situation that must exist. These few data should be used to assure maximum effectiveness in our comprehensive exploration of the entire atmosphere.

5.1 Mesospheric Circulation

The circulation of the mesosphere is an integral part of the stratospheric circulation system, forming the upper half of that central atmospheric monsoonal circulation originating in meridional temperature gradients produced by ozone absorption of solar ultraviolet radiation. These meridional temperature gradients have their greatest strength in the 40- to 50-km altitude region and for a variety of reasons are not symmetrical in the meridional plane, but vary in a complex fashion with latitude. The driving energy for the stratospheric circulation is thus located in the upper stratosphere and becomes of diminishing importance with height in the mesosphere. The energy exchange processes which are basic to the stratospheric circulation are principally concerned with ozone and the energy contained in the 2000- to 3000-angstrom region of the solar spectrum. Since the ozone concentration diminishes with altitude in the mesosphere to a negligible amount at the mesopause, the contribution of these processes to the upper atmospheric heat balance is of decreasing consequence in upper mesospheric altitudes. Absorption of shorter ultraviolet energy by oxygen and other atomic species becomes the dominant mode of upper atmospheric heating at higher levels, and the mesopause is formed by the lack of heat between the ozone-warmed stratopause and the oxygen-warmed ionosphere.

Data on the structure of the stratosphere are now obtained on a synoptic basis over the northwestern quadrant of the globe by the Meteorological Rocket Network. One may confidently expect that the MRN will be expanded in space and time to provide for a comprehensive global synoptic system based on current and improved small rocket systems and special sensitive sensors. Ideally these systems would be extended upward to include all of the stratospheric circulation system to provide a comprehensive observational system on an economically sound basis. There are limits to the range over which sounding techniques can be applied, and attainment of the 80-km level will require additional development in all systems. The first difficulty concerns the temperature measurement, where reduced coupling with the environment introduces lag, and high fall rates result in dynamic heating of the sensor and smooth the remaining sensitivity until only nonrepresentative "nice" smoothed data are obtainable. Faster response can be obtained through use of thin film techniques at the expense of certain other losses in accuracy and sensitivity of the measuring system. It is likely that such techniques will assume a role in the MRN with continued development, but at the present time measurements in the mesosphere with contact-type temperature-sensing devices is limited to 70-km altitude, and a large portion of the data which have been obtained during these first five years' operation of the MRN is limited to 60 km. Currently, the principal source of temperature data in the upper mesosphere is the grenade experiment, which was initiated by the U. S. Army Signal Corps (Weisner, 1956) during the late 1940's and early 1950's and continued by the National Aeronautics and Space Administration as a part of the IGY program and subsequently in a systematic exploration of the mesosphere.

Acquisition of wind data by the more desirable MRN wind sensors is also in trouble in the mesosphere in that fall rates of currently available rocket-borne parachutes are excessive above the 70-km level, although properly applied data reduction techniques can be incorporated to reconstruct the wind profile with reasonable accuracy to slightly higher altitudes. As can be noted in Fig. 2.12, the fall rate increases rapidly with height in this region and it is difficult to construct a sampling configuration with an acceptably small fall rate at the mesopause (approximately 80 km) that accomplishes the entire observational job in a reasonable period. Systems for wind determination having satisfactory characteristics in the mesopause region are in use. These sensors are passive clouds of small cylinders of conducting materials which are adjusted in length to act as antennas to absorb and re-radiate electromagnetic energy at the particular wavelength of a radar tracking device which can then obtain a position-time plot of the cloud as it falls. The fall rate of such a cloud can be reduced to acceptable values at 80 km, but the cloud expands and disperses with resulting inaccuracies in wind measurement with time so that a complete wind profile cannot generally be obtained.

In addition to the problems associated with sensing the environment, which may well be overcome with additional development, current rocket systems limit the upper altitude at which synoptic data are produced by the MRN. It is desirable to deploy the rocket sensing system at a minimum velocity, since any excessive speed will result in a reduction in sensitivity to variations as well as introducing undesirable side effects. Optimally, then, the rocket will reach peak altitude just above the level at which measurements are to be taken and ejection will occur at that point. In practice it is difficult to achieve precision in inexpensive systems, so that all parameters are not generally optimum. Combining all of the above effects results in the current limitations of maximum altitude to which the MRN synoptically obtains data.

The grenade experiment obtains temperature and wind measurements in the upper atmosphere to peak altitudes of the order of 100 km. The data are mean values over the altitude intervals between grenade bursts, which range from 3 to 10 km according to the specific experiment. A principal problem with application of the grenade experiment to synoptic analysis of the mesosphere is the expense of each sounding, so the amount of data obtained with this system is limited to sporadic soundings at carefully selected times and places to achieve maximum information yield. These relatively few data points constitute the reservoir of data on temperature and wind structure of the upper portions of the mesosphere and are used here to extend the general features of our knowledge of the stratospheric circulation up to the mesopause.

The fact that the stratospheric circulation extends upward into and dominates the wind structure of the mesosphere is illustrated by comparison of Figs. 5.3 and 4.7. The data presented in Fig. 5.3 are based on an analysis technique similar to that used in the Stratospheric Circulation Index (SCI) for measuring the stratospheric circulation in general; that is, the vertical wind profile data in components are used to obtain average speeds in 10-km-thick layers, in this case centered at 70- and 80-km altitude instead of the 50-km altitude of the SCI determination. The data used here were obtained during the years 1960 through 1964 and are plotted here primarily to depict the annual cyclic variation that oc-

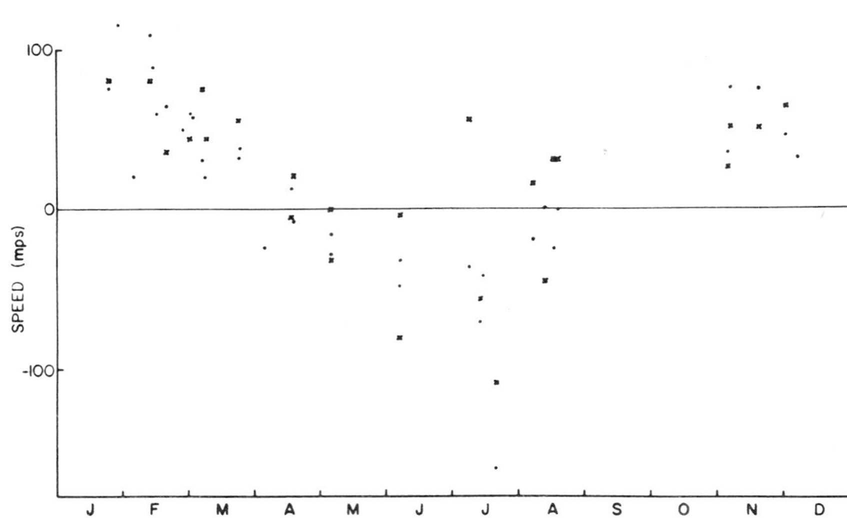

Fig. 5.3. Zonal mean flow in 10-km-thick layers centered at 70 (dots) and 80 (crosses) km in the mesosphere over Wallops Island, Virginia. The data were obtained by use of the grenade experiment since 1960 by the National Aeronautics and Space Administration.

curs in the mesosphere. As shown by these data, the zonal flow in the mesosphere is westerly in winter and easterly in summer, as is the case in the general stratospheric circulation. These midlatitude data indicate that the mesospheric flow follows approximately the same annual course as that observed at the stratopause, with a spring reversal after the spring equinox and an early fall reversal into the winter circulation. The spread of data points in this limited sample indicates a considerable variability in the summer season, a characteristic which was not apparent in the SCI data of Fig. 4.7. It would appear that the winter storm period influences the circulation of the summer mesosphere, which can be taken to mean that the mass transport between hemispheres that has been postulated in the interactions between the winter westerlies and summer easterlies of the stratospheric circulation alters the summer mesosphere.

Zonal winds of the winter season are noticeably lighter in the mesosphere than at the stratopause. This was expected from climatological analyses (Webb, 1965) of MRN vertical profiles, which showed the peak in zonal wind speed to be near the stratopause, with decreasing wind speeds above that level. It is also in agreement with the concept of the circulation disturbances composing the winter storm period interactions and the sudden warming disturbances beginning in tropical regions and working downward toward the winter pole. The first notice of these events (noted in Section 4.2) was in the 15- to 20-degree latitude belt of the winter hemisphere where the principal interaction between the hemispheric circulations occurs, but the data presented here for mesospheric altitudes, particularly in the summer, imply that the resulting circulation disturbances not only propagate downward toward the winter pole but also work upward into the mesospheric easterly circulation. This applies to both the small-scale disturbances which produce the winter storm period in general and the major events such as sudden warmings which produce dramatic changes in the winter polar vortex. If these deductions should prove to be correct, these data represent our first evidence of sudden warming events in the Southern Hemisphere, since the large excursions noted in the July data of the Northern Hemisphere correspond to the time of occurrence of the major stratospheric circulation disturbances in the Southern Hemisphere. The extreme data point of almost 60 meters/sec from the west at 80 km occurred on 8 July 1960, while the extreme easterly wind of more than 160 meters/sec at 70 km occurred on 20 July 1961. These particular measurements could be in error, but the general spreading of the data in the mesospheric summer substantiates the hypothesis that these disturbances actually do move into the summer mesosphere, or even originate there, to at least the extent of middle latitudes.

Meridional components of the data presented in Fig. 5.3 are presented in Fig. 5.4. A most obvious difference in the data presented here and the corresponding SCI data for White Sands Missile Range in Fig. 4.4 is the gross increase in scale of the meridional variability. Instead of a range of a few meters per second, characteristic of the stratopause data, the upper mesosphere evidences a general meridional variability of several tens of meters per second, with extreme values of over 100 meters/sec. While the data are inadequate to establish means with great confidence, it appears likely that there is a southerly component of the meridional flow in the mesosphere also, of the same magnitude or larger than that which had been observed at the stratopause in Figs. 4.8 and 4.24. The largest deviations in meridional flow occurred in July in these data, at the time of the gross changes in zonal circulation which may be correlated with the winter storm period of the opposite hemisphere. This is in opposition to this feature in the stratopause data, where meridional flows are strongest in the winter storm period, and the summer season is characterized by the very small variations that are observed.

As has been noted in Chapter 4, there are considerable differences

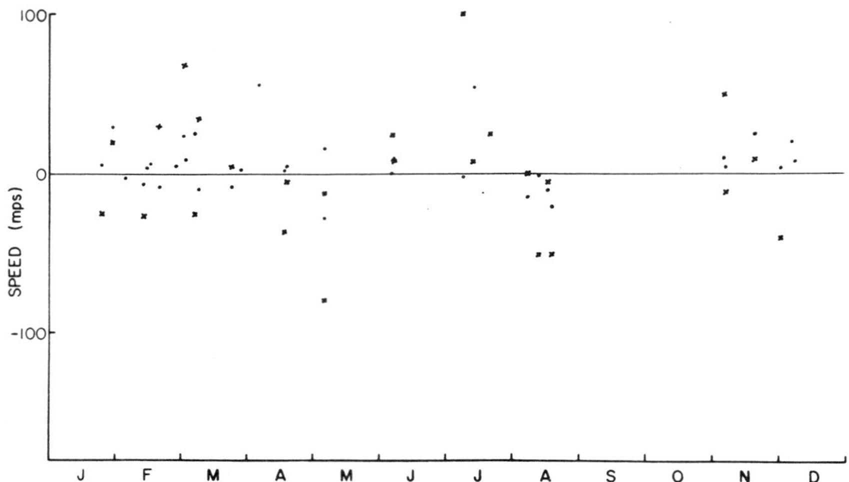

Fig. 5.4. Meridional mean flow in 10-km-thick layers centered at 70 (dots) and 80 (crosses) km in the mesosphere over Wallops Island, Virginia. The data were obtained by use of the grenade experiment since 1960 by the National Aeronautics and Space Administration.

in the stratospheric circulation with latitude in both the winter and summer seasons. These differences are detail features which can only be discerned from considerable amounts of data, but with an indication of the nature of this variability it is possible to detect some symptom of their presence in the mesosphere even with our restricted sample. Available grenade data obtained at Fort Churchill, Canada, are plotted in Figs. 5.5 and 5.6. The zonal data of Fig. 5.5 illustrate the usual strong annual cycle of zonal wind reversal which pictures the stratospheric circulation monsoon. In this case it is apparent that the strong variability which is so much in evidence in the stratopause data is present in the mesosphere. In all cases the mesospheric circulation is weaker than that of the stratosphere. The weaker winter circulation and stronger summer circulation noted in the Wallops Island data (Fig. 5.3) are not apparent in the Fort Churchill data; there is a definite indication that the winter vortex is quite strong and variable and that the summer anticyclone is weakened and steady at the Fort Churchill location. In terms of the circulation features that were discussed in Section 4.4 in connection with the spring reversal, these data would suggest that the upflowing which appears to occur in the mesosphere over the summer pole must begin at latitudes lower than that of Fort Churchill. It would be expected that such divergence would reduce the zonal flow at higher latitudes as is observed here.

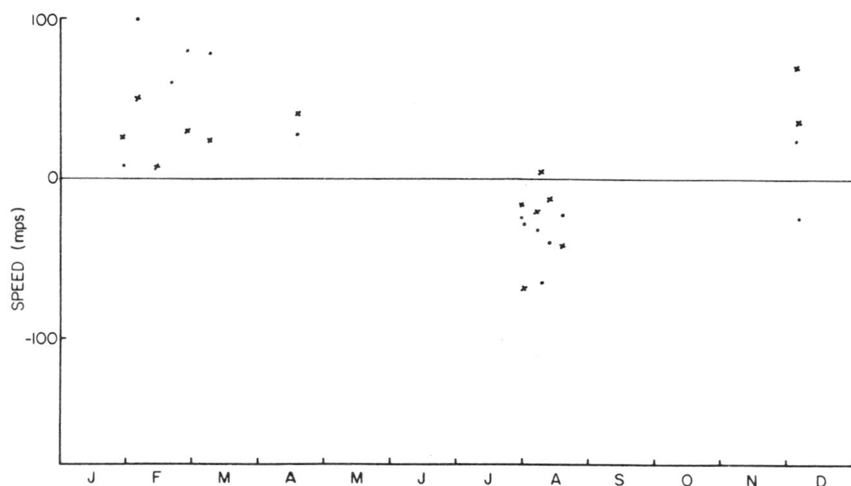

Fig. 5.5. Zonal mean flow in 10-km-thick layers centered at 70 (dots) and 80 (crosses) km in the mesosphere over Fort Churchill, Canada. The data were obtained by use of the grenade experiment since 1958 by the National Aeronautics and Space Administration.

The data of Fig. 5.6 illustrating the meridional flow in the mesosphere over Fort Churchill at the 70- and 80-km levels exhibit certain features that are quite similar to those observed in the stratopause data of high latitudes. The most important of these are the northerly winds of considerable variability in the winter season. In this case the variability is only of approximately the same magnitude as that observed in the stratopause data, rather than the greatly increased values observed in the meridional components of the mesosphere in midlatitudes over their lower-altitude correlaries. There is a difference in the mesospheric meridional flow data in the summer season, however. A flow from the pole toward the equator is in evidence here, with increasing speed with greater height. These data would indicate that the stratopause current sheet has made the turn and headed back for the lower latitudes in the upper portions of the mesosphere. It is probable that the main portion of this current has its position of maximum vertical flow at higher latitudes and that the principal northerly return current is located at higher altitudes in the lower ionosphere.

All of these data would then indicate that the mesospheric region of the upper atmosphere is an integral part of the stratospheric circulation system. One of the most marked features of vertical profiles of mesospheric parameters is the gross variability observed. The detailed structure has been shown to increase in scale with height in the stratosphere,

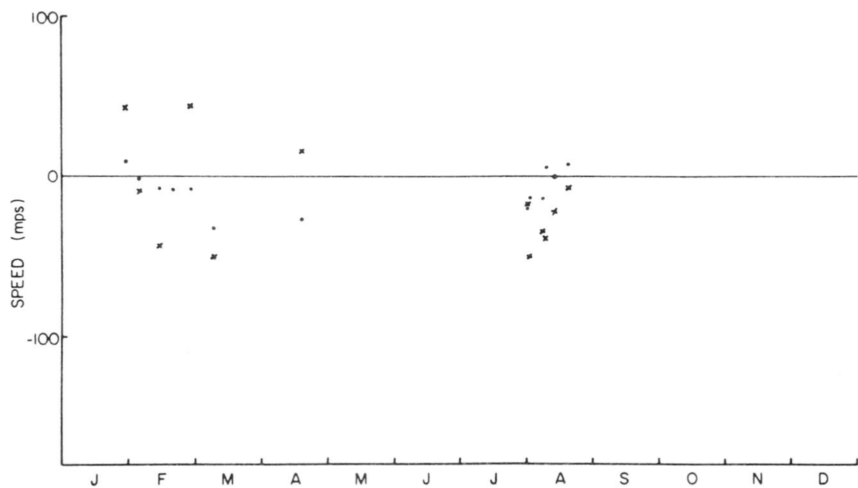

Fig. 5.6. Meridional mean flow in 10-km-thick layers centered at 70 (dots) and 80 (crosses) km in the mesosphere over Fort Churchill, Canada. The data were obtained by use of the grenade experiment since 1958 by the National Aeronautics and Space Administration.

and these data show that this trend is continued throughout the mesosphere, although the sensitivity of the data leaves much to be desired. To establish the vertical profiles of the stratospheric and mesospheric portions of the stratospheric circulation during the several seasonal periods which have been established from the MRN data, it is desirable to average even these small amounts of data for the seasonal periods to determine the mesospheric contribution to these mean profile conditions in the upper atmosphere. A major portion of the mesospheric data now available was obtained at Wallops Island by the National Aeronautics and Space Administration. These data through 1963 were summarized for the noted periods, and a general summary of the zonal wind data is presented in Fig. 5.7. These profiles show that in midlatitudes the peak summer easterlies occur in the middle mesosphere, although they are quite strong at the mesopause, and die out in the lower ionosphere, even reversing to westerly in these data. During the winter seasons the westerly winds decrease with height in the lower mesosphere, but in the winter storm period case a second maximum is indicated in the 70- to 80-km altitude region. In all of the data, the zonal wind component shows a minimum immediately above the mesopause.

Very strong values of zonal winds are indicated for mesospheric altitudes in midlatitudes. Westerly wind extremes of 150 meters/sec are exceeded on occasion near the stratopause peak in the stratospheric

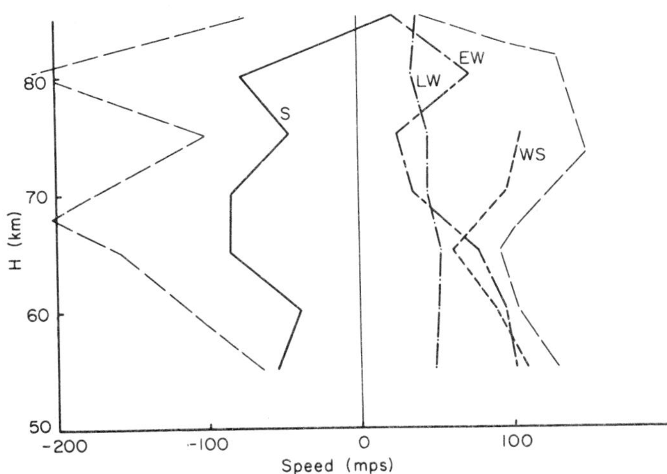

Fig. 5.7. Mean zonal winds of the mesosphere over Wallops Island for each season of the stratospheric circulation. The light dashed lines indicate the extreme values thus far observed. Curves are for summer (S), early winter (EW), winter storm (WS), and late winter (LW) seasons.

circulation, but the over 200 meters/sec that has been measured in the easterly flows of summer over Wallops Island exceeds by far the maximum easterly winds which have thus far been observed in the MRN data. These data for the early and late winter seasons point to some differences, although the limited amount of data makes any conclusions of limited certainty. It would appear, however, that the late winter season is more stable in vertical profile as well as in time. A secondary peak in the westerly zonal speed appears in the early winter data at the 80-km level, and apparently during the winter storm period this upper mesospheric jet increases in vertical extent downward and strengthens to become a well-developed system. This is during the period in which the principal stratospheric circulation system is being dragged down to lower speeds from lower-latitude interaction with the summer hemispheric circulation. These data would suggest that the downward progression of the sudden warming event is not just a move to encompass the entire stratospheric circulation in its changes, but that as it settles into the polar region a circumpolar vortex in the upper portion of the mesosphere may well be intensified.

Such a result could easily be obtained if the descending air which produces the sudden warming event were to be horizontally advected into the subsidence zone at stratospheric altitudes of the stratopause and below. In that case radiational cooling would take care of the continuation of this high-level circumpolar vortex, while adiabatic heating of the descending air in stratospheric level would provide the observed heating in that region.

Mean meridional wind profiles of the mesosphere over Wallops Island for each of the stratospheric circulation seasons (Chapter 4) are illustrated in Fig. 5.8. These data show the same southerly compo-

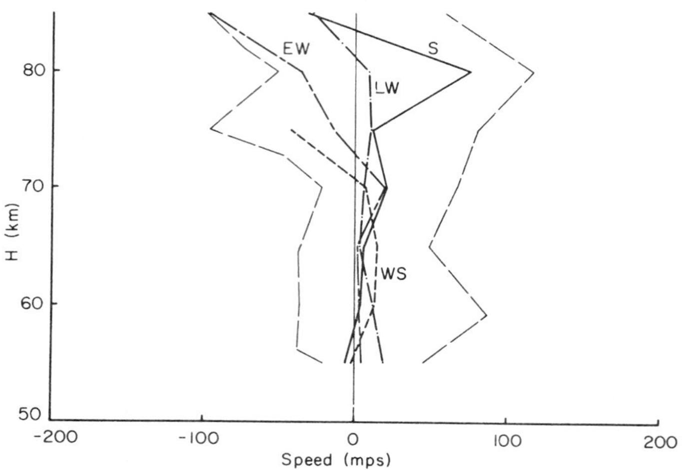

Fig. 5.8. Mean meridional winds of the mesosphere over Wallops Island for each season of the stratospheric circulation. The light dashed lines indicate the extreme values thus far observed.

nent of flow in the lower portion of the mesosphere that was evidenced in the upper stratosphere. Speeds of up to 20 meters/sec occur in these mean data, and thus individual measurements may be expected to exceed these values significantly, as is illustrated by the extremes which have been observed (light dashed lines). These data indicate that southerly components of between 50 and 100 meters/sec are not too unusual in the lower mesosphere, and northerly components in the 30- to 40-meters/sec range occur. The values represent decided increases in range of meridional components over that observed in the upper stratosphere, although the trends established from the Wallops Island MRN data are essentially continued upward through the 70-km level.

At 70-km altitude the situation changes dramatically. These upper portions of the curves are based on too few data to be certain of the validity of the mean values, but they illustrate the fact that the meridional components of the flows in this region characteristically are strong and variable, approaching values as large as the zonal components above the mesopause. It should be noted that there is some disagreement in the data presented here and those obtained by Kochanski (Fig. 5.1). Kochanski's analysis showed a meridional monsoon in the region immediately above the mesopause which consisted of northerly winds in the summer and southerly winds in the winter, with reversals in the spring and fall during April and September, respectively. His source of data was the meteor trail measurements made at Adelaide, Australia (35° S and 138° E) and Jodrell Bank (53° N and 2° W), and consisted of a much larger source of data for these particular locations. While the global applicability is questionable owing to the small spatial distribution, the data represent the most extensive sample of the 80- to 100-km region thus far available.

A principal conclusion that may be drawn from the data of Figs. 5.7 and 5.8 is that the circulation of the mesosphere becomes increasingly variable with height and that this variability is particularly enhanced in the meridional component. These Wallops Island data would then indicate the appearance of a new circulation system with its base in the mesopause region; the system retains some of the features of the stratospheric monsoon, but superimposed on it there is a new element which modifies the basic stratospheric circulation's dominant zonal flow into a partial meridional flow. As was stated in the introduction of this chapter, incorporation of frictional losses in the stratospheric circulation system would provide an adequate explanation of the observed data.

5.2 Temperature Structure

Temperature structure of the upper atmosphere is, in general, controlled by radiational processes; therefore, the geographical distribution and annual variation are closely keyed to the solar input. The stratospheric circulation provides an example of this since the general features of that circulation are accurately forecastable from known solar relationships. Even in this model case, however, there are certain deviations from simple solar control. Most notable of these is the sudden warming phenomenon of the stratospheric winter, where gross alterations occur in the general circulation without obvious relationship to extraterrestrial events. On a different scale, it is possibly more surprising that the large amount of detail structure which is a major characteristic of upper atmospheric profiles must almost certainly be ascribed to dynamic effects apart from the basic energy input process that converts solar radiant energy into the kinetic energy of the circulation.

As was shown in Chapter 4, local changes in temperature in the high-latitude stratonull region on occasion equal the annual range, with the variation accruing over a few days' time. It is generally agreed that these temperature changes have a dynamic origin, specifically the result of descending motions in the lower stratosphere and lower portions of the upper stratosphere. As has been discussed in Section 4.2, there is evidence that the circulation disturbances which cause these descending motions originate in interaction between the gross hemispheric circulations at high levels in tropical regions and that they produce these observed effects only when the disturbances have propagated into high latitudes and lower stratospheric altitudes. It is also postulated that these obviously major events are simply the largest of an entire spectrum of interactions that occur, the smaller scales of which are quite prolific and are too small to delineate with the resolution provided by current MRN observations. The effects of these smaller interactions are shown in the data by the general reduction of the winter stratospheric circulation during what is called the winter storm period. This strong interruption of the smooth annual stratospheric monsoon is the most important case of a failure of solar control of the upper atmosphere.

Small-scale detail features in stratospheric wind and temperature profiles and ionospheric wind and electron density profiles are of such character that an internal wave mode of formation is implied. The exact physical mechanism involved in generation of these disturbances is not clear, nor is the manner in which the observed profiles are altered certain. Again, however, it is almost beyond doubt that these features are not the direct result of inhomogeneities in solar input. The data and previous experience have led to the belief that these special circulations, both large and small, are turbulent eddies in the flow which have the specific function of weakening the general circulation through relaxation of the driving meridional temperature gradient and viscous decay into thermal energy.

Dynamic effects of the type discussed above are familiar items in the troposphere and stratosphere. The great stability of the stratospheric regions makes the large events very special in the stratosphere and limits the development of smaller-scale phenomena, at least in the vertical. The mesosphere, on the other hand, is far less inhibited relative to vertical motions as a result of its positive lapse rate, and the amplification of any wave phenomenon should result from the very important reduction in density in this region. The nature of the thermal structure of the mesosphere is illustrated in Fig. 1.1. These data are for a lower middle latitude of approximately 30 degrees that is within the tropical and subtropical belt, which, owing to only small variations in the annual solar flux incident angle, exhibits little in the way of an annual cycle of temperature. Nordberg and Stroud (1961) accomplished nine rocket grenade soundings over Guam (13.5° N, 145° E) during November 1958 and found a very stable mesosphere (Fig. 5.9). These data indicate the presence of a low (below 50 km) tropical stratopause and a moderately cold mesopause, with an average mesospheric lapse rate of approximately 2.6°C per kilometer.

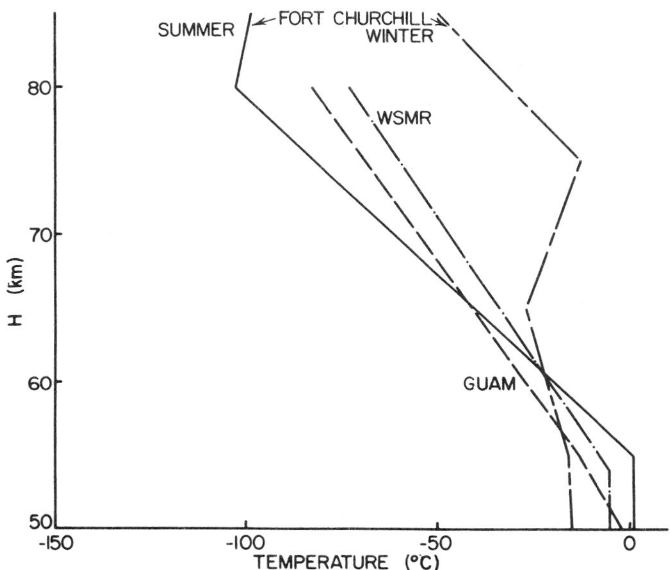

Fig. 5.9. Average temperature profiles obtained from grenade soundings at Guam, Fort Churchill (winter and summer), and White Sands Missile Range by Nordberg and Stroud (1961).

One of the interesting deductions in that report is that there is little in the way of annual cycle in temperature structure in the tropical stratosphere and mesosphere but that there are large day-to-day variations (evidenced by the wind data). It is probable that this limited set of data was providing an initial indication of the turbulent interaction between the hemispheric stratospheric circulations, which in December pushes far into the winter westerly flow to produce the winter storm period (Section 4.1). Data obtained at White Sands Missile Range during the development phases of the grenade experiment are also presented in Fig. 5.9. The stratopause appears to have a slightly different construction at these two stations, appearing very sharp in the Guam mean data, and more in the form of an isothermal layer in the White Sands Missile Range data. Additional data have indicated that this apparent difference is probably due to fluctuations in height of the stratopause as well as some differences in structure. Within the limited amount of low-latitude data available there appears to be little in the way of a latitudinal temperature gradient.

A considerable amount of data has become available in recent years (W. Smith *et al.*, 1964) concerning the temperature structure of the mesosphere over Wallops Island and Fort Churchill. These data are summarized at 5-km intervals in Fig. 5.10 for the seasonal breakdown

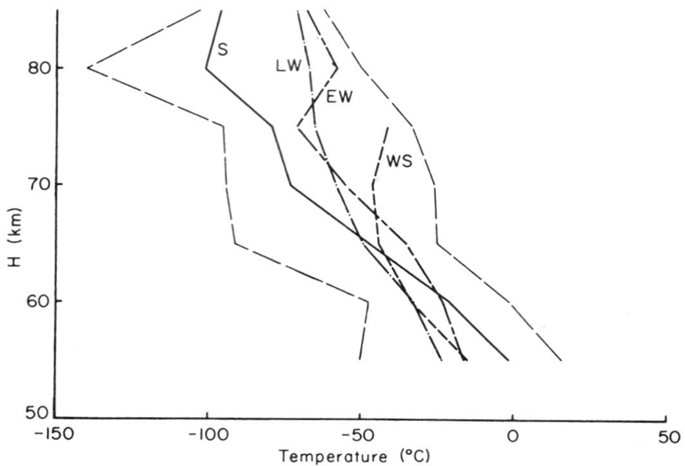

Fig. 5.10. Mean temperature profiles of the mesosphere over Wallops Island for each season of the stratospheric circulation. The light dashed lines indicate the extreme values thus far observed.

of the stratospheric circulation presented in Chapter 4, along with the extremes in the entire body of data. Twenty-two observations were available in the 55- to 70-km region, and 18, 13, and 8 data points were used at the 75-, 80-, and 85-km levels, respectively. These data indicate that at Wallops Island the lapse rate of the lower mesosphere is slightly less than that observed at low latitudes during the winter seasons and is greater at high latitudes during the summer season. This difference becomes pronounced in the upper portions of the mesosphere during the grossly different periods.

The lapse rate during winter over Wallops Island is less than 1°C per kilometer above 70-km altitude. Possibly the effect is due to inadequate data, but the level of the mesopause appears to be higher during the winter period than the familiar 80 km. This significant break in mesospheric lapse rate during the winter season at middle latitudes is difficult to ascribe to radiational effects and again points toward a dynamic control. The attendant circulation could begin in the lower ionosphere in low-latitude regions and work downward while advancing poleward. Compressional heating would then modify the radiationally prescribed temperature distribution of a cold mesopause (approximately —80°C) to the warmer —60° to —70° indicated here. The center of this subsiding motion would then appear to be at the mesopause over Wallops Island. The result is to push the level of the ionospheric inversion to higher altitudes.

The summer thermal structure over Wallops Island is decidedly different from the winter case. The positive lapse rate remains strong to the mesopause in these mean data, resulting in colder mesopause temperatures at midlatitudes than are observed in tropical regions. This effect contradicts the fact that the solar heat input is enhanced during this period. These data then show that there is an annual temperature cycle at the mesopause with an amplitude of approximately 40°C in midlatitudes that is inversely related to the solar input (180° out of phase). This cyclic variation couples with the stratopause cycle, which is in phase with the solar input, to produce an annual oscillation in the mean mesospheric lapse rate having an amplitude of a factor of 2 and varying from slightly over 3°C per kilometer in summer to approximately 1.5°C per kilometer in winter. The pivot point for these oppositely directed variations appears to be located at the 65-km level, which must then be a conservative point in the upper atmospheric structure, at least in middle latitudes.

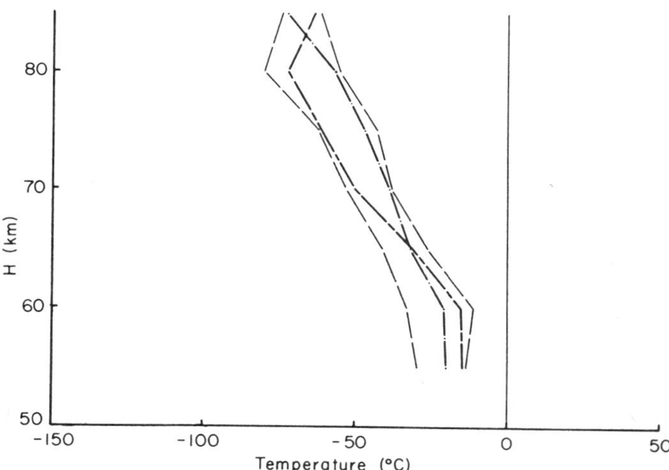

Fig. 5.11. Mean temperature profiles of the mesosphere over Fort Churchill for the early and late winter seasons of the stratospheric circulation. The light dashed lines indicate the extreme values thus far observed during these two seasons.

Mean temperature data for Fort Churchill through 1963 are limited to the early winter and late winter seasons illustrated with their combined extremes in Fig. 5.11. There are five observations included in these curves to 75-km altitude, and two above that level. These data are very similar to the temperature profiles over Wallops Island for the same periods. A significant difference appears in the 55- to 60-km region, where the beginning of the mesospheric positive lapse rate is higher than is evidenced in the Guam or Wallops Island data. It would appear, then, that there is an upslope in the tropopause surface of approximately 10 km between the equator and the winter pole. This higher level of the tropopause in high-latitude winter results in a slightly warmer mesosphere at high latitudes, although the difference is a matter of a very few degrees in these mean data.

Data on the high-latitude summer temperature structure of the mesosphere have been obtained by the National Aeronautics and Space Administration in cooperation with atmospheric research scientists in Sweden in attempts to define environmental conditions associated with formation of noctilucent clouds. Three observations of the temperature were obtained in August 1962, at Kronogård, Sweden (66° N, 19° E) and were averaged to produce the curves presented in Fig. 5.12. The lower portion of the mesosphere exhibits a temperature profile that is quite similar to corresponding ones for Fort Churchill (Fig. 5.9) and Wallops Island (Fig. 5.10). Meridional temperature gradients in the lower meso-

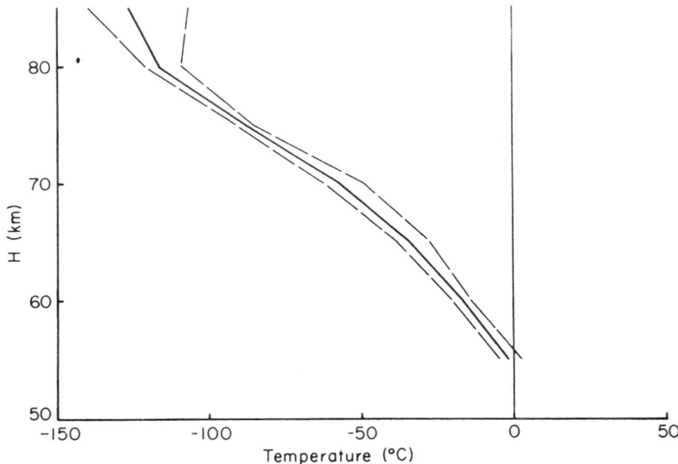

Fig. 5.12. Mean (solid line) and extreme temperature values indicated by three rocket grenade soundings obtained during the noctilucent cloud season (summer) over Sweden.

spheric summer thus appear to be positive, with a value of the order of 0.2°C per degree latitude.

High-latitude summers in the upper mesosphere are significantly different from the middle- and low-latitude ones. A mean positive lapse rate of 6°C per kilometer is indicated by these data in the altitude range from 70 to 80 km. Lowest temperatures are found at 85 km, with a minimum of —140°C. A rather weak lapse rate is maintained in the lower portion of the mesosphere with a change at about 70 km where the steeper lapse rate begins. If, as is discussed in Section 5.3, noctilucent clouds are supported by the vertical air current of the stratospheric tidal jet, the temperature lapse rate illustrated here would be established by adiabatic expansion of the ascending air. It is clear that static approximations will not suffice for an understanding of such a circulation.

The meridional temperature gradient has a mesospheric configuration which is idealized by the latitudinal cross section presented in Fig. 5.13.

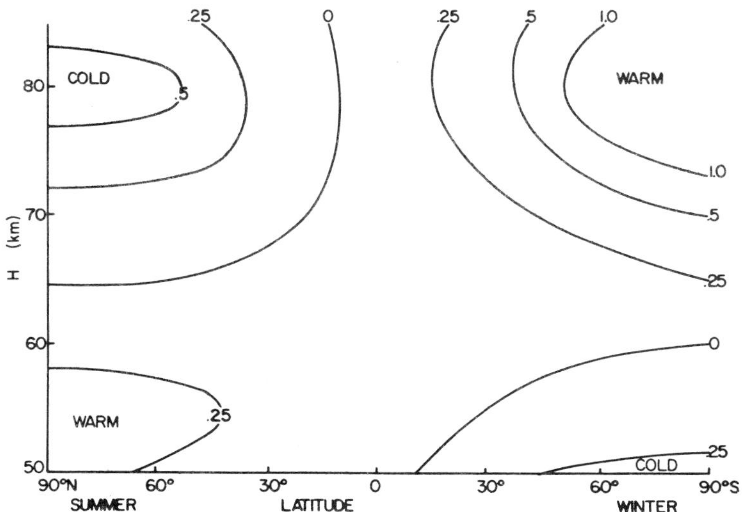

Fig. 5.13. Postulated meridional temperature gradients of the mesosphere. Contours are degrees Celsius per degree latitude.

While the data on temperature structure in this region are relatively meager, the contours illustrated here are designed to cover the characteristics of which we are relatively certain. Basic of these special features is that radiational equilibrium is of minor importance in establishing the thermal structure of the mesosphere, at least in the static sense. This is demonstrated principally by the fact that the temperature gradient is oppositely directed to that which is to be expected from radiation absorption effects. The warm winter mesosphere is most likely the result of subsidence initiated in the shrinking of stratospheric regions in the long winter night. As a result of the atmospheric density distribution, wind values in a circulation system which could supply such a downward motion would have to be strong in the upper portion of the flow, although increased density below the mesopause would possibly preclude the detection of the return current at lower altitudes.

In the case of the cold polar mesopause of summer, the situation is more susceptible to analysis as a result of MRN data. As has been discussed in detail at the beginning of Chapter 4 and in Section 4.8, the special circulation established by diurnal variations in heat input to the stratosphere has the characteristic of forced upward motion in the high-latitude nighttime sector of the upper mesosphere. We know that upward motions must exist to some degree in summer polar areas due to expansion of stratospheric regions during this period. Initial phases of upward motions of this meridional current at the stratopause could well be supported by ozone heating in that region. As the air lifts toward the mesopause, adiabatic cooling would become a very strong effect. It is obvious, however, that the flow will be nongeostrophic if it executes such a motion, at least in the upper mesosphere.

5.3 Noctilucent Clouds

Particulate material in the upper atmosphere has been observed only under very restricted conditions. This results in part because of the very difficult observational problem but is primarily due to the fact that the upper atmosphere is clean in comparison with the standards with which we are experienced in the troposphere. If one accepts the order-of-magnitude reductions in concentration and size of the particles that could have significant resident times in the mesosphere, it is immediately apparent that observation of these sparse clouds of small particles by standard ground optical systems would be extremely unlikely. This does not hold for other vantage points, such as a balloon, aircraft, or satellite platform operating above the contamination of the troposphere.

A marked exception to this rule of a clean mesosphere is the very special "noctilucent clouds" (Figs. 5.14–5.17) of the high-latitude night sky. Measurements made before the turn of the century located these thin clouds at altitudes of the order of the mesopause, and this fact sparked a considerable interest in the circumstances surrounding the appearance of these clouds. Speculation has ranged far and wide in

Fig. 5.14. Noctilucent clouds north-northeast of College, Alaska on the night of 12–13 August, 1963. (Courtesy of B. Fogle, Geophysical Institute, University of Alaska)

Fig. 5.15. Noctilucent clouds northeast of College, Alaska on the night of 11–12 August 1963. (Courtesy of B. Fogle, Geophysical Institute, University of Alaska)

the absence of definitive data on the origin and composition of these somewhat illusive items of the mesosphere.

Perhaps the title "noctilucent clouds," or "night clouds," is a misnomer, since there is no real evidence that they occur only at night. Actually, these phenomena can generally be observed only during the

Fig. 5.16. Noctilucent clouds north of Anchorage, Alaska on the night of 28–29 July 1964. (Courtesy of B. Fogle, Geophysical Institute, University of Alaska)

Fig. 5.17. Noctilucent clouds north of Watson Lake, Yukon Territory, Canada on the night of 12–13 July 1964. (Courtesy of B. Fogle, Geophysical Institute, University of Alaska)

period in which the sun is at some 5° to 8° elevation angle below the horizon. This is illustrated by the diagram presented in Fig. 5.18, which outlines the geometry of a typical noctilucent cloud observation. It is

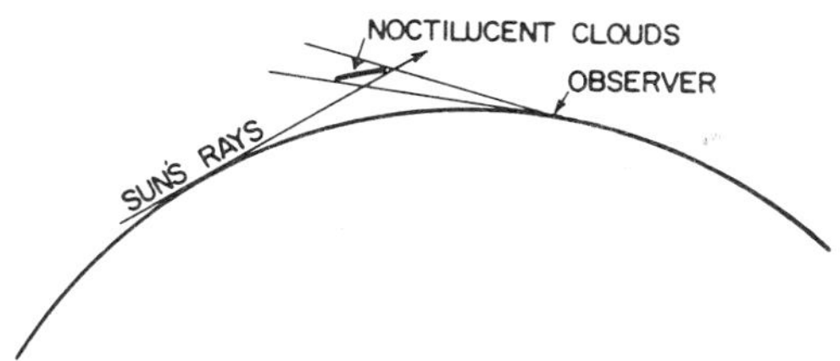

Fig. 5.18. Geometric layout of the very special noctilucent cloud observation problem. The observer must be in darkness while the sun still shines on the clouds. At middle and low latitudes these conditions are realized for only a few minutes each morning and evening, while at high latitudes the observational period may last for several hours.

obvious that an ideal observation point for searching for such clouds would be at high latitudes where the sun's relative position falls within these limits for considerable periods of time. Suspicions that the occurrence data may have been biased by this optimum window for observation have been inspected in the field, and to date, at least, the deduction that these clouds frequent high latitudes appears to be well founded.

A diurnal variation in noctilucent cloud occurrence is less certain. Sky brightness from molecular and particulate scattering of sunlight in the troposphere and stratosphere reduces the visual contrast which can be obtained, and at the same time the scattering angle occupied by the cloud particles is less favorable for maximum illumination in the observer's direction. It seems quite likely that this possible observational bias is of significant import to require investigation of the longitudinal distribution of noctilucent clouds with observational systems that are more sensitive than the human eye.

Noctilucent clouds exhibit a well-defined annual variation, with almost all sightings confined to the summer season between 15 June and 15 August. An early summary of noctilucent cloud observations which were made largely in Scandinavia and Russia reported that a maximum occurred very early in July, but more recent work in the North American area has produced a later maximum, falling in late July. These data are illustrated in Fig. 5.19. It is not known whether real differences are recorded here, or if the variation is observational in nature. The North American data were obtained by a network of well-trained observers, while the earlier data were obtained over a number of years in a more

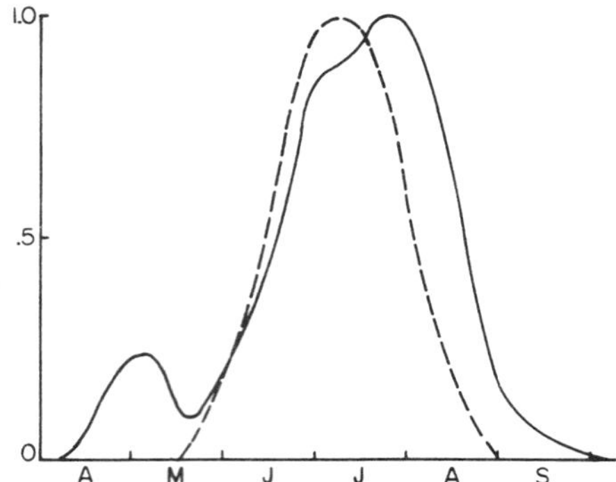

Fig. 5.19. Noctilucent cloud frequency of occurrence distributions according to Vestine (1934) (dashed curve) and Fogle (1964a). These smoothed data show a range of from mid-May to the end of August with a peak in early July from Vestine's study, while Fogle's work indicates a later peak and a wider range, from early April through September.

sporadic fashion. Considerably more data will be required to define the distribution of occurrences accurately, but the data leave little doubt of the existence of a marked summer peak incidence.

A point in favor of the concept of a diurnal variation of noctilucent clouds lies in the fact that they are most frequently observed near midnight local time. The observational circumstances are usually optimum near midnight, but the observational window is not sharp at the latitudes of frequent observation, and the data are suggestive of a diurnal variation. If such is correct, this is an extremely enlightening bit of information relative to the dynamics of the phenomenon which supports the formation of the clouds.

In connection with the gross variability evidenced by noctilucent clouds, there is a tremendous amount of smaller-scale detail structure apparent in the internal structure of the clouds. As is illustrated in the examples in Figs. 5.14 through 5.17, the clouds are characterized by wave motions of many different dimensions that are horizontally oriented. Sequence photographs indicate that these wave features are rapidly varying in addition to the wind motion on which they are superimposed. Proper analysis of these data will yield information on the scale of detail processes which take place in the upper mesospheric structure, and broaden the view of this difficult-to-observe region which has been obtained by isolated soundings. Before these data can be most useful, it is necessary that we understand the processes which govern the appearance of these very special clouds.

The source of particulate materials in the mesopause region that could accumulate in concentrations sufficient to form the observed noctilucent clouds has been a matter of speculation. A dry upper atmosphere concept required that the particles be dust, and thus of either terrestrial or meteoroid origin. It is very difficult to illustrate mechanics which will transport tropospheric particulate materials upward through the stable stratosphere into the noctilucent cloud occurrence zone, although such particles are known to reside in abundance just below the tropopause. If this source is considered improbable, the remaining obvious source of such particulate material is accumulation of small meteorites or debris from larger meteors traversing the upper atmosphere. Principal objections to the latter source have concerned a supposed insufficiency of particles and the absence of an asymmetry in the influx of meteoroids which could account for the rather curious distribution of noctilucent cloud occurrence.

Space probes in the vicinity of the earth during the past few years have extended our knowledge of the concentration and size distribution of meteoroid material. Large numbers of particles (with a flux of 0.1 particle per square centimeter per second) exist that are too small to generate atmospheric interactions that can be observed by ground-based systems. These particles have thus far been observed only by momentum-sensing devices, so the density of the material composing these particles is therefore undefined. If a common value of composition density is assumed for these small particles the data indicate a cumulative influx of meteoroid particles as is illustrated in Fig. 5.20.

Now the density assumed here is not at all certain. The particles in this small size range could be condensed gaseous material from the solar atmosphere or from cometary passages through the solar system. If so, they could be composed of very light condensates similar to snowflakes or hoarfrost. In that case, the mass of the individual particles will be significantly smaller than that found for the assumed 2.5 gm/cm^3 used in calculating the data in Fig. 5.20. If the density of each particle is lower than assumed, the particles must then be larger, since the measured data are the relative momentum. This crowding of the smaller particles into the larger size ranges will straighten out the curve of Fig. 5.20 in the upper portion and indicate the presence of greater numbers of very small particles than is indicated here. Data plotted in the upper portion of the curve are questionable because of the small quantity available and the failing sensitivity of the sensors. The expectation that the curve would begin to flatten out is valid if the particles are of common earth's surface materials ($\rho = 2.5$ gm/cm^3) owing to the large

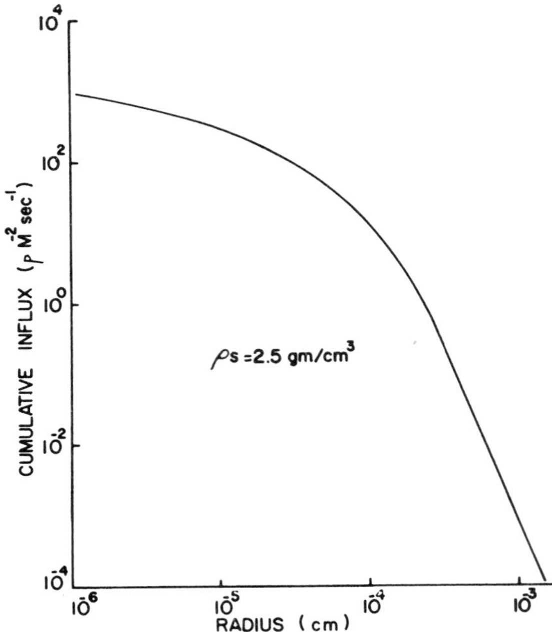

Fig. 5.20. Size distribution of meteoroid particles in the earth orbit vicinity, assuming a composition density of 2.5 gm/cm^3, as observed by radar (large particles) and by satellites (small particles) according to F. L. Whipple (1963).

amount of work required to grind such small particles. It is not valid, however, in the case of particles formed of condensing gases, where the lower size limit of a stable particle could be much smaller.

Another interesting aspect of the problem concerns the actual presence of these particles. Solar radiation pressure becomes of significance to particles of 10^{-5}-cm radius even when they are formed of material as dense as 2.5 gm per cubic centimeter. They should be literally forced out of the solar system by the "sail" effect on their large surface-to-mass ratio, and there is reason to wonder at their continued presence in the earth's vicinity. Information on the origin of these small particles could be one of the more important results of our space program.

In addition to the questions concerning the origin of small meteoroid particles in the solar system, the physics of small-particle orbits about the sun provides information on their interaction with the earth. These particles would assume nearly circular orbits about the sun as they are "blown" away from the sun. Since the earth's orbit is nearly circular this means that the small meteoroids would have about the same speed as the earth and thus the closing velocity would be either very small or very large according to whether the two bodies are rotating about the sun in the same or opposite directions. Without information on the source of these particles we cannot be sure of the direction of rotation, but we do know that a large fraction of the particles that have been observed through visual and radar (greater than 10^{-3}-cm radius) techniques is rotating about the sun in the same sense as the earth.

The earth acts as the center of a central force field to any particle that approaches it. If the particle is traveling at high speed to one side of the earth's centroid, its orbit will simply be perturbed into a parabola whose shape will vary with the exact velocity relationship. The trajectories of these particles are illustrated in Fig. 5.21 by those particles

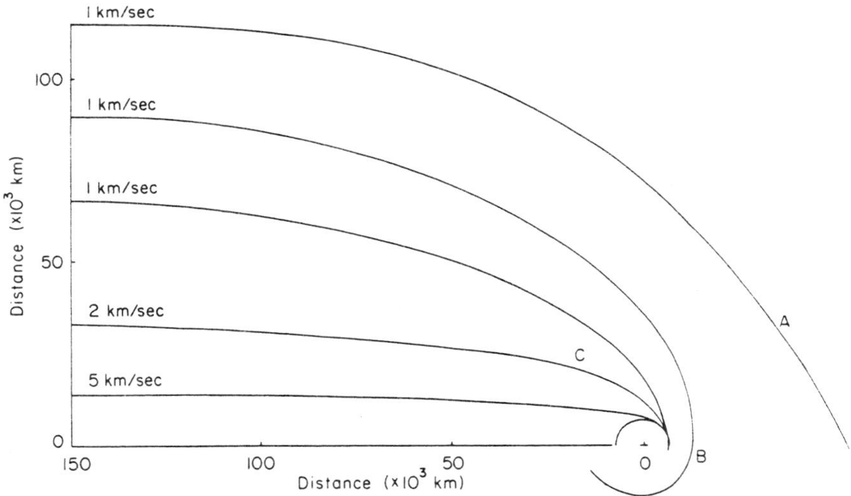

Fig. 5.21. Perturbations of meteoroid orbits in the earth's central force field for the special case where the meteoroid is being overtaken by the earth along parallel vectors at low speeds. Those meteoroids in group A *are not collected and simply enhance the concentration in a focal zone behind the earth's velocity vector. Category* B *meteoroids are collected by the earth's atmosphere after some orbital time, while those meteoroids in category* C *are collected on the first approach.*

that fall in the sector labeled *A*. Trajectories of this type will be generated by those meteoroids that have sufficiently high relative velocities and those that approach at sufficient distances from the center of the central field. It is obvious that there will be an enhancement of the number of particles per unit volume in selected regions in the earth's near space (behind the orbital velocity vector) if there is a preferred mode of approach between the particles and the earth.

Those meteoroids having insufficient momentum to escape the earth's "gravitational well" will be captured. Some of these particles enter earth orbits, owing to small energy losses through collisions with atmospheric molecules in the air of high altitudes. Such cases are illustrated by the trajectories falling in the region labeled *B* in Fig. 5.21. After additional earth orbits, further collisions will cause these particles to be collected by the earth's atmosphere, but the possible modes of accomplishing this are sufficiently complex that the final disposition can be assumed to be random. These class *B* meteoroid encounters can be assumed to generate a dust cloud of particles in the earth's gravitational well which execute geo-orbital motions and thus are entrained into the earth and moon orbital cycle about the sun.

From our point of view the most interesting group of meteoroid encounters to be considered is that labeled *C* in Fig. 5.21. These are the particles that are collected by the earth during the first trajectory of the encounter. Obviously all meteoroids with a relative velocity directed toward the earth's capture cross section will be collected on the first pass regardless of the speed of approach. If a particular meteoroid is aimed to by-pass the earth in an encounter, it will be accelerated radially toward the earth's center in the central field and the probability of capture is then a strong function of the relative velocity. The slower particles will be under the influence of the field for a greater period and their paths will be affected most significantly, introducing considerable curvature in the trajectories as they approach the capture region. This focusing will result in an asymmetric influx of these particles into the earth's atmosphere and thus could have a bearing on the accumulation rate of such particles in the noctilucent cloud zone at the time of occurrence of these clouds.

All of the above discussions hinge on the presence of a preferred relative velocity between the earth and small meteoroid materials. Measurements have demonstrated the presence of an enhanced meteoroid impact in the visual and radar sizes on the side of the earth facing along its orbital velocity vector. These data indicate that these particles are in orbits about the sun which intersect the earth's orbit, and further that at the time of collection the earth has a greater velocity about its own orbital path than does the meteoroid. Both of these factors are satisfied by the presence of slightly more elliptical orbits on the part of the meteoroids than that evidenced by the earth. In addition to these central field concepts of why these small particles should be captured by the earth in an asymmetric fashion, it is also now known that the solar wind approaches the earth at an angle of about 40 degrees sunward from the earth's orbital velocity vector. The force exerted on the high surface-to-mass ratio small particles would tend to decelerate them as they are pushed away from the sun, producing an additional reason for their encounters to be concentrated on the forward side of the earth.

Assuming the presence of an efficient "gravitational lens" associated with the encounters between small meteoroids and the earth, it is possible to calculate the latitude distribution of impacts which will occur due to these trajectory perturbations and to obtain estimates of the influx rates in the capture zone (assumed to be at 100-km altitude). Considering only those particles that have a relative velocity of approach to the earth parallel to the earth's velocity vector but in the opposite direction, we can evaluate the latitude of intercept for various relative speeds and lateral distances of entry into the gravitational sphere of the earth through the following relation:

$$\frac{r_0}{r} = \frac{1 - \cos\theta}{\alpha \sin^2\gamma} + \frac{1}{\sin\gamma}, \tag{5.1}$$

where r_0 is the initial distance of the particle from the earth's center; r is the earth's capture radius (6451 km); θ is latitude measured for the front along a meridian through the earth's velocity vector; α is equal to $r_0 v_0^2/\mathrm{GM}$; $\gamma = \theta + 90°$; v_0 is the relative speed between the meteoroid and the earth's center; and M is the mass of the earth. Evaluation of this equation within the stated limitations shows that there is a very large enhancement of the earth's capture cross section when velocities are low in this particular geometry. These effects result in increased concentrations of the incoming particles by a factor of 30 for lower speeds near 1 km/sec as the meteoroid enters the earth's gravitational field. The effect is very sensitive to closure speed, becoming essentially negligible at a relative speed of 5 km/sec. It should be noted that the impact speed of capture is the initial relative speed plus the speed gained by accelerating in the earth's potential field, and this amounts to somewhat over 11 km/sec for the cases discussed here.

Equally important to the general enhancement of meteoroid influx which results from the gross gravitational lens produced by relative motions is the latitudinal distribution which is produced. Under the assumptions of the above computations we obtain a very broad band of strong enhancement concentric about the earth's orbital velocity vector, extending from midlatitudes on the forward side all of the way to the antiorbital point on the back side. This information indicates that there will be far fewer particles entering the mesopause region at low latitudes around noon and midnight and that the influx will be quite strong at sunrise and sunset. At high latitudes the influx will remain strong at all hours, with little variation resulting from the earth's rotation.

Now the case used here as an example is extreme, and it is likely that the actual circumstances fall intermediately between the data observed for slightly larger particles obtained by radar techniques and the above results. In the radar case there are order-of-magnitude enhancements in the influx on the elliptical plane at angles of approximately 60° relative to the direction of the earth's orbital velocity vector. In the absence of actual data an estimated enhancement of a factor of 3 of the smaller particles considered here would appear reasonable. Thus, the earth's cross-sectional plane passing through the orbital velocity vector and tilted approximately 45° to the elliptic plane into the daytime summer hemisphere should have the distribution of influx illustrated in Fig. 5.22. This estimate assumes that the particles of interest will be spiraling slowly outward from the sun.

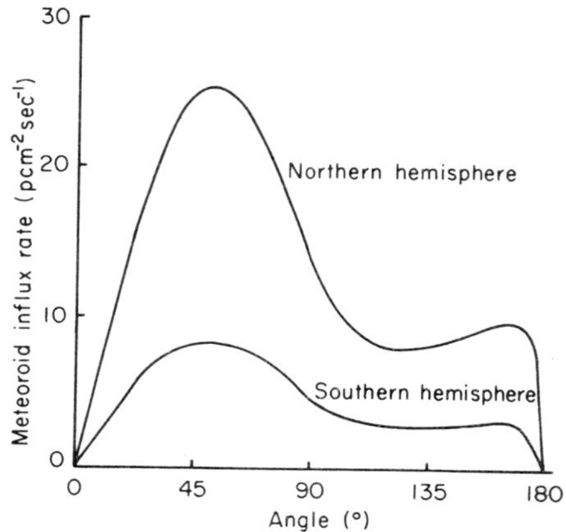

Fig. 5.22. Meteoroid influx enhancement at the earth's capture radius over that observed in free space along a cross-sectional plane through the earth's velocity vector and tilted 45 degrees into the summer daytime hemisphere. Angular coordinates are measured from the earth's center and are zero in the direction of motion.

It would appear then, that the maximum influx of meteoroid particles in the 10^{-6} to 5×10^{-5} cm radius size range will be found in the high-latitude daytime sky of the Northern Hemisphere summer. The minimum influx should be at the point immediately under the point of the orbital vector. If our best estimate of 0.1 particle per square centimeter per second in the near space is used for this minimum value of small-particle influx, the value at the maximum in noctilucent cloud latitudes would be approximately 24 particles per square centimeter per second, and on the nighttime side of the summer hemisphere where noctilucent clouds have actually been observed, the influx should be approximately 8 particles per square centimeter per second. These numbers would appear to be sufficient to assure a dirty mesosphere unless some cleansing processes are active.

It is essential at this point that we have some idea of the fall rate of the noctilucent cloud particles. The clouds have been observed over an altitude range of from 65 to 95 km, so we require terminal fall velocity data from the captive altitude of 100 km to the stratopause at 50 km. The problem must be treated on the basis of kinetic energy exchange between individual molecules of the resisting gas and the falling particle. In the absence of more sophisticated data, we shall assume that the particles are elastic spheres of a material density of 2½ gm/cm^3. We can then calculate the terminal fall rates of various sized particles from the equation

$$v_w = \frac{3g(M_s - M)}{16N(R + R_s)^2} \left[\frac{2(M + M_s)}{kT_\pi MM_s} \right]^{1/2} \quad (\text{cm sec}^{-1}), \tag{5.2}$$

where g is gravitational acceleration (cm sec^{-2}); M_s is the mass of a particle (grams); M is average mass of air molecules (gram); N is the number density of air (p cm^{-3}); R is the average collision radius of air molecules (cm); R_s is the radius of a particle; k is the Boltzmann constant (ergs deg^{-1}C); and T is absolute temperature (°K). Using values of $k = 1.38 \times 10^{-16}$ erg deg^{-1}C, $R = 3.82 \times 10^{-8}$ cm, and the variable atmospheric parameters from the "U. S. Standard Atmosphere 1962," the curves of Fig. 5.23 are obtained. The data presented here are probably conservative relative to the particles of interest. For instance, if the particles are less dense than assumed they will fall slower. Also, if they are nonspherical they will probably fall slower. Thus these curves represent the maximum fall rate allowable within the guidance of the experimental data.

It is clear from these data that the meteoroid particles will have lost their orbital speeds and will be falling quite slowly when they reach the mesopause. Fall rates of the order of tens of centimeters per second accrue at the 82-km level of most noctilucent clouds for all particles smaller than 10^{-5}-cm radius. At lower mesospheric levels the fall rates drop to a few centimeters per second. This parameter is inversely dependent on the density scale height of the region, and it thus has a scale height of approximately 10 km. The particulate concentration varies inversely with the fall rate and thus has a scale height similar to that of the density distribution. The concentration thus increases with decreasing height to produce a concentration distribution, if all particles are 10^{-6} cm in radius, of 0.02 particle per cubic centimeter at an altitude

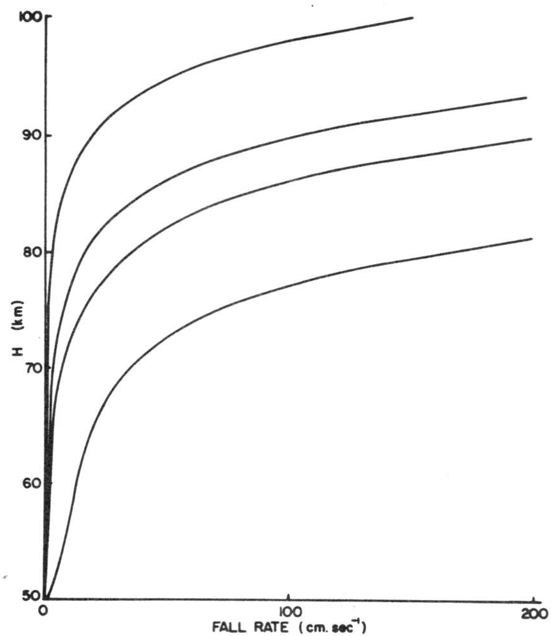

Fig. 5.23. Fall rates of spherical particles of density 2.5 gm/cm³ and radii 10^{-6}, 5×10^{-6}, 10^{-5}, and 5×10^{-5} for the curves from top to bottom, respectively.

of 100 km, 0.2 at 90, 2.0 at 80, and 30 at 70 km. If the incoming particles should be of 10^{-5} cm radius, the distribution would be 0.015 particle per cubic centimeter at 90, 3.0 at 80, and 30 at 70 km. Now estimates of the visibility of such particles indicate that concentrations of such particles would be visible in our observational window at all times under the conditions in which noctilucent clouds are observed. This is not the case, as is demonstrated by observation.

This discrepancy could be due to erroneous estimates of the concentrations. It could, however, be the result of alternate processes which result in reduction in particulate concentration. One of the approximations involved in obtaining these results was that the environment remains stationary. The approximation is probably erroneous, since we know that the mesosphere is a turbulent region, and thus the concentration profile is likely to be governed by eddy diffusion rather than the molecular process employed in the above calculations. Vertical wind speeds of several centimeters per second are to be expected in convective overturnings of mesospheric air, particularly in the strong lapse rates of the summer mesosphere.

Probably of equal importance is the process of coagulation which is an inherent factor in colloidal suspensions, particularly in the case of gaseous suspension. The rate at which coagulation proceeds is a strong function of the concentration and mean free path of the environment, since these factors establish the frequency of contact between particles. The mean free path has values of approximately 0.1 cm at 70-km altitude, 0.4 at 80, and 2.5 at 90 km. It has been demonstrated that the particulate concentration varies with a curve governed by the density scale height. The same is true of the mean free path, except that the exponential function is negative in this case. If we neglect the complications introduced by turbulent factors, these curves will combine to establish a vertical profile of particulate concentration (under the conditions assumed here).

Measurements have been made in the lower stratosphere to maximum altitudes of 30 km with a resulting concentration of one particle per cubic centimeter being a characteristic value. Such a value for the uniform concentration of the upper stratosphere and mesosphere is in accord with other facts, such as the general absence of optically detectable concentrations of such particles in the upper atmosphere. In the absence of additional information, it will be assumed that there is an ample number of meteoritic particles entering the mesosphere at most locations, but that under general conditions the concentration is controlled by destructive processes so that concentrations are insufficient to be optically observable from our vantage point. The question then remains as to how this equilibrium situation can be altered to permit formation of the observed noctilucent clouds.

Such a physical situation could be generated by convective motion. An upward current of air will sharply reduce the particulate concentration scale height, even changing its sign if the vertical wind speed is in excess of the particulate fall rate. Thus, a 50-cm/sec upward air flow would stop the Eulerian descent of all particles smaller than 10^{-5}-cm radius at an altitude of 82 km and would push the smaller particles upward to higher altitudes. Although the coagulation rate would be higher owing to an increased mean free path at these higher equilibrium altitudes, the differences in scale heights of these two parameters would foster an increase in concentration of the meteoroid particulate material.

These considerations lead to the conclusion that the most likely cause of noctilucent cloud formation is the presence of a vertical flow of mesospheric air. The source and physical characteristics of such a flow were discussed at the beginning of Chapter 5 and of Chapter 4, and in Section 4.8. A "stratospheric tidal jet" is postulated to be formed by the diurnal heat input from the sun to the ozonosphere during the summer season, and this jet of stratospheric air has been shown to occur at the time and place of noctilucent cloud occurrence. Observational systems have not yet produced adequate data for a complete description of this very special circulation system, so the noctilucent clouds provide a most important source of information on this important aspect of the upper atmospheric circulation.

Inspection of noctilucent cloud observations by Ludlam (1957) yielded the fact that observers persistently report the individual cloud elements to be moving from the northeast at a speed of the order of 80 meters/sec. These observations are suspect in that the detail features of the clouds which provide the means of displacement measurement are rapidly changing elements, and the motions could be the result of alterations in the clouds' small-scale structure. If that were so, the phenomenon that produces these detailed features must then move with the above general velocity, which does not seem likely. It is likely that these observations do represent actual winds in the noctilucent cloud zone with a reasonable accuracy. In that event, the observations provide us with an estimate of the flows required for their support.

The observational data is best satisfied if it is assumed that vertical motions are sufficient to lift the particles so that they are carried away from the polar region in the lower ionosphere by the return flow of our diurnally driven stratospheric tidal circulation. As is indicated by the curves of Fig. 5.23, these vertical speeds must be of the order of a meter per second at the mesopause. It was shown at the beginning of Chapter 5 that such a vertical flow would amount to only a few percent of the mass involved in the stratospheric tidal jet. Divergence of the stratospheric tidal jet into the lower ionosphere is then postulated to supply the mass required to support the flows detected by Kochanski (1963). This meridional circulation then has its origin in low latitudes at the stratopause level, and moves poleward with a significant ascending motion at high latitudes. In the nighttime upper mesosphere the air involved in this meridional circulation is injected through momentum effects across the mesopause and begins to flow away from the polar regions in the stable lower ionosphere.

Another most important feature of noctilucent cloud occurrences is the very general great variability they exhibit. Uniformity is a very unusual quality in noctilucent clouds, and the rule is that gross changes occur over intervals of the order of minutes. The divergent phase of the stratospheric tidal jet is then highly variable, as could reasonably be expected in the case of such an unusual circulation system. If continued research should show that meteoroid particles are always adequate for production of the clouds, as is inferred by the current data, then the noctilucent clouds clearly delineate the events which comprise this important phase of the global meridional circulation which ties together the stratospheric, mesospheric, and ionospheric regions of the upper atmosphere. The input to the ionosphere portion of this circulation thus appears to occur principally in the nighttime sector of the mesopause region and is additionally a highly variable series of pulses as instabilities alter the stratospheric tidal jet.

An imperative task of upper atmospheric researchers is adequate measurement of the diurnal circulation, which includes the stratospheric tidal jet. MRN data obtained to date are responsible for the picture deduced above, with adequate observations only available at middle and low latitudes. This diurnal circulation is located near the stratopause in these locations, but rises into the mesosphere in the tidal jet portion of summer high latitudes. The data of Fig. 4.79 indicate that the altitude of the tidal circulation has lifted from the stratopause to mesospheric levels at Fort Greely. This matter has been inspected further as is illustrated by the data illustrated in Fig. 5.24. Here the SCI data

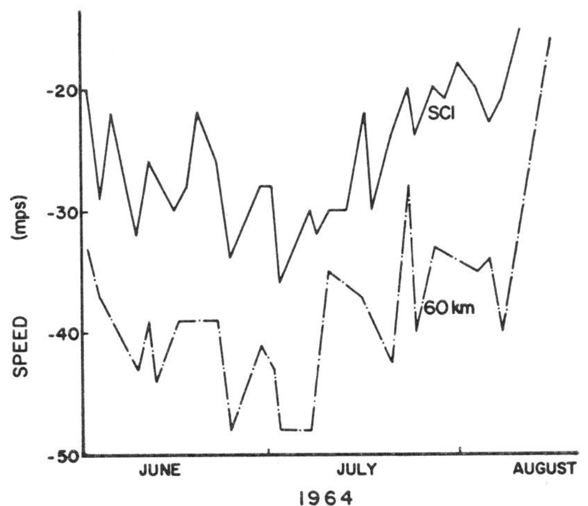

Fig. 5.24. Comparison of zonal SCI and 55- to 65-km mean flows at Fort Greely during the summer of 1964.

are compared with a similarly derived measure of mesospheric circulation; namely, the mean zonal flow in the layer between 55 and 65 km. During the summer season of interest the zonal easterlies are stronger at this higher altitude, which is interpreted as supporting the concept that the level of the diurnal circulation does indeed lift with increasing latitude.

It is desirable to consider again the structure of such a circulation system as was discussed in Fig. 4.1. MRN data are obtained near local noon, so it is representative of the stratospheric structure along the ridge of high pressure and temperature. As has been noted before, the meridional temperature gradient of the summer season is a minimum along the ridge line, and thus the MRN data on zonal flow provide an underestimate of the average easterly winds. The maximum meridional temperature gradient is located in the nighttime sector, so peak zonal flows will be located there. Very importantly, this maximum gradient is also located in the upper mesosphere. The data presented in Fig. 5.24 are then an underestimate on all counts, and it is to be expected that a series of observations through a summer season which include the region to 90 km would illustrate a much more pronounced effect. An interesting problem in the noctilucent cloud picture is clarified by the data of Fig. 5.24 in that the flow in the 60-km region does not break with the SCI as the summer easterlies begin to wane, but holds strong until August and breaks sharply with the decrease of noctilucent clouds obtained by Fogle (1964a) as is illustrated in Fig. 5.19. The development of this tidal circulation into a circumpolar circulation of the summer season has been discussed in Sections 4.4 and 4.5, and comparison of these data with the annual variation in noctilucent clouds (Fig. 5.19) provides a very satisfactory correlation between these parameters.

The very strong reservations with which initial noctilucent cloud observations were accepted by the atmospheric scientific community have been largely dispelled by the data obtained in recent years by Fogle in a truly synoptic network of observational stations. It is comparatively certain that the night clouds and the circulation that supports them are general in the high-latitude high-altitude nighttime sky over the Eurasian and North American continental regions. The data do indicate and may be adequate for proof that these clouds are a truly diurnal phenomenon. An important aspect of the similarity of the circulation structures of the two hemispheres may well be established by careful comparison of observational data obtained at comparable latitudes. Noctilucent clouds are already known to exist in the Southern Hemisphere, and we already have reason to expect (Fig. 5.22) that the events will be less plentiful if the circulation systems are identical. All of these observations point toward these special clouds being a most important means of exploring the upper atmosphere.

5.4 Electrical Characteristics

The mesosphere is a transition zone between the relatively neutral lower atmosphere and the electrically active ionosphere. It is the region where tropospheric sensing techniques begin to lose their effectiveness as atmospheric observation tools, but where electrical measuring systems are hampered by excessive neutral atmosphere deteriorating effects. In electrical terms, the mesosphere constitutes the lower half of the ionospheric D region, which extends roughly from 60- to 100-km altitude. The number density of the total gas varies from approximately 6.4×10^{21} molecules per cubic centimeter at the base to approximately 4.2×10^{20} molecules per cubic centimeter at the mesopause. Charged particles constitute only a small fraction of the total population, but as is frequently true, these minor elements exert an apparent dominant influence on the physical state of the medium.

The mesosphere is thus located in the near field of the upper condenser plate which sustains the earth's atmospheric electric field. A current of negative ions of the order of 2000 amperes' intensity is forced toward the ionosphere from below by the fair-weather electric field. On the average, approximately 2000 negative ions per square centimeter per second approach the ionosphere through the mesosphere. Much of the neutralization of this current is accomplished by physical processes in the mesosphere. The carriers involved in this current flow are probably of atomic size in the mesosphere, and owing to the large mean free path (tenths of centimeters) in that region their mobility should be orders of magnitude greater than at the surface. A positive space charge immediately above the earth's surface results from the reduction in mobility with decreasing altitude, and the presence of impurities such as condensation nuclei which further reduce the mobility. Such a space charge should be of negligible importance in the mesopause region, however, since mobility-reducing mechanisms should be ineffective except in noctilucent cloud regions.

No great degree of uniformity should be expected, however, since the discharge current flowing into the ionospheric condenser owing to the fair-weather field must necessarily be replaced by a return current in the opposite direction. This return flow has been postulated to be related to thunderstorm activity, and if so it is probably characterized by intense local upward flows of positive ions above thunderstorm activity. While thunderstorms occur occasionally at almost all locations, they are most frequent at low latitudes in the summer hemisphere in the PM sector. The return current, then, is not likely to be symmetrical, as is the case of the fair-weather electric field current. This is a matter of rather small consequence, since the ionosphere has a high conductivity and can easily redistribute the charge to maintain an equipotential surface. These processes require, however, that the electric field of at least the lower portion of the mesosphere be reversed over these centers of the return current. Because of the high mobility of charged particles in the upper atmosphere these fields will be small, but this reversal of electric fields over thunderstorm activity will complicate the mesospheric structure.

A further complexity of the distribution of electrically charged particles in the mesosphere is introduced by the presence of the earth's magnetic field. The magnetic field is a vector quantity with an intensity of approximately 0.5 gauss, oriented generally along meridian lines, with a nearly horizontal direction in equatorial regions and becoming nearly vertical at high latitudes. Electrical conductivity is significantly higher along the magnetic lines of force than it is perpendicular to these lines. A horizontal gradient in charge concentration is thus relaxed with greater facility in equatorial regions than at high latitudes, if the gradient is in the meridional direction. In general, the charge composition of the mesosphere is principally ions, which are only slightly sensitive to magnetic effects as a result of their large mass, but are generally entrained into ambient motions. Internally, then, the magnetic field which pervades the mesosphere has a minor impact on the physical situation. Externally, however, the earth's magnetic field has a strong influence on mesospheric structure.

This is demonstrated most clearly by auroral activity. Principal auroral activity is found in a high-latitude belt symmetrical about the geomagnetic poles, with a peak intensity in the 20- to 25-degree colatitude region. The visible aurora is confined to the upper atmosphere above the mesopause, in general, but associated effects such as electron and ion enhancements extend well down into the mesosphere. There is evidence that the auroral emissions are produced by atomic and molecular reactions which are activated by bombardment of environmental gases by high-speed charged particles which are guided into the particular auroral zone location by the earth's magnetic field. These incoming particles, principally protons and electrons, appear to have a solar source. Evidence on the participation of these particles in the radiation belt activity about the earth is contradictory, and it appears probable that the auroral-generating particles represent a direct precipitation of the solar wind into the earth's atmosphere. The geomagnetic radiation belts may be involved in a secondary capacity such as storage, but the considerably greater energies of the auroral particles than those observed in trapped orbits of the radiation belts makes the direct dumping of trapped particles an unlikely source of auroral particles.

Auroral activity is observable only about 5% of the nights in a year. Its occurrence follows the sunspot cycle with an amplitude of approximately a factor of 5 and with the maximum activity following the peak in sunspot activity by some two tenths of a cycle. In addition, there is a well-defined maximum in spring and fall. Auroral light has been observed only at night, and the question of daytime occurrence is a matter of some controversy. Observation of daytime auroral light has proved difficult owing to the high background light level. There is little reason to expect the incidence of solar wind particles to stop on the daylight side of the earth, although their number and strength may well be reduced at auroral altitudes as a result of the asymmetric shape of the geomagnetic cavity.

Auroral activity occurs in a great variety of geometric forms. They range from faint diffuse glows high up in the ionosphere to clearly defined arcs, rays, and bands which sometimes extend down the D region to mesospheric levels.

Color spectra of the aurora result from line emissions of the excited atmospheric constituents, both atomic and molecular, which cover the entire range of the visible spectrum. In many cases little color is evident, but this appears to be generally due to low intensities which fall below the color threshold of the eye.

Since auroral activity is constrained by the earth's magnetic field, it can offer little information relative to the upper atmospheric circulation. It can, however, yield data on the composition and electrical characteristics of the D region of the ionosphere and possibly on the structure of the mesosphere. In particular, the general ionospheric configuration is well represented by horizontal stratification, while it is clear that any auroral influences are likely to exhibit a nearly vertical stratification. It is known from electromagnetic reflection and absorption characteristics that an enhanced quantity of free electrons attends auroral occurrences. It is then to be expected that considerable latitudinal variability will be introduced into the electrical characteristics of the mesospheric regions if the auroral phenomenon should extend to low altitudes.

A most important electrical characteristic of the D region of the ionosphere is that it introduces considerable absorption of a traversing electromagnetic signal, increasing in importance with decreasing altitude, increased electron concentration, and decreased frequency of the probing radiation. The principal mechanism for absorption of the radiant energy is the collision of participating electrons oscillating with the wave with neutral molecules, in which case a portion of the incident electromagnetic energy is converted into kinetic motion of the gas (that is, heating). Ambient heating is generally of small consequence to the general ionospheric structure, but loss of energy by the traversing radiation may be of great significance. Now the mesosphere offers an ideal environment for conduct of this physical process in that the ambient density is sufficient to assure collision of the electrons during their wave-induced motions. A major factor in the efficiency of the mechanism is the concentration of electrons in the mesosphere.

Electrons are freed in profusion during daytime hours down to the stratopause. Close proximity of neighboring molecules sets a limit on the lifetime of those free electrons, with the destructive process centering principally on electron attachment on a neutral molecule to form a negative ion although the electron may become associated with a positive ion to form a neutral particle or attach to a dust particle. All of these destructive processes limit the electron concentration at mesospheric altitudes to a few hundred per cubic centimeter in the general case as is illustrated in Fig. 5.25. Even this small number of electrons is sufficient to have a significant damping effect on very low-frequency electromagnetic signals.

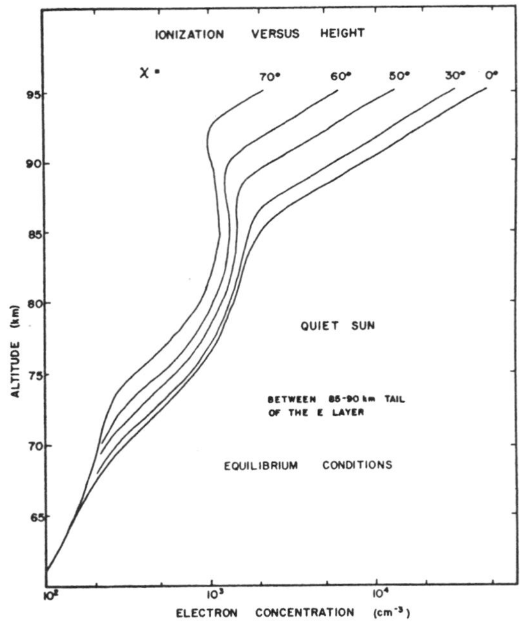

Fig. 5.25. Variation of electron concentration with height as a function of solar zenith angle (χ) according to Nicolet and Aikin (1960). (Courtesy Journal of Geophysical Research)

The concentration of free electrons in the mesosphere has a strong diurnal cycle. A great majority is liberated by ionization processes which involve absorption of solar ultraviolet radiation; so, as this energy source is denied during nighttime hours, the electron concentration of the ionosphere will decrease to near zero, sustained only by cosmic radiation particle release of a small number of additional electrons on a more or less uniform basis. Since opportunity exists for destruction of the free electron population, lifetime of electrons in the mesosphere is short and the concentration curve follows the generation curve very closely. In the diurnal cycle a few hundred electrons per cubic centimeter represent the maximum concentration, near noon, with a concentration of the order of 100 electrons per cubic centimeter during nighttime hours. In fact, this diurnal cycle represents an advancing wave of D-region electron concentration that progresses through middle and low latitudes at a speed of approximately 0.5 km/sec, traveling from east to west.

There is also a marked latitudinal variation in mesospheric electron density. Because of the longer path lengths traversed by solar radiation, its intensity is reduced by absorption, so that at a given level in polar D regions, ionization will be reduced at all times and will exhibit a strong annual cycle, with maximum concentrations in the summer season. The polar night shadow of the earth will preclude production of electrons by solar radiation to altitudes well up into the ionosphere and thus leave the winter polar mesosphere effectively void of free electrons. This has a marked effect on radio propagation in that the entire electron content of the polar winter ionosphere is diminished, becoming less than that concentration required to refract the higher-frequency radio waves. This phenomenon gives rise to the "polar cap blackout" of communications systems.

Auroral activity enhances the electron population of the lower ionosphere locally in winter polar regions and provides a more favorable environment for conduct of long-range communications. Lifting of the winter polar ionosphere decreases the fair-weather electric field in that region, and modifies the physical processes through which destruction of the fair-weather field current ions is accomplished. Lowering of the ionospheric highly conducting layer in a characteristic circumpolar ring of about 20–25 degrees co-latitude will thus complicate the electrical structure of the mesosphere, providing horizontal gradients, particularly along latitude lines.

This matter of electron density in the high-latitude winter during auroral activity has been analyzed by Lerfald *et al.* (1964) with the resulting estimates illustrated in Fig. 5.26. The symbol n refers to an absorption coefficient which is a negative power of the effective frequency, and is directly related to the energy spectrum of the incident auroral electrons. A smaller value of n infers higher energies in the incoming spiraling radiations, and thus results in greater electron densi-

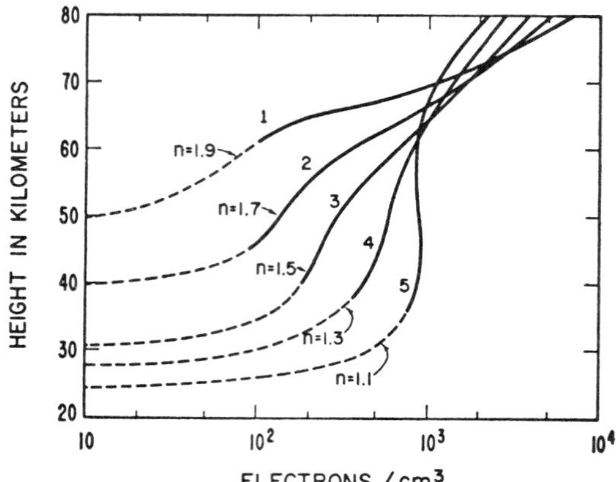

Fig. 5.26. Estimated electron density profiles for various auroral situations. (Courtesy Journal of Geophysical Research)

ties in the lower mesosphere and the upper portions of the upper stratosphere. The experimental data on which that study was based indicated a variability in the value of n from case to case and with time during a particular event, and thus would indicate a variability of the electron density with time during auroral events amounting to about an order of magnitude above the natural level induced by cosmic rays at the stratopause level.

There is then indication that the auroral zones of each hemisphere are the sites of enhanced electron concentrations amounting to an order of magnitude or more increase during the periods in which auroral activity is present. Most auroral activity occurs in the fall and spring periods, and thus these "horns" of ionospheric intrusion into the lower atmosphere have their maximum occurrence during those periods. It is in these vertically oriented bands about the polar regions at about 60 degrees latitude that gross absorption of radio frequency energy is experienced. The high ionosphere which is produced by lack of solar ultraviolet ionizing radiation in the polar region during winter forces the use of low frequencies in order to obtain refractions, and these walls of enhanced electron density act to absorb these low frequencies quite effectively. The ionosphere does not offer the smooth spherical shell of reflecting surface which would be desirable from a communications standpoint, but evidences considerable variations in space on a global basis as well as locally and with time.

Nicolet and Aikin (1960) have also considered the general electron density in the mesosphere as a function of the state of the solar emission. Auroral associated enhancements of the D-region electron density are the result of focusing of particles from the solar wind by the earth's magnetic field as they are precipitated. In general, the intensity of the solar wind and thus of the incident auroral particles is related to the state of the solar surface, but solar emission enhancements of ionizing radiations other than electrons and protons are known to be coincident phenomena. These additional radiations are in the short wavelengths of the continuum, in the region of the spectrum which is dominated by the Lyman alpha radiation of hydrogen. These radiations are not subjected to the deflecting effects of the earth's magnetic field and thus have a general effect over the sunlight portion of the globe in enhancing the electron concentration of the D region. The magnitudes of these effects are illustrated by the curves of Fig. 5.27, which were constructed by

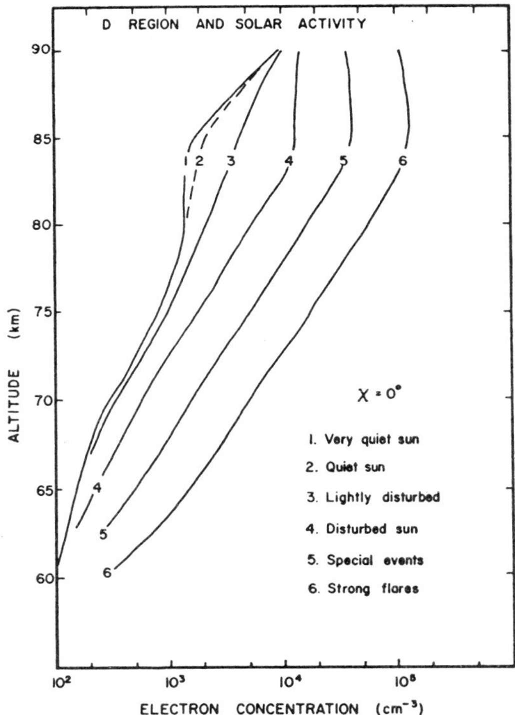

Fig. 5.27. Variation of the electron concentration according to solar conditions, from very quiet sun to strong solar flares. (Courtesy Journal of Geophysical Research)

Nicolet and Aikin (1960) for the case in which the sun's radiations are vertically incident. There are, then, variations of two orders of magnitude in the electron density at the mesopause over the 11-year sunspot cycle if the short-term effects of large solar flares are included.

As has been indicated, the particulate input into the auroral zones will also be enhanced during peak solar activity, so the gradients on the sunlight side of the earth might well be relatively stable. Since the electron density of the night side of the earth is fixed by cosmic ray ionization, the known reductions in cosmic ray abundances during intense solar activity should result in a reduction in electron density in that quarter, and thus result in maximum gradients in electron concentration in the mesosphere, both latitudinally owing to the auroral zone enrichments and longitudinally by the strong changes associated with the sunset and sunrise lines. All of these considerations point toward an electrical situation in the mesosphere characterized by free electron concentrations of approximately 200 electrons per cubic centimeter generally in the lower mesosphere, occasionally enhanced to 1000 or more per cubic centimeter during auroral events and strong solar activity. In general the electron density of the mesosphere exhibits a negative lapse rate, with concentrations at the mesopause approximately an order of magnitude higher than those observed at the stratopause.

As is illustrated in Fig. 5.25, there is some indication of a weak peak in concentration of electrons at the mesopause level. The possibility exists that the presence of meteoroid debris in this region has an equivalent electrical effect to that exhibited by atmospheric impurities in the lower troposphere, in that the mobility of charges is reduced by association with these particles, and the reactive processes are altered. It would normally be assumed that the presence of these particles would result in a reduction in electron density. It must be remembered that the ionosphere is comparatively stable, and that small meteoroid particles decelerating in the region immediately above the mesopause would tend to fall at their terminal velocity in this viscous medium with little effect by transport in wind current. The mesosphere, however, appears to be a turbulent medium, quite probably characterized by considerable convective motions. Eddy diffusion of the particulate material downward would suffice to reduce the density of the meteoroid material in the region immediately below the mesopause. The proper view, then, is that the peak in electron density indicated in Fig. 5.25 is caused by a reduction in electron density in the layer above the mesopause caused by presence of particulate material.

The D region of the ionosphere, which roughly corresponds to the mesosphere, is then one of the most interesting regions of the atmosphere. It occupies the transition sphere separating the more or less neutral troposphere and stratosphere from the electrical ionosphere, and thus has some of the characteristics of both regions. Sensing systems that have proved efficient in the lower or upper environments are usually difficult to apply in the mesosphere, so the data are generally inadequate at this time. It is also certain that a complete picture of the mechanics and thermodynamics of the earth's atmosphere will not be obtained until adequate knowledge is available for this isolating region. The synoptic approach of the meteorologist is a must for this region, since the data already available clearly indicate that circulation effects are of prime importance in establishing the physical character of the region. The most important reason for embarking on this new extension of atmospheric synoptic study rests directly on the compelling scientific necessity for a complete understanding of our atmospheric environment.

Early estimates of electron and ion distributions were based on considerations concerning a static atmosphere. This was in large part the result of a lack of data on upper atmospheric motions, a situation which has gradually altered for specific altitude regions. First realization of important large-scale motions came from sound wave propagation of the stratospheric region and the first hints of a turbulent environment were obtained from radio sounding of ionized trails formed by meteors. The MRN data represent the first effort at synoptic exploration of the higher levels, and while the data do not yet provide a global sample, they do offer a preliminary look at the dynamic structure of the stratospheric circulation almost to the mesopause. Reaction rates of the several chemical and physical processes are relatively slow in mesospheric and lower ionospheric altitudes, so advective and convective motions of an active circulation could have pronounced effect on the electrical and other characteristics of the medium. Static considerations are apt to yield a comparatively simple picture of any phenomena, and the introduction of a perturbing factor such as the meridional and zonal circulations that have been discussed in this book can be expected to profoundly modify some currently accepted concepts. As the data become more applicable and the mating of the fields of meteorology and aeronomy becomes more secure it is probable that some of the confusing aspects of upper atmospheric physics will be clarified.

REFERENCES

Aiken, A. C. (1960). A preliminary study of sunrise effects on the D-region. Sci. Report No. 133, AF19 (604)-4663, 1.6.60, AD 240 289.

aufm Kampe, H. J., M. E. Smith, and R. M. Brown. (1962). Winds between 60 and 110 kilometers. *J. Geophys. Res.* **67**, 4243–4258.

Bandeen, W. R., R. M. Griffith, and W. Nordberg. (1959). Measurement of temperatures, densities, and winds over Ft. Churchill, Canada, by means of rocket grenade experiment. U. S. Army Signal Research and Development Laboratory, Fort Monmouth, Tech. Report 2076, 71 pp.

Barth, C. A. (1961). Nitrogen and oxygen atomic reactions in the chemosphere. Jet Propulsion Laboratories, TR 32–63, 14.4.

Bates, D. R. (1960). The airglow. *In* "The Physics of the Upper Atmosphere" (J. A. Ratcliffe, ed.), pp. 219–267. Academic Press, New York.

Bates, D. R., and M. Nicolet. (1950). The photochemistry of atmospheric water vapour. *J. Geophys. Res.* **55**, 301–327.

Batten, E. S. (1961). Wind systems in the mesosphere and lower ionosphere. *J. Meteorol.* **18**, 283–291.

Berning, W. W. (1958). IGY upper air research at the Ballistic Research Laboratories. Experimental Results of the U. S. Rocket Program for the International Geophysical Year to July 1, 1958. IGY Rocket Report Series, Nat. Acad. Sci., Washington, D.C.

Beth, E. W. (1947). Discussion of a method for the investigation of pressure, temperature, average molecular weight, wind velocity and dissociation in the upper atmosphere. U. S. Air Force, Air Materiel Command, Report No. 5–10, 39 pp.

Bigg, E. K. (1956). The detection of atmospheric dust and temperature inversions by twilight scattering. *J. Meteorol.* **13**, 262.

Biriukova, L. A. (1959). Opyt opredeleniia iarkostl neba do vysoty 60 km. *Tr. Tsentr. Aerol. Observ.* **25**, 77–84.

Blamont, J. E. (1958). The sodium twilight airglow 1955–1957. I. *Ann. Geophys.* **14**, 253–281; II. *ibid.*, pp. 282–304.

Blamont, J. E. (1959). Nuages artificiels de sodium: vitesse du vent, turbulence et densité de la haute atmosphere. *Compt. Rend.* **249**, 1248–1250.

Blamont, J. E., and M. L. Long. (1963). New direct measurements of ionospheric temperature. *In* "First International Symposium on Rocket and Satellite Meterology" (H. Wexler and J. E. Caskey, Jr., eds.), pp. 71–75. Wiley, New York.

Bourdeau, R. E. (1959). Analytic and experimental electrical conductivity between the stratosphere and the ionosphere. *J. Geophys. Res.* **64**, 1363–1370.

Bowen, E. G. (1953). The influence of meteoritic dust on rainfall. *Australian J. Phys.* **6**, 490.

Boyd, R. L. F., and M. J. Seaton. (1954). "Rocket Exploration of the Upper Atmosphere," 376 pp. Pergamon Press, Oxford.

Brasefield, C. J. (1954). Winds at altitudes up to 80 km. *J. Geophys. Res.* **59**, 233–237.

Bruch, A., and G. M. Morgan, Jr. (1961). Wind variability in the mesosphere as determined by the tracking of falling objects. New York University, AFCRL 660. Final Report Contract No. AF19(604)–6193.

Byram, E. T., T. A. Chubb, and H. Friedman. (1955). Dissociation of oxygen in the upper atmosphere. *Phys. Rev.* **98**, 1594–1597.

Byram, E. T., T. A. Chubb, and H. Friedman. (1956). The study of extreme ultraviolet radiation from the sun with rocketborne photon counters. *In* "Rocket Exploration of the Upper Atmosphere" (R. L. F. Boyd and M. J. Seaton, eds.), pp. 276–278. Pergamon Press, Oxford.

Cahill, L. J., Jr. (1959). Investigation of the equatorial electrojet by rocket magnetometer. *J. Geophys. Res.*, **64**, 489–503.

Campbell, W. H., and J. M. Young. (1963). Auroral-zone observations of infrasonic pressure waves related to ionospheric disturbances and geomagnetic activity. *J. Geophys. Res.* **68**, 5909–5916.

Crackett, K. F., F. A. Paneth, and E. J. Wilson. (1949). Chemical composition of the stratosphere at 70 km height. *Nature* **164**, 128–129.

Champion, K. W. (1959). Atmospheric densities from satellites and rocket observations. *Planetary Space Sci.* **1**, 259–264.

Chapman, S. (1930). A theory of upper atmospheric ozone. *Mem. Roy. Meteorol.* **3,** 103–125.

Chapman, S. (1951). Photochemical processes in the upper atmosphere and resultant composition. *Compendium Meteorol.* pp. 262–274.

Chapman, S. (1955). Rockets and the magnetic exploration of the ionosphere. *In* "Rocket Exploration of the Upper Atmosphere" (R. L. F. Boyd and M. J. Seaton, eds.), pp. 292–305. Pergamon Press, Oxford.

Charney, J. G., and P. G. Drazin. (1961). Propagation of planetary-scale disturbances from the lower into the upper atmosphere. *J. Geophys. Res.* **66,** 83–109.

Cheers, B. (1959). Ultraviolet detector tests at high altitudes using gun fired projectiles. Canadian Armament Research and Development Establishment, Interval Memorandum Report No. 0101–02.

Chubb, T. A. (1957). Rocket observations of X-ray emission in a solar flare. *Nature* **179,** 861–862.

Chubb, T. A., E. T. Byram, H. Friedman, and J. E. Kupperian. (1958). The use of radiation absorption and luminescence in upper air density measurements. *Ann. Geophys.* **14,** 109–116.

Cole, A. E., and A. J. Kantor. (1963). Horizontal and vertical distributions of atmospheric density, up to 90 km. U. S. Air Force, Cambridge Research Laboratories, Bedford, Massachusetts, Surveys in Geophysics, No. 157, 22 pp.

Cooper, H. W., I. S. Gulledge, M. J. Koomen, and D. M. Packer. (1960). Altitude distribution of night airglow emissions by rocket-borne photometers. *J. Geophys. Res.* **65,** 2484.

Court, A., A. J. Kantor, and A. E. Cole. (1962). Supplemental atmospheres. U. S. Air Force, Cambridge Research Laboratories, L. G. Hanscom Field, Massachusetts, Meteorological Development Laboratory Project 8624, AFCRL-62-899.

Curtis, A. R., and R. M. Goody. (1956). Thermal radiation in the upper atmosphere. *Proc. Roy. Soc.* **A236,** 193–206.

Curtis, J. P., W. A. Rense, and B. Todd. (1954). Extreme ultraviolet research. Colorado State University, Engineering Experimental Station, Fort Collins, Colorado, Sci. Report No. 1, 28 pp.

Dalgarno, A. (1958). The latitudes and excitation mechanisms of the night airglow. *Ann. Geophys.* **14,** 241–252.

Davies, J. G. (1957). Radio observations of meteors. *Advan. Electron. Electron Phys.* **9,** 95–128.

Deirmendjian, D., and E. H. Vestine. (1959). Some remarks on the nature and origin of noctilucent cloud particles. *Planetary Space Sci.* **1,** 146–153.

Ditchburn, R. W. (1955). Ultraviolet radiation absorption cross sections. *In* "Rocket Exploration of the Upper Atmosphere" (R. L. F. Boyd and M. J. Seaton, eds.), pp. 313–326. Pergamon Press, Oxford.

Ditchburn, R. W. (1956). Absorption of ultraviolet radiation by the atmospheric gases. *Proc. Roy. Soc.* **A236,** 216–226.

Ditchburn, R. W., and P. A. Young. (1962). The absorption of molecular oxygen between 1850 and 2500 A. *J. Atmospheric Terrest. Phys.* **24,** 127–139.

Ditchburn, R. W., J. E. S. Bradley, C. G. Cannon, and G. Munday. (1957). Absorption cross sections for Lyman and neighboring lines. *In* "Rocket Exploration of the Upper Atmosphere" (R. L. F. Boyd and M. J. Seaton, eds.), pp. 327–334. Pergamon Press, Oxford.

Dobrovol'skii, S. P. (1964). The nature of noctilucent clouds. *Priroda* **53,** 87–94.

Dole, S. H. (1962). The gravitational concentration of particles in space near the earth. *Planetary Space Sci.* **9,** 541–553.

Dubin, M. (1954). Meteor Impacts by acoustical techniques. *In* "Rocket Exploration of the Upper Atmosphere" (R. L. F. Boyd and M. J. Seaton, eds.), pp. 26–27. Pergamon Press, Oxford.

Dubin, M., and C. W. McCracken. (1962). Measurements of distribution of interplanetary dust. *Astron. J.* **67,** 248–256.

Elford, W. G. (1959). Winds in the upper atmosphere. *J. Atmospheric Terrest. Phys.* **15,** 132–136.

Elvey, C. T. (1965). Morphology of auroral displays. *In* "Auroral Phenomena" (M. Walt, ed.), pp. 1–15. Stanford Univ. Press, Stanford, California.

Ference, M., W. G. Stroud, and J. R. Walsh. (1956). Measurements of temperatures at elevations of 30 to 80 km by the rocket-grenade experiments. *J. Meteorol.* **13,** 5–12.

Fleagle, R. G. (1958). Inferences concerning the dynamics of the mesosphere. *J. Geophys. Res.* **63,** 137–146.

Fogle, B. (1964a). Results of the study of noctilucent clouds over North America during 1963. University of Alaska, Geophysical Institute, NSF Grant GP 1759, Cloud Report No. 2.

Fogle, B. (1964b). Noctilucent clouds in the Southern Hemisphere. *Nature* **204,** 14–18.

Franklin Institute, Philadelphia. (1950). Development, design, instrumentation. Blossom 4 series. Contract AF19 (122)-33, Final Report F-2106.

Friedland, S. S. (1956). Pulsed searchlighting the atmosphere. *J. Geophys. Res.* **61,** 415–434.

Friedman, H. (1960). Survey of observations of solar ultraviolet and X rays. *J. Geophys. Res.* **65,** 2491.

Friedman, H., T. A. Chubb, J. E. Kupperian, Jr., and J. C. Lindsay. (1958). X-ray and ultraviolet emission of solar flares. Experimental Results of the U. S. Rocket Program for the International Geophysical Year to July 1, 1958. IGY Rocket Report Series, Natl. Acad. Sci., Washington, D. C.

Gibbons, J. J., and A. H. Waynick. (1959). The normal D-region of the ionosphere. *Proc. IRE* **47,** 160–161.

Gossard, E. E. (1962). Vertical flux of energy into the lower ionosphere from internal gravity waves generated in the troposphere. *J. Geophys. Res.* **67,** 745–757.

Gossard, E. E., and W. Munk. (1954). On gravity waves in the atmosphere. *J. Meteorol.* **11,** 259–269.

Greenhow, J. S. (1952). Characteristics of radio echoes from meteor trails: III. The behavior of the electron trails after formation. *Proc. Phys. Soc.* (*London*) **B65,** 169–181.

Greenhow, J. S., and E. L. Neufeld. (1955). Diurnal and seasonal wind variations in the upper atmosphere. *Phil. Mag.* [7] **46,** 549–562.

Greenhow, J. S., and E. L. Neufeld. (1956). The height variation of the upper atmosphere winds. *Phil. Mag.* [8] **1,** 1157–1171.

Greenhow, J. S., and E. L. Neufeld. (1959). Measurements of turbulence in the upper atmosphere. *Proc. Roy. Soc.* **A74,** 1–10.

Greenhow, J. S., and A. C. B. Lovell. (1960). The upper atmosphere and meteors. *In* "Physics of the Upper Atmosphere" (J. A. Ratcliffe, ed.), pp. 513–549. Academic Press, New York.

Greenhow, J. S., and E. L. Neufeld. (1961). Winds in the upper atmosphere. *Quart. J. Roy. Meteorol. Soc.* **87**, 472–489.

Gregory, J. B. (1960). Radiowave reflection of the mesosphere. *Trans. Am. Geophys. Union* **61**, 620.

Gregory, J. B. (1961). Radiowave reflections from the mesosphere. *J. Geophys. Res.* **66**, 429–445.

Grishin, N. I. (1961). Wave motion and meteorological conditions for the appearance of noctilucent clouds. *Ann. Intern. Geophys. Yr.* **11**, 20–22.

Gromov, L. F. (1963). Nekotorye dannye o chastote poiavleniia serebristykh, Leningrad University, 1960. *NASA (Natl. Aero. Space Admin.), Tech. Transl.* **F84.**

Groves, G. V. (1963). U. K. Meteorological rocket grenade studies. *In* "First International Symposium on Rocket and Satellite Meteorology" (H. Wexler and J. E. Caskey, Jr., eds.), pp. 42–59. Wiley, New York.

Gutenberg, B. (1946). Physical properties of the atmosphere up to 100 km. *J. Meteorol.* **3**, 27–30.

Hanson, A. M. (1963). Noctilucent clouds at 76.3° North. *Weather* **18**, 142–144.

Hanson, W. B. (1961). Structure of the ionosphere. *In* "Satellite Environment Handbook" (F. S. Johnson, ed.), pp. 27–46. Stanford University Press, Stanford, California.

Haurwitz, B. (1957). Solar activity and atmospheric tides. *J. Geophys. Res.* **62**, 489–491.

Haurwitz, B. (1960). The winds in the upper mesosphere. *Trans. Am. Geophys Union* **41**, 620.

Haurwitz, B. (1961a). Frictional effects and the meridional circulation in the mesosphere. *J. Geophys. Res.* **66**, 2381–2391.

Haurwitz, B. (1961b). Wave formations in noctilucent clouds. *Planetary Space Sci.* **5**, 92–98.

Haurwitz, B. (1964). Comment on wave forms in noctilucent clouds. University of Alaska, Geophysical Institute, NSF Grant GP 1759.

Havens, R. J., R. T. Koll, and H. E. LaGow. (1952). The pressure, density and temperature of the earth's atmosphere to 160 km. *J. Geophys. Res.* **57**, 59–72.

Hawkins, G. S. (1956). Variation in the occurrence rate of meteors. *Astron. J.* **61**, 386–391.

Hemenway, C. L., R. K. Soberman, and G. Witt. (1963). Particle sampling from noctilucent clouds. *Nature* **199**, 269–270.

Hemenway, C. L., R. K. Soberman, and G. Witt. (1964). Sampling of noctilucent cloud particles. *Tellus* **16**, 84–88.

Heppner, J. P., and L. H. Meredith. (1958). Nightglow emission altitudes from rocket measurements. *J. Geophys. Res.* **63**, 51–65.

Hesstvedt, E. (1961). Note on the nature of noctilucent clouds. *J. Geophys. Res.* **66**, 1985–1987.

Hines, C. O. (1959a). An interpretation of certain ionosphere motions in terms of atmospheric waves. *J. Geophys. Res.* **64**, 2210–2211.

Hines, C. O. (1959b). Turbulence at meteor height. *J. Geophys. Res.* **64**, 939–940.

Hines, C. O. (1960). Internal atmospheric gravity waves at ionospheric heights. *Can. J. Phys.* **38**, 1441–1481.

Hines, C. O. (1963). The upper atmosphere in motion. *Quart. J. Roy. Meteorol. Soc.* **89**, 1–42.

Hinteregger, H. E. (1961). Preliminary data on solar extreme ultraviolet radiation in the upper atmosphere. *J. Geophys. Res.* **66**, 2367–2380.

Hultquist, B., and J. Ornter. (1959). Strongly absorbing layers below 50 km. *Planetary Space Sci.* **1,** 193–204.

Humphreys, W. J. (1933). Nacreous and noctilucent clouds. *Monthly Weather Rev.* **61,** 228–229.

Ivanov-Kholodnyi, G. S. (1959). O raketnykh issledovaniiakh korotkovolnovoi radiatsii solntsa. *Izv. Akad. Nauk SSSR Ser. Geofiz.* **1,** 108–121.

Jacchia, L. G., and Z. Kopal. (1952). Atmospheric oscillations and the temperature profile of the upper atmosphere. *J. Meteorol.* **9,** 13–23.

Jacchia, L. G. (1957). A preliminary analysis of atmospheric densities from meteor decelerations for solar, lunar and yearly oscillations. *J. Meteorol.* **14,** 34–37.

Jastrow, R. (1959). Density and temperature of the upper atmosphere. *Astronautics* **4,** 24.

Jesse, O. (1896). Die Höhe der leuchtenden Nachtwolken. *Astron. Nachr.* **140,** 161.

Johnson, F. S., J. D. Purcell, R. Tousey, and K. Watanabe. (1952). Direct measurements of the vertical distribution of atmospheric ozone to 70 km altitude. *J. Geophys. Res.* **57,** 157–177.

Johnson, F. S., J. D. Purcell, R. Tousey, and N. Wilson. (1954). The ultraviolet spectrum of the sun. *In* "Rocket Exploration of the Upper Atmosphere" (R. L. F. Boyd and M. J. Seaton, eds.), pp. 279–288. Pergamon Press, Oxford.

Jones, L. M., and J. W. Peterson. (1961). Upper air densities and temperatures measured by the falling sphere method. Univ. of Michigan Report 03558-5-T, AF19(604)-6185.

Jones, L. M., F. F. Fischback, and J. W. Peterson. (1958). Seasonal and latitude variations in upper-air density. Experimental Results of the U. S. Rocket Program for the International Geophysical Year to July 1, 1958. IGY Rocket Report Series, Natl. Acad. Sci., Washington, D. C.

Jursa, A. S., F. J. Le Blanc, and Y. Tanaka. (1955). Results of a recent attempt to record the solar spectrum in the region of 900-3000 A. *J. Opt. Soc. Am.* **45,** 1085–1086.

Jursa, A. S. (1959). Nitric oxide and molecular oxygen in the earth's upper atmosphere. *Planetary Space Sci.* **1,** 161–172.

Kaiser, T. R., and S. Evans. (1955). Upper atmospheric data from meteors. *Ann. Geophys.* **11,** 148–151.

Kalitin, N. N. (1944). Cosmic dust according to actinometric measurements. *Compt. Rend. Acad. Sci. URSS* **45,** 375.

Kallmann, H. K. (1958). Recent results of high altitude research by means of rockets and satellites. Rand Corp., Santa Monica, California, RM 2275, AD 207 200.

Kane, H. A. (1959). Arctic measurements of electron collision frequency in the D-region of the ionosphere. *J. Geophys. Res.* **64,** 133–140.

Kaplan, J., and H. K. Kallmann. (1957). Upper atmosphere research. Univ. of California Institute of Geophysics Final Report, AF19 (604)-111, AD 133 688, 250 pp.

Kaplan, L. D. (1960). The influence of carbon dioxide variations on the atmospheric heat balance. *Tellus* ***12,*** 204–208.

Kaufman, F., and J. R. Kelso. (1961). The homogeneous recombination of atomic oxygen. *In* "Proceedings of an International Symposium at Stanford Research Institute," pp. 255–268. (Interscience), Wiley, New York.

Kellogg, W. W. (1952). Temperatures and motions of the upper atmosphere. "Physics and Medicine of the Upper Atmosphere," Albuquerque, pp. 54–74.

Kellogg, W. W. (1959a). IGY rockets and satellites: A report on the Moscow meetings, August, 1958. *Planetary Space Sci.* **1,** 71–84.

Kellogg, W. W. (1959b). Review of IGY upper air results. Rand Corp., Santa Monica, California, Papers, 21 pp.

Kellogg, W. W. (1960a). The dynamics of the polar mesosphere in winter. *Trans. Am. Geophys. Union* **41,** 620.

Kellogg, W. W. (1960b). Upper atmosphere studies. *Trans. Am. Geophys. Union* **41,** 179–183.

Kellogg, W. W. (1961). Chemical heating above the polar mesopause in winter. *J. Meteorol.* **18,** 373–381.

Kellogg, W. W. (1963). Report on symposium on meteorological rockets. *In* "First International Symposium on Rocket and Satellite Meteorology" (H. Wexler and J. E. Caskey, Jr., eds.), pp. 3–14. Wiley, New York.

Kellogg, W. W. (1964). Pollution of the upper atmosphere by rockets. Rand Corp., Santa Monica, California, RM-3961-PR., 86 pp.

Khvostikov, I. A. (1952). Silvery clouds. *Priroda* **5,** 49.

Kochanski, A. (1963). Circulation and temperatures at 70–100 kilometer height. *J. Geophys. Res.* **68,** 213–226.

Kroening, J. L. (1960). Ion-density measurements in the stratosphere. *J. Geophys. Res.* **65,** 145–151.

Kupperian, J. E., Jr., E. T. Byram, T. A. Chubb, and H. Friedman. (1958). Far ultraviolet radiation in the night sky. Experimental Results of the U. S. Rocket Program for the International Geophysical Year, 1957/1958. I. G. Y. Rocket Report Series, pp. 186–189. Natl. Acad. Sci., Washington, D. C.

Kupperian, J. E., Jr., E. T. Byram, and H. Friedman. (1959). Molecular oxygen densities in the mesosphere over Ft. Churchill. *J. Atmospheric Terrest. Phys.* **16,** 174–178.

LaGow, H. E. (1954). Physical properties of the atmosphere up into the F-layer. *In* "Rocket Exploration of the Upper Atmosphere" (R. L. F. Boyd and M. J. Seaton, eds.), pp. 73–81. Pergamon Press, Oxford.

LaGow, H. E., R. Horowitz, and J. Ainsworth. (1958). Arctic atmosphere structure to 250 km. Experimental Results of the U. S. Rocket Program for the International Geophysical Year to July 1, 1958. IGY Rocket Report Series, Natl. Acad. Sci., Washington, D. C.

Lear, J. (1957). Miners of the sky. *Saturday Rev.* **30,** 42–46.

Lerfald, G. M., C. G. Little, and R. Parthasarathy. (1964). D-Region electron density profiles during auroras. *J. Geophys. Res.* **69,** 2857–2860.

Ley, W. (1951). Upper atmosphere, its exploration and exploitation. *Aero. Eng. Rev.* **10,** 20–24.

Liller, W., and F. L. Whipple. (1954). High-altitude winds by meteor-train photography. *In* "Rocket Exploration of the Upper Atmosphere" (R. L. F. Boyd and M. J. Seaton, eds.), pp. 112–130. Pergamon Press, Oxford.

Lindley, W. B. (1957). Noctilucent clouds in Alaska, July 27–28, 1957. *Monthly Weather Rev.* **85,** 272–281.

Link, F. (1950). Couche de poussieres meteoriques dans une atmosphere planetaire. *Bull. Astron. Inst. Czech.* **2,** 1.

London, J. (1956). Radiative properties of the stratosphere. Final Report, AF19 (604)-1285, October, AD 110 220.

London, J. (1959). Dynamics of the mesosphere. Final Report, AF 19(604)-1738, December, AD 232 548.

Lovell, A. C. B. (1954). "Meteor Astronomy." Oxford Univ. Press (Clarendon), London and New York.

Lovell, A. C. B., and J. A. Clegg. (1952). "Radio Astronomy." Chapman & Hall, London.

Ludlam, F. H. (1956). The forms of ice clouds, II. *Quart. J. Roy. Meteorol. Soc.* **82**, 257.

Ludlam, F. H. (1957). Noctilucent clouds. *Tellus* **9**, 341–364.

McKinley, D. W. R. (1961). "Meteor Science and Engineering," 309 pp. McGraw-Hill, New York.

Maeda, K. (1963). On the heating of polar night mesosphere. *Proc. Intern. Symp. Stratospheric Mesospheric Circulation* pp. 451–506. Inst. Meteorol. Geophys. Freien Univ. Berlin.

Malzev, V. (1926). Luminous night-clouds. *Nature* **118**, 14.

Manring, E. R. (1962). Study of winds, diffusion. and expansion of gases in the upper atmosphere. Quarterly Progress Report, Geophys Corp. of America, NASA/GSFC Contract No. NASW-396.

Manring, E. R., J. F. Bedinger, and H. B. Pettit. (1959). Some wind determinations in the upper atmosphere using artificially generated sodium clouds. *J. Geophys. Res.* **64**, 584–592.

Manring, E. R., J. Bedinger, H. B. Knaflinch, and R. Lynch. (1960). Measurements of upper atmospheric winds and diffusion coefficients. *J. Geophys. Res.* **65**, 2509.

Marmo, F. F., L. M. Aschenbrand, and J. Pressman. (1959). Artificial electron clouds. *Planetary Space Sci.* **1**, 227–237.

Martin, G. R. (1954). The composition of the atmosphere above 60 km. *In* "Rocket Exploration of the Upper Atmosphere" (R. L. F. Boyd and M. J. Seaton, eds.), pp. 161–168. Pergamon Press, Oxford.

Meadows, E. B., and J. W. Townsend. (1956). Neutral gas composition of the upper atmosphere by a rocket-borne mass spectrometer. *J. Geophys. Res.* **61**, 576–577.

Meinel, A. B., and C. P. Meinel. (1964). Low-latitude noctilucent cloud of 2 November 1963. *Science* **143**, 38–39.

Michigan University, Engineering Research Institute. (1950a). Atmospheric phenomena at high altitudes. Final Progress Report for period July 15, 1946 to Aug. 31, 1950, Contract W-30-039 sc-32307, Project DA 3-99-07-022.

Michigan University, Engineering Research Institute. (1950b). Pressure and temperature measurements in the upper atmosphere. Final Report, Contract W-33-038-ac-14050.

Michigan University, Dept. of Aeronautical Engineering. (1951). Atmospheric phenomena at high altitudes. Final Report, Sept. 1, 1950–Dec. 3, 1951, Contract DA 36-039-sc-125, Progress Report No. 5.

Michigan University, Upper Atmosphere Research Project. (1954). The measurement of diffusive separation in the upper atmosphere. *In* "Rocket Exploration of the Upper Atmosphere" (R. L. F. Boyd and M. J. Seaton, eds.), pp. 143–156. Pergamon Press, Oxford.

Mikhnevich, V. V. (1957). Izmerenie davleniia v verkhnei atmosfere. *Usp. Fiz. Nauk* **63**, 197–204.

Millman, P. M. (1959). Visual and photographic observations of meteors and noctilucent clouds. *J. Geophys. Res.* **64**, 2122–2128.

Mirtov, B. A. (1957). Raketnye issledovania sostava atmosfery na bol'shikh vysotakh. *Usp. Fiz. Nauk* **63,** 181–196.

Mitra, A. P., and E. E. Jones. (1954). Recombination in the lower ionosphere. *J. Geophys. Res.* **59,** 391–406.

Mitra, S. K. (1952). "The Upper Atmosphere," 2nd ed., 713 pp. Asiatic Society, Calcutta.

Murcray, D. G., F. H. Murcray, W. J. Williams, and F. E. Leslie. (1960). Water-vapour distribution above 90,000 ft. *J. Geophys. Res.* **65,** 3641–3649

Murgatroyd, R. J. (1957). Winds and temperatures between 20 and 100 km—a review. *Quart. J. Roy. Meteorol. Soc.* **83,** 417–458.

Murgatroyd, R. J., and R. M. Goody. (1958). Sources and sinks of radiative energy from 30 to 90 km. *Quart. J. Roy. Meteorol. Soc.* **84,** 225–234.

Murgatroyd, R. J., and R. Singleton. (1961). Possible meridional circulation in the stratosphere and mesosphere. *Quart. J. Roy. Meteorol. Soc.* **87,** 125–137.

Nazarek, A. (1950). The temperature distribution of the upper atmosphere over New Mexico. *Bull. Am. Meteorol. Soc.* **31,** 44–50.

Newell, H. E., Jr. (1950). A review of upper atmosphere research from rockets. *Trans. Am. Geophys. Union* **31,** 25–34.

Newell, H. E., Jr. (1951). Temperatures and pressures in the upper atmosphere. *Compendium Meteorol.* pp. 303–310.

Newell, H. E., Jr. (1954). Rockets and the upper atmosphere. *Sci. Monthly* **128,** 30–36.

Newell, H. E., Jr. (1955). Rocket data on atmospheric pressure, temperature, density and winds. *Ann. Geophys.* **11,** 115–144.

Newell, H. E., Jr. (1958). The rocket research program. *Astronautics* **3,** 116–118.

Newell, H. E., Jr., and J. W. Siry. (1953). Rocket upper air research. U. S. Naval Res. Lab. Report No. 12-53, 20 pp.

Nicolet, M. (1959). The constitution and composition of the upper atmosphere. *Proc. IRE* **47,** 142–147.

Nicolet, M. (1958). Aeronomic conditions in the mesosphere and lower thermosphere. 89 pp. Pennsylvania State Univ. Sci. Report 102, AF19(604)-1304, 1.4.58, AD 152 550.

Nicolet, M., and A. C. Aikin. (1960). The formation of the D-region of the ionosphere. *J. Geophys. Res.* **65,** 1469–1483.

Nicolet, M., and P. Mange. (1954). The dissociation of oxygen in the high atmosphere. *J. Geophys. Res.* **59,** 15–45.

Nordberg, W., and W. Smith. (1963). Grenade and sodium rocket experiments at Wallops Island, Virginia. *In* "First International Symposium on Rocket and Satellite Meteorology" (H. Wexler and J. E. Caskey, Jr., eds.), pp. 119–134. Wiley, New York.

Nordberg, W., and W. G. Stroud. (1961). Results of IGY rocket-grenade experiments to measure temperatures and winds above the island of Guam. *J. Geophys. Res.* **66,** 455–464.

Paneth, F. A. (1954). The chemical analysis of atmospheric air. *In* "Rocket Exploration of the Upper Atmosphere" (R. L. F. Boyd and M. J. Seaton, eds.), pp. 157–158. Pergamon Press, Oxford.

Paton, J. (1949). Luminous night clouds. *Meteorol. Mag.* **78,** 354.

Paton, J. (1950). Aurorae and luminous night clouds. *Proc. Phys. Soc.* (*London*) **B63,** 1039.

Paton, J. (1951). Simultaneous occurrence of aurora and noctilucent clouds. *Meteorol. Mag.* **80,** 145.

Paton, J. (1954). Direct evidence of vertical motion at about 80 km provided by photgraphs of noctilucent clouds. *Proc. Toronto Meteorol. Conf., 1953* p. 31. Roy. Meteorol. Soc., London.

Paton, J. (1961). Noctilucent clouds. *Ann. Intern. Geophys. Yr.* **11,** 4–6.

Paton, J. (1964). Noctilucent clouds. *Meteorol. Mag.* **93,** 161–179.

Pearson, P. H. O. (1963). Measurements of atmospheric density, temperature and pressure at Woomera on 29 March, 1962 by the falling sphere method. Department of Supply, Salisbury, South Australia, TN 5AD 121.

Penndorf, R. (1949). The vertical distribution of atomic oxygen in the upper atmosphere. *J. Geophys. Res.* **54,** 7–38.

Petteway, M. L. V. (1962). Wave-guide propagation inside elongated irregularities in the ionosphere. *J. Geophys. Res.* **67,** 5107–5118.

Pressman, J. (1960). Artificial electron clouds—VI, low altitude study, release of cesium at 69, 82, and 91 km. *Planetary Space Sci.* **2,** 228–237.

Quiroz, R. S. (1961a). Seasonal and latitudinal variations of air density in the mesosphere. *J. Geophys. Res.* **66,** 2129–2139.

Quiroz, R. S. (1961b). Air density profiles for the atmosphere between 30 and 80 km. U. S. Air Force, Air Weather Service, TR 150, AD 254 659.

Quiroz, R. S. (1964). On the origin and climatology of noctilucent clouds. U. S. Air Force, Air Weather Service, Technical Report 181, 29 pp.

Ragsdale, G. C., and P. E. Wasko. (1963). Wind flow in the 80–400 km altitude region of the atmosphere. *NASA (Natl. Aeron. Space Admin.), Tech. Note* **TN D-1573.**

Rakipova, L. P. (1947). Possible effect of dust on vertical air movements and on isothermy in the stratosphere. *Izv. Akad. Nauk SSSR* **11,** 15.

Rapp, R. R. (1960). Accumulation of extraterrestrial particulate matter near the mesopause. *Trans. Am. Geophys. Union* **41,** 620.

Ratcliffe, J. A., ed. (1960). "Physics of the Upper Atmosphere," 586 pp. Academic Press, New York.

Rawer, K. (1952). "The Ionosphere," 202 pp. Frederick Ungar, New York.

Reasbeck, P., and B. S. Wiborg. (1954). Chemical analysis of upper atmosphere air samples from 50 to 93 km. *In* Rocket Exploration of the Upper Atmosphere" (R. L. F. Boyd and M. J. Seaton, eds.), pp. 158–161. Pergamon Press, Oxford.

Rocket Panel, Harvard College Observatory. (1952). Pressure, densities, and temperatures in the upper atmosphere. *Phys. Rev.* **88,** 1027–1032.

Rofe, B. (1963). Australian Sounding rocket experiments. Department of Supply, Salisbury, Australia, TN5 AD 127.

Scultetus, H. R. (1949). Altere Beobachtungen von Leuchtstreifen. *Z. Meteorol.* **3,** 272.

Seddon, J. C. (1960). Rocket observations of high-electron density gradients in the ionosphere. *Trans. Am. Geophys. Union,* 113–118.

Singer, S. F. (1956). Research on the upper atmosphere with high-altitude sounding rockets. *J. Atmospheric Terrest. Phys.* **4,** 878–912.

Smith, L. B. (1960a). Monthly observations of winds between 150 000 and 300 000 ft. over a one-year period. *J. Geophys. Res.* **65,** 2524.

Smith, L. B. (1960b). The measurement of winds between 100,000 and 300,000 ft. by use of chaff rockets. *J. Meteorol.* **17,** 296–310.

Smith, L. B. (1962). Monthly wind measurements in the meso-decline over a one-year period. *J. Geophys. Res.* **67**, 4653–4672.

Smith, W., L. Katchen, P. Sacher, P. Swartz, and J. Theon. (1964). Temperature, pressure density and wind measurements with the rocket grenade experiment 1960–1963. Goddard Space Flight Center, NASA, Greenbelt, Maryland, X-651-64-106, 55 pp.

Soberman, R. K. (1963). Noctilucent clouds. *Sci. Am.* **208**, 50–59.

Spangenberg, W. W. (1949). Über die leuchtenden Nachtwolken, 1932–41. *Wetter Klima* **2**, 15–23.

Spencer, N. W., R. L. Boggess, and D. Taeusch. (1958). Pressure, temperature and density to 90 km. over Ft. Churchill. Experimental Results of the U. S. Rocket Program for the International Geophysical Year to July 1, 1958. IGY Rocket Report Series, Natl. Acad. Sci., Washington, D. C.

Stormer, C. (1933). Height and velocity of luminous night clouds observed in Norway, 1932. Publ. 6, Univ. Obs., Oslo, 21 pp.

Stormer, C. (1935a). Measurements of luminous night clouds in Norway, 1933 and 1934. *Astrophys. Norvegica* **1**, 87–114.

Stormer, C. (1935b). Luminous night clouds over Norway in 1933 and 1934. *Nature* **135**, 103–104.

Stroud, W. G. (1958). Meteorological rocket soundings in the Arctic. *Trans. Am. Geophys. Union* **39**, 789–794.

Stroud, W. G., W. Nordberg, and J. R. Walsh. (1956). Atmospheric temperature and winds between 30 and 80 km. *J. Geophys. Res.* **61**, 45–56.

Stroud, W. G., W. R. Bandeen, W. Nordberg, F. L. Bartman, J. Otterman, and P. Titus. (1958). Temperature and winds in the Arctic as obtained by the rocket-grenade experiments. World Data Center A, IGY Rocket Report No, 1, pp. 58–79.

Stroud, W. G., W. Nordberg, and W. R. Bandeen. (1959). Rocket grenade observations of atmospheric heating in the Arctic. *J. Geophys. Res.* **64**, 1342–1344.

Stroud, W. G., W. Nordberg, W. R. Bandeen, F. L. Bartman, and P. Titus. (1960). Rocket grenade measurements of temperatures and winds in the mesosphere over Churchill, Canada. *J. Geophys. Res.* **65**, 2307–2323.

Surtees, W. J. (1964). An outline of some characteristics of the upper atmosphere. Canadian Armament Research and Development Establishment, Valcartier, Quebec, Report TR 470/64, 124 pp.

Temple University Research Institute. (1952). Research in the physical properties of the upper atmosphere with V-2 rockets. Final Report, April 1, 1948–Feb. 29, 1952, Contract W19-122-ac-12.

Thompson, W. E. (1955). The mean molecular weight of the upper atmosphere. U. S. Air Force, Cambridge Research Center, Geophysical Research Paper No. 36.

Thomsen, J. W. (1953). The annual deposit of meteoritic dust. *Sky Telescope* **12**, 147.

Tousey, R. (1953). Rocket spectroscopy. *J. Opt. Soc. Am.* **43**, 245–251.

Tousey, R. (1954). Observations of the solar ultraviolet spectrum from rockets. *8th Rept. Comm. Study Solar Terrest. Relationships, Paris,* 1953. pp. 49–57. Intern. Council Sci. Unions, Brussels.

Tousey, R. (1958). Rocket measurements of the night airglow. *Ann. Geophys.* **14**, 186 and 195.

Townsend, J. W. (1960). Composition of the upper atmosphere. World Data Center A, IGY Rocket Report No. 6, pp. 61–82.

Townsend, J. W., Jr., E. B. Meadows, and E. C. Pressly. (1954). A mass spectrometric study of the atmosphere. *In* "Rocket Exploration of the Upper Atmosphere" (R. L. F. Boyd and M. J. Seaton, eds.), pp. 169–188. Pergamon Press, Oxford.

Upper Atmosphere Research Project, University of Michigan. (1954). The measurement of diffusive separation in the upper atmosphere. *In* "Rocket Exploration of the Upper Atmosphere" (R. L. F. Boyd and M. J. Seaton, eds.), pp. 143–156. Pergamon Press, Oxford.

Van Allen, J. A. (1950). Use of rockets in upper atmosphere research. IUGG. Association of Terrestrial Magnetism and Electricity. *IATME Bull.* **13,** 531–536.

Van Mieghem, J. (1963). New aspects of the general circulation of the stratosphere and mesosphere. *Proc. Intern. Symp. Stratospheric Mesospheric Circulation,* pp. 5–63. Inst. Meteorol. Geophys. Freien Univ., Berlin.

Vassy, E. (1954). La Haute atmosphere. *Meteorologie* [4] **35,** pp. 185–194.

Vassy, E. (1963). Revue sur la physique de la haute atmosphere. *In* "First International Symposium on Rocket and Satellite Meteorology" (H. Wexler and J. E. Caskey, Jr., eds.), pp. 154–163. Wiley, New York.

Vestine, E. H. (1934). Noctilucent clouds. *J. Roy. Astron. Soc. Can.* **28,** 249–272 and 303–317.

Watanabe, K. (1958). Ultraviolet absorption processes in the upper atmosphere. *Advan. Geophys.* **5,** 153–221.

Webb, W. L. (1965). Morphology of noctilucent clouds. *J. Geophys. Res.* **70,** 4463–4475.

Webb, W. L., and K. R. Jenkins. (1961). Sonic structure of the mesosphere. U. S. Army Signal Missile Support Agency, White Sands Missile Range, New Mexico. Special Report 50, April, AD 255 767.

Weisner, A. G. (1954). The determination of temperatures and winds above thirty kilometers. *In "Rocket Exploration of the Upper Atmosphere"* (R. L. F. Boyd and M. J. Seaton, eds.), pp. 133–142. Pergamon Press, Oxford.

Weisner, A. G. (1956). Measurements of winds at elevations of 30 to 80 km by the rocket-grenade experiment. *J. Meteorol.* **13,** 30–39.

Wenzel, E. A., L. T. Loh, M. H. Nichols, and L. M. Jones. (1958). The measurement of diffusive separation in the upper atmosphere. World Data Center A, IGY Rocket Report No. 1, pp. 91–106.

Whipple, E. C. (1960). Direct measurements of ion-density and conductivity in the D-region. *NASA (Natl. Aeron. Space Admin.), Tech. Note* **TN D-567,** AD 244 348.

Whipple, F. L. (1951). Meteors as probes of the upper atmosphere. *Compendium Meteorol.* pp. 356–365.

Whipple, F. L. (1952). Exploration of the upper atmosphere by meteoritic technique. *Advan. Geophys.* **1,** 119–154.

Whipple, F. L. (1952). Results of rocket and meteor research. *Bull. Am. Meteorol. Soc.* **33,** 13–25.

Whipple, F. L. (1954). Density, pressure, and temperature data above 30 kilometers. *In* "The Earth As A Planet" (G. P. Kuiper, ed.), pp. 491–513. Chicago Press, Chicago, Illinois.

Whipple, F. L. (1961). The earth's dust belt. *Astron. Sci. Rev.* **3,** 17–20.

Whipple, F. L. (1963). On meteoroids and penetration. *J. Geophys. Res.* **68,** 4929–4939.

White, M. L. (1960). Atmospheric tides and ionospheric electrodynamics. *J. Geophys. Res.* **65,** 153–171.

Witt, G. (1960). Polarization of light from noctilucent clouds. *J. Geophys. Res.* **65,** 3.

Young, D., and E. S. Epstein. (1962). Atomic oxygen in the polar winter mesosphere. *J. Atmospheric Sci.* **19,** 435–443.

Zhekulin, L. (1960). Radio wave propagation characteristics of a simple ionospheric model based on rocket data. *Planetary Space Sci.* **2,** 110–120.

6

Summary

The discussions of Chapter 4 and 5 raise important questions relative to governing physical processes in the stratospheric circulation. Most obvious of these special results of MRN data analysis is the solstice time invasion of the winter hemisphere by easterlies of the summer hemisphere (see Section 4.1). Competing for first place is the diurnal variation which is induced by solar heat input into the earth's rotating system (see Section 4.8). The tidal motions produced by this diurnal heat cycle of the stratosphere are effective in mixing the stratosphere and mesosphere, both on a hemispheric scale and locally in the strong vertical and horizontal shears associated with this restricted circulation. These two special circulation systems provide the possibility of intelligently testing the stratospheric circulation for adherence to geostrophic conditions and/or deviations from uniformity in composition. The principal inferences are summarized here as a guide for future research studies which will attempt to clarify the many aspects of the problem which are hazy at this time.

A most important point is the fact that maximum strengths of the easterlies and westerlies which make up the global monsoonal circulation of the upper stratosphere and mesosphere are located at different altitudes (Fig. 1.2). Peak speeds in the winter westerlies are generally found near the stratopause (50 km altitude), with a negative wind lapse rate below that level and a positive lapse rate above. In general, the positive mesospheric wind lapse rate has a magnitude of approximately one-half the negative wind lapse rate of the upper stratosphere. This means, then, that the mesopause region does not exhibit a strong minimum in the zonal circulation such as that found at the stratonull level, and that the stratospheric monsoonal circulation extends into the lower ionosphere. The easterlies of the summer stratospheric circulation have a completely different character, increasing rather steadily with height (negative lapse rate) throughout the upper stratosphere and the mesosphere (Fig. 5.7). Peak easterly wind speeds are generally found in the middle or upper mesosphere, and only in the upper portion of the mesosphere near the mesopause does the wind lapse rate become positive during summer. It is interesting to note that the maximum speeds in these different circulations are of approximately the same magnitude, differing only in that the westerlies operate in the relatively dense air of the stratopause whereas the easterlies concentrate their principal flows in the low densities of the mesosphere. In general, the winter westerlies must be assumed to be characterized by a considerable momentum when compared with the summer easterlies.

On the other hand, the summer easterlies permeate a much larger portion of the stratospheric region than that occupied by the winter westerlies. Easterly winds first appear at a latitude of about 30 degrees, and, with development, the latitudinal maximum moves to lower latitudes, so that at the peak of the summer season (late July) approximately 70% of the stratospheric circulation volume is incorporated in the easterly flow. The total momentum of the easterly circulation is thus not small, even though its maximum speeds are located in a less favorable density environment. The anticyclonic curvature of the easterly flow around the polar high-pressure region provides a considerable stability in the general circulation so that day-to-day variations in the stratospheric hemispheric circulations are of relative unimportance in the summer easterlies when compared with the rather highly asymmetric character of the winter westerly circulation.

As was discussed in Section 4.1, for a short period at the time of the equinox, stratospheric circulation winds are westerly over practically the entire globe. The buildup of westerly flow in the winter hemisphere is very rapid so that about two months after the equinox, zonal winds with mean speeds of the order of 100 meters/sec are in evidence in middle latitudes. Encroachment of the summer easterly circulation into the winter hemisphere is accompanied by retreat of the circumpolar westerly circulation of that region to higher latitudes (Figs. 4.9–4.20). At the peak of the easterly invasion the westerlies are confined to the region above approximately twenty degrees latitude (Fig. 4.15). It should be remembered that this frontal zone of interaction between these major circulation systems is not a smooth narrow band symmetrical in latitude as is illustrated by these mean data, but is rather a broad meandering zone, generally penetrating poleward over oceanic regions and equatorward over continental areas. The usual latitudinal range of this interaction region is from about 20 to about 40 degrees latitude, although on occasion of specially strong thrust of easterly circulation, effects may be observed at very high latitudes of the winter hemisphere.

The nature of the diurnal perturbation of the stratospheric circulation has been outlined, the principal points being that peak intensity of the circulation is located at the stratopause level, although a lifting of altitude in sunrise-sunset regions is indicated by general considerations of heat input geometry. This effect should evidence some importance in the summer polar regions where the maximum impressed diurnal temperature variations will be largely meridional and thus will result in a strong zonal diurnal oscillation with a peak in the nighttime sector. While such an upward flaring of the diurnal heating zone must also occur in the winter hemisphere near the Arctic Circle, the period of impressed heating is short (less than one-half day) so that its influence will be reduced. The end of the diurnal heat wave which terminates in the winter hemisphere should be characterized by a negative meridional temperature gradient at the stratopause level in upper middle latitudes, and thus the thermal wind associated with this diurnal heat wave which the earth's rotation impresses on the stratospheric circulation is divided at the equatorial plane and thus is most efficient in mixing the summer hemisphere, since there it exhibits the maximum latitudinal extent and also appears to traverse the maximum altitude range.

The general structure of these circulation systems is illustrated for a meridional plane through the subsolar point at the time of the Northern Hemisphere summer solstice in Fig. 6.1. This picture of the summer easterly invasion and upper mesospheric tidal circulation is based on somewhat fragmentary data, and details of these features will have to be verified by special experimental data. General features are in good agreement with available data, however, and are probably depicted here in at least the proper general configuration. The diurnal tidal flow of Fig. 6.1 (shaded area) is not symmetrical in the meridional plane, except possibly during the equinox periods, since the direction of the flow is opposite in each hemisphere, and this division occurs at the rotational equator. The path of tidal motions is short in the winter hemisphere, while in the summer hemisphere the tidal circulation loops around the pole and rises significantly into the mesosphere. Hemispheric transport and mixing thus appear to be maximum in the summer hemisphere.

Also of great interest are the dynamic phenomena which will be a

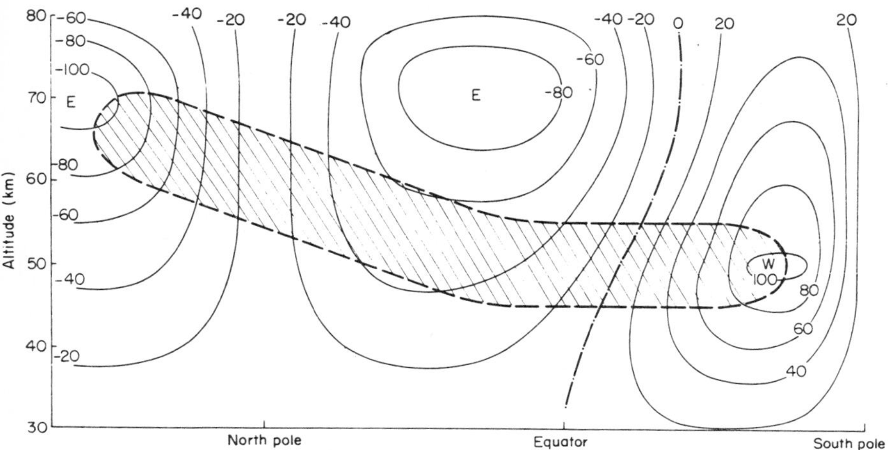

Fig. 6.1. Meridional cross section of the stratospheric circulation at noon during the summer solstice of the Northern Hemisphere. Contours are zonal wind speed in meters per second, with a wind from the west positive. The shaded region illustrates the geometry of the diurnal oscillation, with the stratospheric tidal jet located at the upper left end on the nighttime side of the summer pole.

part of the diurnal circulation in the earth's equatorial vicinity. Expansion of stratospheric air in the diurnal heat wave should be a maximum over the equator at equinox times and should be quite large throughout the year. The meridional flows set up by this heat wave are controlled by rotational effects, however, and they should be zero at the rotational equator and should be oppositely directed in the Northern and Southern Hemispheres. This depletion of equatorial mass during the morning hours and influx of mass during the afternoon and evening could be compensated for by variations in the zonal flow, but it is far more likely that the relatively rigid (strong negative temperature lapse rate) upper and lower boundaries would become involved. The upper boundary (D region) could be expected to be the most obviously affected, with phenomena of the "equatorial electrojet" type a direct result of the physics of this diurnal interplay of solar heating and earth's rotation.

Elementary calculations of the lowering of the mesopause and everything above it which could result from this diurnal equatorial divergence have maximum values of several kilometers, and thus it would produce downward velocities in the tens of centimeters per second range. In addition, if only a small portion of the diverging mass were supplied by vertical currents, such flows could easily be in the meters per second range, and they would surely be located at low latitudes away from the rotational equator due to the nature of the coriolis function. Such a downflow would produce enhancements in electron concentration (Hanson, 1961, see references in Chapter 5) in the strong negative density gradient ($H\rho \sim 7$ km) and strong positive electron density gradients ($H_e \sim 3$ km) and would have an interaction between the earth's magnetic field and these charged particles according to the relation

$$\vec{F} = q(\vec{E} + \vec{v} \times \vec{B}), \tag{6.1}$$

where $\vec{F}$ is the force acting on the charged particle of charge q, $\vec{E}$ is the electric field, $\vec{v}$ is the vertical wind component, and $\vec{B}$ is the earth's magnetic field. In the equatorial case where $\vec{B}$ is near horizontal (oriented north), Eq. (6.1) specifies that this rectifying action will displace electrons and negative ions westward and positive ions eastward. As a result of the gross differences in mobilities of these constituents of the upper atmosphere, a net current will be generated which will flow toward the east.

The current generated by the charge separation process described above will produce a local horizontal magnetic field in the lower ionosphere which will be in the same direction as the earth's permanent field and thus will enhance the field at those altitudes. The vertical gradient of the magnetic field will be reduced. In the same manner, the monsoonal circulation of the upper atmosphere will produce a separation of charge as the negative and positive charges are forced across the magnetic field. Easterly winds will force electrons upward, while westerly winds will force them downward. This current will be vertical and will produce a magnetic field which will diminish the permanent magnetic field in winter and enhance it in summer. Such variations have been observed by Cahill (1959, see references in Chapter 5) in the equatorial electrojets.

All of the MRN data thus far available on the diurnal temperature and wind structure of the stratospheric circulation have been combined with other sources of information to make the structure illustrated in Fig. 6.2. The isotherms and wind fields of the sunlit portion of the stratospheric circulation at the summer solstice are presented, with temperatures indicating the distribution in degrees Celsius above the minimum at 30 degrees latitude (approximately 270°K). In the instance illustrated here, the sun is above the ninetieth meridian, and the crest of the ridge of high temperature (and pressure) lags by some two hours the maximum heat input. In this solstice time situation, highest temperatures of the heated ridge are to be found over the summer pole, and maximum meridional temperature gradients (positive) of the summer hemisphere are located in the nighttime sector of the high latitude

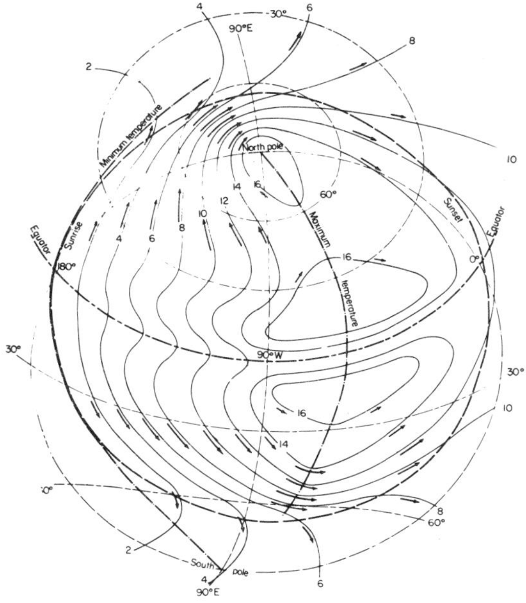

Fig. 6.2. Diurnal temperature and wind structure of the sunlit hemisphere at the Northern Hemisphere summer solstice. The temperature contours (solid lines) are in degrees Celsius above the diurnal minimum at 30 degrees latitude. The temperature and wind fields are projected on the stratopause level, which is portrayed through an equidistance projection centered at 42 degrees latitude and 103 degrees west longitude.

mesosphere. Of almost equal prominence in this picture is the structure of the heat wave in equatorial regions. These features are developed because the coriolis parameter reverses sign for horizontal motions between the hemispheres, becoming zero at the equator, and thus making the geostrophic relationship meaningless. From other considerations a downward circulation system can be expected to develop on both sides of the equator at low latitudes during the heating period, with adiabatic heating supplementing the radiational heat input during the period in which air is flowing meridionally out of equatorial regions.

Special features of the MRN data which deserve attention concern the gross detail structure which is observed in vertical profiles of any parameter which is measured with sensitivity. It is difficult to conceive of this detail structure being generated in any fashion other than small-scale motions. In an exponential atmosphere, motions which involve vertical and lateral components will produce such structures in any parameter which has relaxation time constants which are long compared with the time constants of the motions. If these are turbulent eddies, we at least have an idea of the vertical scale of these motions, and we know that they are larger at greater heights. It does not follow, nor is it even likely, that they are isotropic. Stability derived from thermal structure would lead one to expect that the eddies should be highly stratified in the horizontal in the upper stratosphere, with vertical components very severely damped. In the mesosphere the situation is much more favorable for a greater degree of symmetry of the horizontal and vertical components of these small-scale motions. These considerations are supported by the geometry of stratospheric tidal motions, which are indicated to be generally confined to a very narrow sheet of the order of 10–15 km thick in the vicinity of the stratopause, with a rather sharp peak in the responsive circulation at that point. The circulation system would seem to be much more confined than would be expected from consideration of heat input geometry. The strong shears associated with such eddy mixing, whether convective or damped, apparently provide an efficient means of transfering energy between circulation systems, and in general converting the kinetic energy of the organized circulation into potential energy (thermal) of the stratopause region. Turbulent eddies associated with the diurnal tidal circulation apparently result in mixing induced by the strong maximum in wind and temperature which occurs near local noon at the stratopause level and thus they should be of maximum intensity during daylight hours, the period in which most MRN observations have been obtained.

Of paramount interest is the apparent important role which diurnal tidal motions play in the dynamics of the stratospheric circulation. Maximum winter westerly winds are found at the same time and place as the maximum poleward location of the tidal circulation in the winter hemisphere, and maximum easterly winds of the summer season are found at the time and place of the zonal traverse of the poleward end of the diurnal heat wave in the summer hemisphere. This result is to be expected from radiational equilibrium considerations for all except the summer polar portion of the upper stratospheric circulation, where the observed positive meridional temperature gradient in the middle mesosphere can be considered unusual. In this same region the coldest temperatures of the earth's atmosphere are found at the mesopause level (approximately 80 km) and cannot be accounted for through radiational considerations. The presence of a supergeostrophic wind in that region with a vertical component of flow could supply the necessary initial conditions, and it is likely that the diurnal tidal circulation discussed here provides just such a circulation at the required time and place. Strong convergence of equator-to-pole winds along the leading edge of the diurnal heat wave into the twilight regions of polar latitudes supplies a jet of middle mesospheric air into the nighttime sector, which then enters a divergence situation in the weak temperature gradients which are characteristics of the trailing edge of the heat wave.

The physical mechanisms through which the diurnal tidal motions are confined to a thin sheet near the stratopause and are decelerated in the nighttime sector of the upper mesosphere are of paramount importance. A physical mechanism which could produce the observed results centers around frictional effects in the flow. Resistance to the flow resulting from transport of D region charged particles across magnetic field lines can be shown to be negligible for our purposes, as can molecular diffusion transport through the mean structure of the mesospheric region. A principal source of question in the last instance is the fact that the rate in which kinetic energy will be converted to potential energy, or will be transported along the gradient, is a strong function of the actual microscale gradients which occur. Molecular diffusion (Eq. (4.3)) is a relatively slow process (at least compared to one day) when one considers the mean gradient from the stratonull to the stratopause. The large amount of detailed structure which has been demonstrated by the sensitive MRN sensors to be a general characteristic of the stratospheric circulation system indicates that actual gradients may be significantly larger than the mean, and places a new emphasis on consideration of the eddy mode of energy transport as a possibly important mechanism of energy redistribution in this region of the atmosphere. A broad spectrum of eddy sizes exists, and there is every reason to expect the classical progression of energy toward relatively smaller sized eddies, which generally means toward stronger local gradients and more active molecular diffusion. Haurwitz (1961, see references of Chapter 5) has considered the importance of eddy diffusion in the stratospheric circulation in the mean case, using typical profile data obtained from the rocket-grenade observational system, which characteristically smooths the vertical profile data over considerable altitude intervals and reduces the maximum gradients purported to exist. It is apparent from the MRN wind and temperature data that small scale gradients do exist in the stratospheric circulation which are at least several times greater than that indicated by the mean data, with possible order-of-magnitude enhancements when smoothing effects which may have been introduced in the MRN data are considered.

When frictional effects result in deviations from geostrophic conditions (Eq. (1.3)) the horizontal equation of motion can be written

$$\frac{d\vec{V}}{dt} = f\vec{V} \times \vec{k} - g\nabla_p h - \vec{F} \tag{6.2}$$

where $\vec{V}$ is horizontal wind and $\vec{F}$ is frictional acceleration, in which the term on the right represents the frictional contribution to the flow.

This acceleration is in the direction of the wind, but is of opposite sign. Thus in all cases frictional effects result in subgeostrophic winds. Such winds will produce divergent circulations around high-pressure centers and convergent circulations around low-pressure centers, and will tend to erase the pressure distribution which generated the original motion. To evaluate Eq. (6.2) it is essential that we have a measure of the kinetic-energy depleting factors which at this point at least appear to have considerable importance. Since an accurate detailed evaluation of this physical parameter seems impossible with the data at hand, it is desirable to find indirect techniques which will at least yield order-of-magnitude estimates of the eddy diffusion rates which are characteristic of the mesosphere.

Such an estimate may be obtained for the meridional flow in the mean case illustrated in Fig. 6.1. If one assumes that the compositions of the upper stratosphere and mesosphere are uniform, the development of an easterly circulation in low latitudes of the winter hemisphere and the withdrawal of the westerlies to higher latitudes can be attributed to viscous effects. Evaluation of the meridional components which must accompany the general zonal flow to produce the observed displacements through frictional effects provides values of approximately 0.3 and 0.5 meter/sec at 50 and 70 km altitude, respectively, directed from the high pressure cell over the summer polar region toward the low pressure center over the winter pole. This result is in very general agreement with the conclusions derived by Haurwitz from typical vertical profile grenade data, and is too small to be detected in the data, even after means are taken, due to system errors and the uncertainties induced in the data analysis by the diurnal tidal component. These considerations then indicate that frictional effects increase by a factor of two from the base to the top of the stratosphere. It must also be noted that these estimates are smoothed over the entire stratospheric circulation and apply only in tropical regions where the horizontal coriolis acceleration is of small magnitude.

As is indicated by the restricted spatial extent of the diurnal oscillation discussed above, the MRN data indicate that local friction losses to this special circulation system may be significantly greater than these mean values. Along the lower margin of the current sheet produced by the tidal circulation, the principal mode of energy transport may be by very small scale eddy motions and by molecular diffusion across very strong gradients produced by internal wave motions in the fluid. The upper surface of the current sheet may be characterized by eddies of greater vertical extent, possibly of a convective nature. The eddy friction forces discussed above will then serve to transport momentum from the tidal circulation to the general zonal flows of each hemisphere, with the principal transfer effected where the tidal circulation is zonal and most likely to be nongeostrophic. Such processes are in agreement with the geometry indicated in Fig. 6.1, where the maximum in zonal circulation of the two hemispheres is found where this mechanism would be most effective.

In the model developed above, the maximum microscale transport of energy from one circulation system to another on a routine basis is to be expected where the diurnal tidal motion converges in high latitudes of the middle mesosphere of the early morning and late nighttime period. The maintenance of a positive wind lapse rate in the summer easterlies to the middle mesosphere and above, which is indicative of a positive meridional temperature gradient in the laminar flow case, may be the result of transport of energy into that region by the circulation of the stratospheric tidal oscillation. In general, one would expect that some of the kinetic energy would be finally dissipated from the tidal circulation through diffusive mixing of ever smaller eddies into the thermal energy content of the environment. The positive meridional temperature gradient thus induced would then drive the strong easterly zonal circulation of that region. Such an explanation of the abnormally cold temperatures of the summer mesopause of high latitudes presents practically the only direction available within the limited data which have been obtained. The cold air of that region is hypothesized to be produced by vertical eddy motions of the stratospheric tidal jet as it operates against the stable base of the lower ionosphere. Instabilities in the tidal jet and turbulent eddies along its margins may produce intermittent transport across the mesopause boundary, in which case the mesospheric air which crosses the mesopause would probably not return to the mesosphere, but would be trapped by thermal stability and would take part in lateral circulations in the D region. The cold mesopause of summer is thus probably not the result of radiational processes, but is a sink for kinetic energy of the stratospheric circulation and could not exhibit a westerly circulation, although a westerly thermal wind does serve to reduce the summer easterlies with height and put a cap on the summer stratospheric circulation. Since the D region should exhibit a positive meridional temperature gradient in summer, the easterlies should again increase with altitude above the mesopause, although if the above deductions concerning frictional effects are correct they should be of major importance in the organized zonal flow of that region so that the zonal flow may well be less intense and the meridional circulation may dominate.

The MRN does not today provide adequate data for a complete description of the circulations outlined above. In particular, the MRN sensors do not operate effectively above the 70-km level and thus fail to obtain data on a significant portion of the region in which the stratospheric monsoonal and tidal circulations operate. Supplementary observation systems exist in the form of radar observation of wind drift of meteor trails in the 80–100 km altitude range and through observation of occurrence and characteristics of noctilucent cloud formations. Noctilucent clouds provide a limited pictorial presentation of the dimensions of wave motions in the mesopause region which is very difficult to obtain by any other observational system and thus can be expected to exert a strong influence on interpretation of detailed structure in vertical profiles obtained by MRN rocket-sounding systems. In addition, they provide a relatively efficient system for observation of at least one aspect of the stratospheric circulation which will permit comparison of hemispheric similarity of this monsoonal flow. Past assumptions of uniformity in composition of the stratosphere and mesosphere seem more questionable in the light of current knowledge of the circulation structure, so that it is also possible that noctilucent cloud occurrence will provide important information on stratospheric and mesospheric composition and the physical processes which govern the transport of energy in those regions. Quite probably there is no more important experimental data required today than a measure of the ozone and water vapor distributions of the stratopause region.

While the application of radar techniques to observation of time displacement of the ionized trails produced by meteors captured by the earth's atmosphere is in some ways difficult in execution and interpretation, it would seem to offer the most efficient means of data gathering related to meteorological parameters which will be available in the foreseeable future for the D region of the atmosphere. Exploratory studies with this technique were informative but were of limited use due to their isolated and thus fragmentary nature. With development of the MRN, this situation has been corrected, and the meteor trail observation technique now offers the most efficient system for extending our knowledge of upper atmospheric circulation to higher altitudes. Meteor trail wind data should, after development of desirable observational techniques, compare favorably with MRN data in accuracy and possibly in sensitivity, and should excel in general coverage since the observations can be obtained regularly in time and thus exclude the bias problems inherent in the MRN observational program. As was pointed out in Section 4.8, the significant diurnal variation induced by periodic solar heating will result in a bias in the data obtained by a network which does not take several samples per day.

The few years of development of the MRN have been years of revolution in meteorological concepts concerning the upper atmosphere. Concepts of a quiescent structure, which had already begun to crumble during the middle 1900's, were completely demolished as far as wind and temperature are concerned by gross variability in the MRN data of all aspects which can be measured adequately. These data force a renewed analysis of the spatial and time distribution of each atmospheric parameter, particularly the composition. There is every indication that the next few years' research on the meteorology of the upper atmosphere (and hence on the entire atmosphere) will be even more fruitful than has been evidenced by these MRN developmental years. Expansion of meteorologists' sphere of interest to include all of the atmospheric system may herald attainment of maturity by the profession, assuming only that results of these new explorations will be incorporated into the framework of meteorological knowledge. Elimination of specific physical processes from complex systems such as the atmosphere should always be viewed with suspicion and should be reviewed with regard to the impact of each new element of data on the over-all system rather than some limited aspect. This book, then, is not aimed at the completion of another phase in our understanding of the meteorology of the earth's atmosphere, but is rather to be looked on as the opening of another door which serves to broaden the horizons of meteorologists to include the entire atmosphere.

AUTHOR INDEX

Numbers in italics refer to pages on which the complete references are listed.

J

K

L

SUBJECT INDEX

D

E

F

G

H

I

O

P

Q

R

S

T

U

V

W